Learning DevOps

The complete guide to accelerate collaboration with Jenkins, Kubernetes, Terraform and Azure DevOps

Mikael Krief

BIRMINGHAM - MUMBAI

Learning DevOps

Commissioning Editor: Vijin Boricha
Acquisition Editor: Meeta Rajani
Content Development Editor: Drashti Panchal
Senior Editor: Arun Nadar
Technical Editor: Prachi Sawant
Copy Editor: Safis Editing
Project Coordinator: Vaidehi Sawant
Proofreader: Safis Editing
Indexer: Tejal Daruwale Soni
Production Designer: Nilesh Mohite

First published: October 2019

Production reference: 1251019

Published by Packt Publishing Ltd.
Livery Place
35 Livery Street
Birmingham
B3 2PB, UK.

ISBN 978-1-83864-273-0

www.packt.com

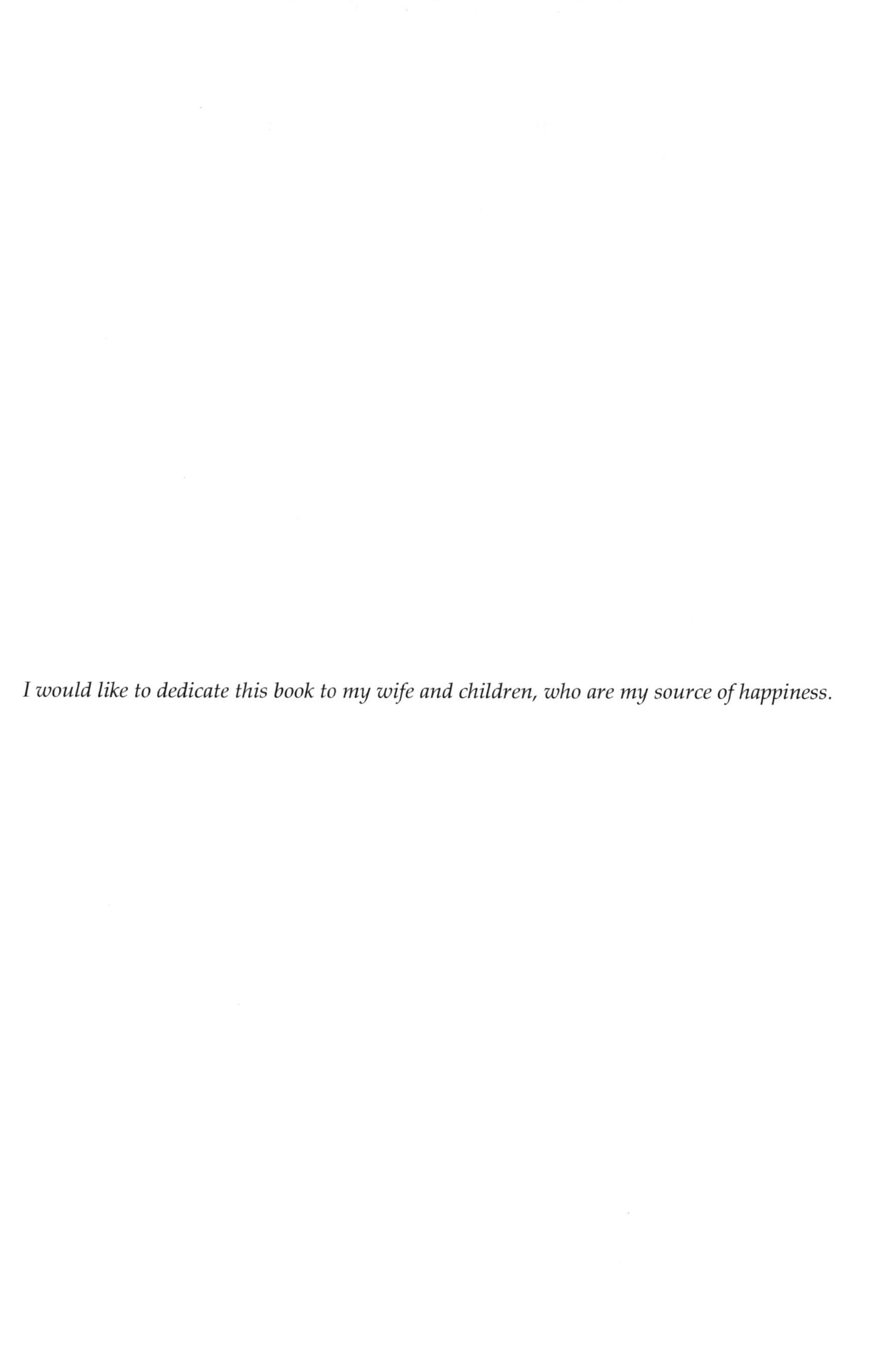

I would like to dedicate this book to my wife and children, who are my source of happiness.

Foreword

Having discussed DevOps with Mikael Krief on several occasions, it is clear that he understands the importance of empowering both Dev and Ops in order to deliver value.

DevOps is the union of people, processes, and products to enable the continuous delivery of value to our end users. Value is the most important word of that definition. DevOps is not about software, automation, shipping a feature, or getting to the bottom of your product backlog. It is about delivering value. To deliver value, you must measure your application while it is running in production and use the telemetry to guide what you deliver next. To deliver value, your team must fully embrace the culture of DevOps.

The hardest part of DevOps is the people part: building the culture that is required to succeed. Learning DevOps does a great job of focusing on the culture behind DevOps. To succeed, you must change the way your team thinks about their roles. Everyone must have a common goal that encourages collaboration. Delivering value to the end user is the responsibility of everyone involved in the application.

Our community tends to spend more time on the Dev side of DevOps. *Learning DevOps,* however, has invested considerable time on Infrastructure as Code. As more workloads move to the cloud, IaC becomes more valuable. The ability to provision and configure your infrastructure as part of your pipeline allows engineers to innovate. IaC can save companies money by shutting down environments when they are no longer in use or simply provisioning them on demand. Once your entire infrastructure is stored in version control and acted upon via your pipeline, recovering from a disaster is simply a deployment.

The time to debate whether you should or should not implement DevOps is over. You either implement DevOps or you lose.

Donovan Brown

Principal Cloud Advocate Manager at Microsoft

Contributors

About the author

Mikael Krief lives in France and works as a DevOps engineer, and for 4 years he has worked as a DevOps consultant and DevOps technical officer at an expert consulting company in Microsoft technologies. He is passionate about DevOps culture and practices, ALM, and Agile methodologies. He loves to share his passion through various communities, such as the ALM | DevOps Rangers community, which he has been a member of since 2015. He also contributes to many open source projects, writes blogs and books, speaks at conferences, and publishes public tools such as extensions for Azure DevOps. For all his contributions and passion in this area, he has received the Microsoft© Most Valuable Professional (MVP) award for the last 4 years.

> *I would like to extend my thanks to my family for accepting that I needed to work long hours on this book during family time. I would like to thank Meeta Rajani for giving me the opportunity to write this book, which was a very enriching experience. Special thanks to Drashti Panchal, Prachi Sawant, Arun Nadar for their valuable input and time reviewing this book and to the entire Packt team for their support during the course of writing this book.*

About the reviewers

Abhinav Krishna Kaiser manages in a leading consulting firm. He is a published author and has penned three books on DevOps, ITIL, and IT communication.

Abhinav has transformed multiple programs into the DevOps ways of working and is one of the leading DevOps architects on the circuit today. He has assumed the role of an Agile Coach to set the course for Agile principles and processes in order to set the stage in development. Apart from DevOps and Agile, Abhinav is an ITIL expert and is a popular name in the field of IT service management.

Abhinav's latest publication, on recasting ITIL with the DevOps processes, came out in 2018. *Reinventing ITIL in the Age of DevOps* transforms the ITIL framework to work in a DevOps project. His earlier publication, *Become ITIL Foundation Certified in 7 Days*, is one of the top guides for IT professionals looking to become ITIL Foundation certified and to those getting into the field of service management.

Abhinav started consulting with clients 15 years ago on IT service management, where he created value by developing robust service management solutions. Moving with the times, he eventually went into DevOps and Agile consulting. He is one of the foremost authorities in the area of configuration management and his solutions have stood the test of time, rigor, and technological advancements.

Abhinav blogs and writes guides and articles on DevOps, Agile, and ITIL on popular sites.

While the life of a consultant is to go where the client is, currently he is based in London, UK. He is from Bangalore, India, and is happily married with a daughter and a son.

Ebru Cucen works as a technical principal consultant at Contino, and is also a public speaker and trainer on Serverless. She has a BSc in mathematics and started her journey as a .NET developer/trainer in 2004. She has over 10 years of experience in digital transformation of financial enterprise companies. She's spent the last 5 years working with the cloud, covering the full life cycle of feature development/deployment and CI/CD pipelines. Being a lifetime student, she loves learning, exploring, and experimenting with technology to understand and use it to make our lives better.

She enjoys living in London with her 7-year-old son and her husband, Tolga Cucen, to whom she is thankful for supporting her during the nights/weekends she has worked on this book.

Packt is searching for authors like you

If you're interested in becoming an author for Packt, please visit `authors.packtpub.com` and apply today. We have worked with thousands of developers and tech professionals, just like you, to help them share their insight with the global tech community. You can make a general application, apply for a specific hot topic that we are recruiting an author for, or submit your own idea.

Table of Contents

Section 2: DevOps CI/CD Pipeline

Section 3: Containerized Applications with Docker and Kubernetes

Section 5: Taking DevOps Further

Preface

Today, with the evolution of technologies and ever-increasing competition, companies are facing a real challenge to design and deliver products faster – all while maintaining user satisfaction.

One of the solutions to this challenge is to introduce (to companies) a culture of collaboration between different teams, such as development and operations, testers, and security. This culture, which has already been proven and is called a DevOps culture, can ensure that teams and certain practices reduce the time to market of companies through this collaboration – with shorter application deployment cycles and by bringing real value to the company's products and applications.

Moreover, with the major shift of companies toward the cloud, application infrastructures are evolving and the DevOps culture will allow better scalability and performance of applications, thus generating a financial gain for a company.

If you want to learn more about the DevOps culture and apply its practices to your projects, this book will introduce the basics of DevOps practices through different tools and labs.

In this book, we will discuss the fundamentals of the DevOps culture and practices, and then we will examine different labs used for the implementation of DevOps practices, such as Infrastructure as Code, using Git and CI/CD pipelines, test automation, code analysis, and DevSecOps, along with the addition of security in your processes. A part of this book is also dedicated to the containerization of applications, with coverage of a simple use of Docker and the management of containers in Kubernetes. It includes downtime reduction topics during deployment and DevOps practices on open source projects. This book ends with a chapter dedicated to some good DevOps practices that can be implemented throughout the life cycle of your projects.

The book aims to guide you through the step-by-step implementation of DevOps practices using different tools that are mostly open source or are leaders in the market.

In writing this book, my goal is to share my daily experience with you; I hope that it will be useful for you and be applied to your projects.

Who this book is for

This book is for anyone who wants to start implementing DevOps practices. No specific knowledge of development or system operations is required.

What this book covers

`Chapter 1`, *DevOps Culture and Practices,* explains the objectives of the DevOps culture and details the different DevOps practices – IaC and CI/CD pipelines – that will be seen throughout this book.

`Chapter 2`, *Provisioning Cloud Infrastructure with Terraform,* details provisioning cloud infrastructure with IaC using Terraform, including its installation, its command line, its life cycle, a practical usage for provisioning a sample of Azure infrastructure, and the protection of tfstate with remote backends.

`Chapter 3`, *Using Ansible for Configuring IaaS Infrastructure,* concerns the configuration of VMs with Ansible, including Ansible's installation, command lines, setting up roles for an inventory and a playbook, its use in configuring VMs in Azure, data protection with Ansible Vault, and the use of a dynamic inventory.

`Chapter 4`, *Optimizing Infrastructure Deployment with Packer,* covers the use of Packer to create VM images, including its installation and how it is used for creating images in Azure.

`Chapter 5`, *Managing Your Source Code with Git,* explores the use of Git, including its installation, its principal command lines, its workflow, an overview of the branch system, and an example of a workflow with GitFlow.

`Chapter 6`, *Continuous Integration and Continuous Delivery,* shows the creation of an end-to-end CI/CD pipeline using three different tools: Jenkins, GitLab CI, and Azure Pipelines. For each of these tools, we will explain their characteristics in detail.

`Chapter 7`, *Containerizing Your Application with Docker,* covers the use of Docker, including its local installation, an overview of the Docker Hub registry, writing a Dockerfile, and a demonstration of how it can be used. An example of an application will be containerized, executed locally, and then deployed in an Azure container instance via a CI/CD pipeline.

`Chapter 8`, *Managing Containers Effectively with Kubernetes,* explains the basic use of Kubernetes, including its local installation and application deployment, and then an example of Kubernetes managed with Azure Kubernetes Services.

`Chapter 9`, *Testing APIs with Postman*, details the use of Postman to test an example of an API, including its local use and automation in a CI/CD pipeline with Newman and Azure Pipelines.

`Chapter 10`, *Static Code Analysis with SonarQube*, explains the use of SonarQube to analyze static code in an application, including its installation, real-time analysis with the SonarLint tool, and the integration of SonarQube into a CI pipeline in Azure Pipelines.

`Chapter 11`, *Security and Performance Tests*, discusses the security and performance of web applications, including demonstrations of how to use the ZAP tool to test OWASP rules, Postman to test API performance, and Azure Plan Tests to perform load tests.

`Chapter 12`, *Security in the DevOps Process with DevSecOps*, explains how to use security integration in the DevOps process through testing the compliance of infrastructure with Inspec, the usage of Vault for protecting sensitive data, and an overview of Azure's Secure DevOps Kit for testing Azure resource compliance.

`Chapter 13`, *Reducing Deployment Downtime*, presents the reduction of downtime deployment with Terraform, the concepts and patterns of blue-green deployment, and how to apply them in Azure. A great focus is also given on the use of feature flags within an application.

`Chapter 14`, *DevOps for Open Source Projects*, is dedicated to open source. It details the tools, processes, and practices for open source projects with collaboration in GitHub, pull requests, changelog files, binary sharing in GitHub releases, and an end-to-end examples of a CI pipeline in Travis CI and in GitHub Actions. Open source code analysis and security are also discussed with SonarCloud and WhiteSource Bolt.

`Chapter 15`, *DevOps Best Practices*, reviews a DevOps list of good practices regarding automation, IaC, CI/CD pipelines, testing, security, monitoring, and project management.

To get the most out of this book

No development knowledge is required to understand this book. The only languages you will see are declarative languages such as JSON or YAML. In addition to this, no specific IDE is required. If you do not have one, you can use Visual Studio Code, which is free and cross-platform. It is available here: `https://code.visualstudio.com/`.

As regards the operating systems you will need, there are no real prerequisites. Most of the tools we will use are cross-platform and compatible with Windows, Linux, and macOS. Their installations will be detailed in their respective chapters.

The cloud provider that serves as an example in this book is Microsoft Azure. If you don't have a subscription, you can create a free account here: `https://azure.microsoft.com/en-us/free/`.

Download the example code files

You can download the example code files for this book from your account at `www.packt.com`. If you purchased this book elsewhere, you can visit `www.packtpub.com/support` and register to have the files emailed directly to you.

You can download the code files by following these steps:

1. Log in or register at `www.packt.com`.
2. Select the **Support** tab.
3. Click on **Code Downloads**.
4. Enter the name of the book in the **Search** box and follow the onscreen instructions.

Once the file is downloaded, please make sure that you unzip or extract the folder using the latest version of:

- WinRAR/7-Zip for Windows
- Zipeg/iZip/UnRarX for Mac
- 7-Zip/PeaZip for Linux

The code bundle for the book is also hosted on GitHub at `https://github.com/PacktPublishing/Learning_DevOps`. In case there's an update to the code, it will be updated on the existing GitHub repository.

We also have other code bundles from our rich catalog of books and videos available at `https://github.com/PacktPublishing/`. Check them out!

Download the color images

We also provide a PDF file that has color images of the screenshots/diagrams used in this book. You can download it here: `https://static.packt-cdn.com/downloads/9781838642730_ColorImages.pdf`.

Code in Action

Visit the following link to check out videos of the code being run:
`http://bit.ly/2ognLdt`

Conventions used

There are a number of text conventions used throughout this book.

`CodeInText`: Indicates code words in text, database table names, folder names, filenames, file extensions, pathnames, dummy URLs, user input, and Twitter handles. Here is an example: "To execute the initialization, run the `init` command."

A block of code is set as follows:

```
resource "azurerm_resource_group" "rg" {
     name = var.resoure_group_name
    location = var.location
     tags {
         environment = "Terraform Azure"
     }
```

When we wish to draw your attention to a particular part of a code block, the relevant lines or items are set in bold:

```
resource "azurerm_resource_group" "rg" {
    name = "bookRg"
    location = "West Europe"
    tags {
        environment = "Terraform Azure"
    }
}
```

Any command-line input or output is written as follows:

```
git push origin master
```

Bold: Indicates a new term, an important word, or words that you see on screen. For example, words in menus or dialog boxes appear in the text like this. Here is an example: "Choose the integration of Git in Windows Explorer by marking the **Windows Explorer integration** checkbox."

Warnings or important notes appear like this.

Tips and tricks appear like this.

Get in touch

Feedback from our readers is always welcome.

General feedback: If you have questions about any aspect of this book, mention the book title in the subject of your message and email us at `customercare@packtpub.com`.

Errata: Although we have taken every care to ensure the accuracy of our content, mistakes do happen. If you have found a mistake in this book, we would be grateful if you would report this to us. Please visit `www.packtpub.com/support/errata`, selecting your book, clicking on the Errata Submission Form link, and entering the details.

Piracy: If you come across any illegal copies of our works in any form on the internet, we would be grateful if you would provide us with the location address or website name. Please contact us at `copyright@packt.com` with a link to the material.

If you are interested in becoming an author: If there is a topic that you have expertise in and you are interested in either writing or contributing to a book, please visit `authors.packtpub.com`.

Reviews

Please leave a review. Once you have read and used this book, why not leave a review on the site that you purchased it from? Potential readers can then see and use your unbiased opinion to make purchase decisions, we at Packt can understand what you think about our products, and our authors can see your feedback on their book. Thank you!

For more information about Packt, please visit `packt.com`.

1
Section 1: DevOps and Infrastructure as Code

The objectives of the first section are to present the DevOps culture and to provide all of the keys for the best Infrastructure as Code practices. This section explains the DevOps application on cloud infrastructure, showing provisioning using Terraform and configuration with Ansible. Then, we improve this by templating this infrastructure with Packer.

We will have the following chapters in this section:

- `Chapter 1`, *DevOps Culture and Practices*
- `Chapter 2`, *Provisioning Cloud Infrastructure with Terraform*
- `Chapter 3`, *Using Ansible for Configuring IaaS Infrastructure*
- `Chapter 4`, *Optimizing Infrastructure Deployment with Packer*

1
DevOps Culture and Practices

DevOps, a term that we hear more and more in enterprises with phrases such as *We do DevOps* or *We use DevOps tools*, is the contraction of the words Development and Operations.

DevOps is a culture different from traditional corporate cultures and requires a change in mindset, processes, and tools. It is often associated with **continuous integration** (**CI**) and **continuous delivery** (**CD**) practices, which are software engineering practices, but also with **Infrastructure as Code** (**IaC**), which consists of *codifying* the structure and configuration of infrastructure.

In this chapter, we will see what DevOps culture is, what DevOps principles are, and the benefits it brings to a company. Then, we will explain CI/CD practices and, finally, we will detail IaC with its patterns and practices.

In this chapter, the following topics will be covered:

- Getting started with DevOps
- Implementing CI/CD and continuous deployment
- Understanding IaC

Getting started with DevOps

The term DevOps was introduced in 2007-2009 by Patrick Debois, Gene Kim, and John Willis, and it represents the combination of **Development** (**Dev**) and **Operations** (**Ops**). It has given rise to a movement that advocates bringing developers and operations together within teams. This is to be able to deliver added business value to users more quickly and hence be more competitive in the market.

DevOps culture is a set of practices that reduce the barriers between developers, who want to innovate and deliver faster, on the one side and, on the other side, operations, who want to guarantee the stability of production systems and the quality of the system changes they make.

DevOps culture is also the extension of agile processes (scrum, XP, and so on), which make it possible to reduce delivery times and already involve developers and business teams, but are often hindered because of the non-inclusion of Ops in the same teams.

The communication and this link between Dev and Ops does, therefore, allow a better follow-up of end-to-end production deployments and more frequent deployments of a better quality, saving money for the company.

To facilitate this collaboration and improve communication between Dev and Ops, there are several key elements in the processes to be put in place, as in the following examples:

- More frequent application deployments with integration and continuous delivery (called **CI/CD**)
- The implementation and automation of unitary and integration tests, with a process focused on **Behavior-Driven Design** (**BDD**) or **Test-Driven Design** (**TDD**)
- The implementation of a means of collecting feedback from users
- Monitoring applications and infrastructure

The DevOps movement is based on three axes:

- **The culture of collaboration**: This is the very essence of DevOps—the fact that teams are no longer separated by silos specialization (one team of developers, one team of Ops, one team of testers, and so on), but, on the contrary, these people are brought together by making multidisciplinary teams that have the same objective: to deliver added value to the product as quickly as possible.
- **Processes**: To expect rapid deployment, these teams must follow development processes from agile methodologies with iterative phases that allow for better functionality quality and rapid feedback. These processes should not only be integrated into the development workflow with continuous integration but also into the deployment workflow with continuous delivery and deployment. The DevOps process is divided into several phases:
    - The planning and prioritization of functionalities
    - Development
    - Continuous integration and delivery
    - Continuous deployment
    - Continuous monitoring

  These phases are carried out cyclically and iteratively throughout the life of the project.

- **Tools**: The choice of tools and products used by teams is very important in DevOps. Indeed, when teams were separated into Dev and Ops, each team used their specific tools—deployment tools for developers and infrastructure tools for Ops—which further widened communication gaps.

With teams that bring development and operations together, and with this culture of unity, the tools used must be usable and exploitable by all members.

Developers need to integrate with monitoring tools used by Ops teams to detect performance problems as early as possible and with security tools provided by Ops to protect access to various resources.

Ops, on the other hand, must automate the creation and updating of the infrastructure and integrate the code into a code manager; this is called **Infrastructure as Code**, but this can only be done in collaboration with developers who know the infrastructure needed for applications. Ops must also be integrated into application release processes and tools.

The following diagram illustrates the three axes of DevOps culture—the collaboration between Dev and Ops, the processes, and the use of tools:

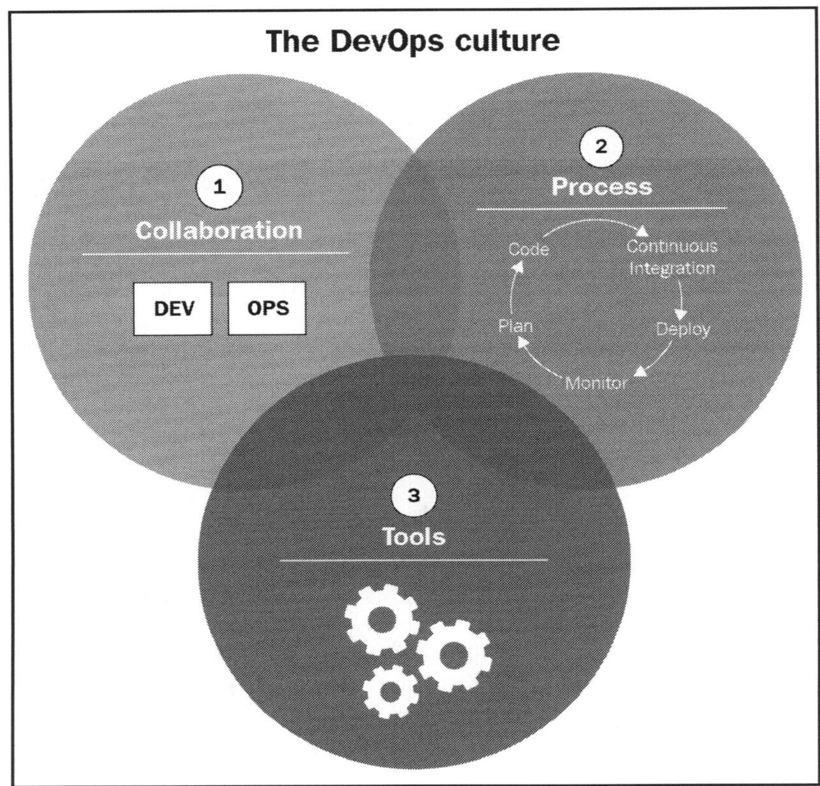

So, we can go back to DevOps culture with Donovan Brown's definition (`http://donovanbrown.com/post/what-is-devops`):

> *"DevOps is the union of people, process, and products to enable continuous delivery of value to our end users."*

The benefits of establishing a DevOps culture within an enterprise are as follows:

- Better collaboration and communication in teams, which has a human and social impact within the company.
- Shorter lead times to production, resulting in better performance and end user satisfaction.

- Reduced infrastructure costs with IaC.
- Significant time saved with iterative cycles that reduce application errors and automation tools that reduce manual tasks, so teams focus more on developing new functionalities with added business value.

For more information about DevOps culture and its impact on and transformation of enterprises, read the book by Gene Kim and Kevin Behr, *The Phoenix Project: A Novel about IT, DevOps, and Helping Your Business Win*, and *The DevOps Handbook: How to Create World-Class Agility, Reliability, and Security in Technology Organizations* by Gene Kim, Jez Humble, Patrick Debois, and John Willis.

Implementing CI/CD and continuous deployment

We saw earlier that one of the key DevOps practices is the process of integration and continuous delivery, also called CI/CD. In fact, behind the acronyms of CI/CD, there are three practices:

- **Continuous integration** (CI)
- **Continuous delivery** (CD)
- **Continuous deployment**

What does each of these practices correspond to? What are their prerequisites and best practices? Are they applicable to all?

Let's look in detail at each of these practices, starting with continuous integration.

Continuous integration (CI)

In the following definition given by Martin Fowler, there are three key things mentioned, *members of a team*, *integrate*, and *as quickly as possible*:

> *"Continuous Integration is a software development practice where members of a team integrate their work frequently... Each integration is verified by an automated build (including test) to detect integration errors as quickly as possible."*

That is, CI is an automatic process that allows you to check the completeness of an application's code every time a team member makes a change. This verification must be done as quickly as possible.

We see DevOps culture in CI very clearly, with the spirit of collaboration and communication, because the execution of CI impacts all members in terms of work methodology and therefore collaboration; moreover, CI requires the implementation of processes (branch, commit, pull request, code review, and so on) with automation that is done with tools adapted to the whole team (Git, Jenkins, Azure DevOps, and so on). And finally, CI must run quickly to collect feedback on code integration as soon as possible and hence be able to deliver new features more quickly to users.

Implementing CI

To set up CI, it is, therefore, necessary to have a **Source Code Manager** (**SCM**) that will allow the centralization of the code of all members. This code manager can be of any type: Git, SVN, or **Team Foundation Source Control** (**TFVC**). It's also important to have an automatic build manager (CI server) that supports continuous integration such as Jenkins, GitLab CI, TeamCity, Azure Pipelines, GitHub Actions, Travis CI, Circle CI, and so on.

In this book, we will use Git as an SCM, and we will look a little more deeply into its concrete uses.

Each team member will work on the application code daily, iteratively and incrementally (such as in agile and scrum methods). Each task or feature must be partitioned from other developments with the use of branches.

Regularly, even several times a day, members archive or commit their code and preferably with small commits (trunks) that can easily be fixed in the event of an error. This will, therefore, be integrated into the rest of the code of the application with all of the other commits of the other members.

This integration of all the commits is the starting point of the CI process.

This process, executed by the CI server, must be automated and triggered at each commit. The server will retrieve the code and then do the following:

- Build the application package—compilation, file transformation, and so on.
- Perform unit tests (with code coverage).

It is also possible to enrich the process with static code and vulnerability analysis, which we will look at in `Chapter 10`, *Static Code Analysis with SonarQube*, which is dedicated to testing.

This CI process must be optimized as soon as possible so that it can run fast and developers can have quick feedback on the integration of their code. For example, code that is archived and does not compile or whose test execution fails can impact and block the entire team.

Sometimes, bad practices can result in the failure of tests in the CI, deactivating the test execution, taking as arguments: *it is not serious, it is necessary to deliver quickly*, or *the code that compiles it is essential*.

On the contrary, this practice can have serious consequences when the errors detected by the tests are revealed in production. The time saved during CI will be lost on fixing errors with hotfixes and redeploying them quickly with stress. This is the opposite of DevOps culture with poor application quality for end users and no real feedback, and, instead of developing new features, we spend time correcting errors.

With an optimized and complete CI process, the developer can quickly fix their problem and improve their code or discuss it with the rest of the team and commit their code for a new integration:

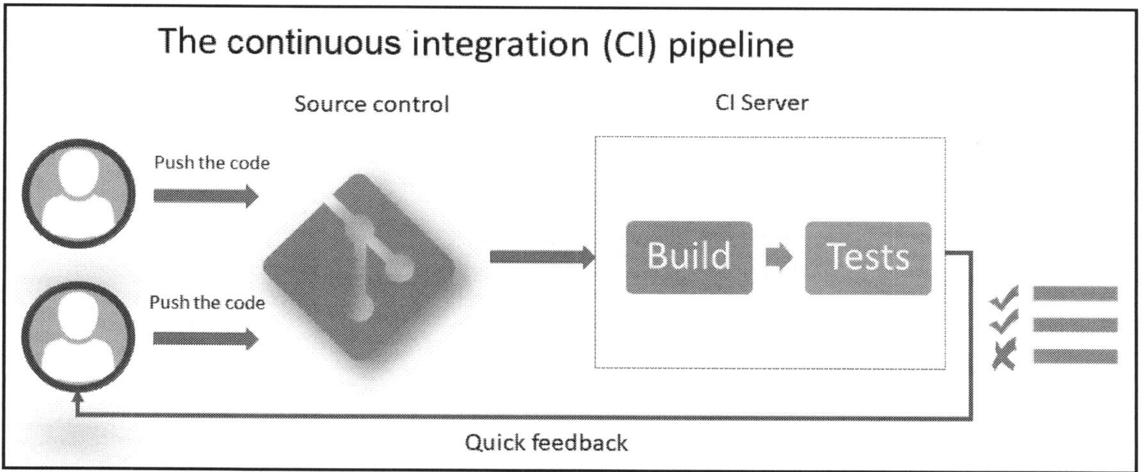

This diagram shows the cyclical steps of continuous integration that include the code being pushed into the SCM by the team members and the execution of the build and test by the CI server. And the purpose of this fast process is to provide rapid feedback to members.

We have just seen what continuous integration is, so now let's look at continuous delivery practices.

Continuous delivery (CD)

Once continuous integration has been successfully completed, the next step is to deploy the application automatically in one or more non-production environments, which is called **staging**. This process is called **continuous delivery** (**CD**).

CD often starts with an application package prepared by CI, which will be installed according to a list of automated tasks. These tasks can be of any type: unzip, stop and restart service, copy files, replace configuration, and so on. The execution of functional and acceptance tests can also be performed during the CD process.

Unlike CI, CD aims to test the entire application with all of its dependencies. This is very visible in microservices applications composed of several services and APIs; CI will only test the microservice under development while, once deployed in a staging environment, it will be possible to test and validate the entire application as well as the APIs and microservices that it is composed of.

In practice, today, it is very common to link CI with CD in an *integration* environment; that is, CI deploys at the same time in an environment. It is indeed necessary so that developers can have at each commit not only the execution of unit tests but also a verification of the application as a whole (UI and functional), with the integration of the developments of the other team members.

It is very important that the package generated during CI and that will be deployed during CD is the same one that will be installed on all environments, and this should be the case until production. However, there may be configuration file transformations that differ depending on the environment, but the application code (binaries, DLL, and JAR) must remain unchanged.

This *immutable*, unchangeable character of the code is the only guarantee that the application verified in an environment will be of the same quality as the version deployed in the previous environment and the same one that will be deployed in the next environment. If changes (improvements or bug fixes) are to be made to the code following verification in one of the environments, once done, the modification will have to go through the CI and CD cycle again.

The tools set up for CI/CD are often completed with others solutions, which are as follows:

- **A package manager**: This constitutes the storage space of the packages generated by CI and recovered by CD. These managers must support feeds, versioning, and different types of packages. There are several on the market, such as Nexus, ProGet, Artifactory, and Azure Artifacts.
- **A configuration manager**: This allows you to manage configuration changes during CD; most CD tools include a configuration mechanism with a system of variables.

In CD, the deployment of the application in each staging environment is triggered as follows:

- It can be triggered automatically, following a successful execution on a previous environment. For example, we can imagine a case where the deployment in the pre-production environment is automatically triggered when the integration tests have been successfully performed in a dedicated environment.
- It can be triggered manually, for sensitive environments such as the production environment, following a manual approval by a person responsible for validating the proper functioning of the application in an environment.

What is important in a CD process is that the deployment to the production environment, that is, to the end user, is triggered manually by approved users:

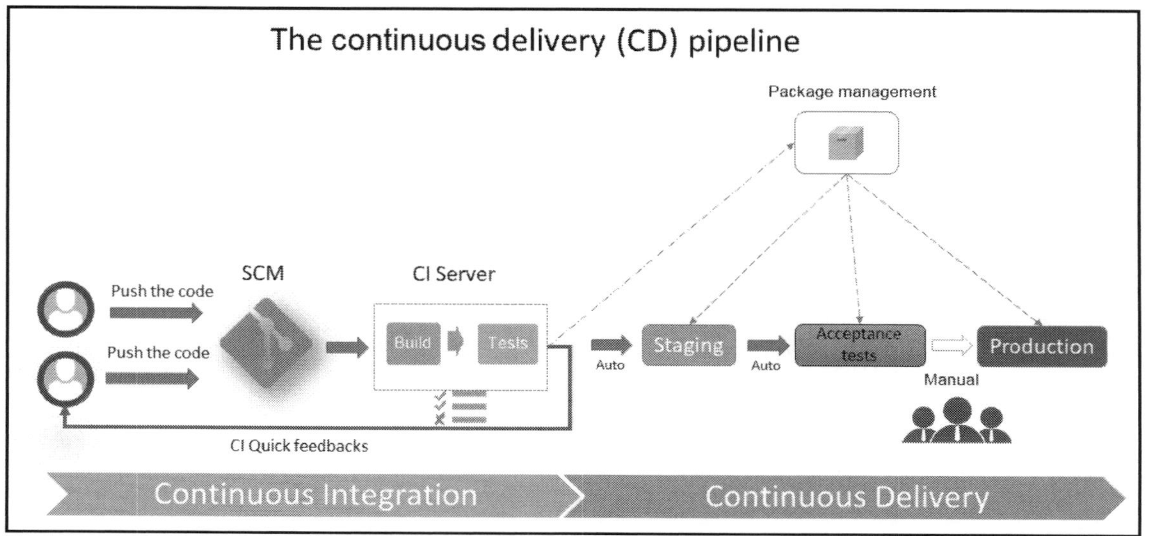

This diagram clearly shows that the CD process is a continuation of the CI process. It represents the chain of CD steps, which are automatic for staging environments but manual for production deployments. It also shows that the package is generated by CI and is stored in a package manager and that it is the same package that is deployed in different environments.

Now that we've looked at CD, let's look at continuous deployment practices.

Continuous deployment

Continuous deployment is an extension of CD, but this time, with a process that automates the entire CI/CD pipeline from the moment the developer commits their code to deployment in production through all of the verification steps.

This practice is rarely implemented in enterprises because it requires a wide coverage of tests (unit, functional, integration, performance, and so on) for the application, and the successful execution of these tests is sufficient to validate the proper functioning of the application with all of these dependencies, but also automated deployment to a production environment without any approval action.

The continuous deployment process must also take into account all of the steps to restore the application in the event of a production problem.

Continuous deployment can be implemented with the use and implementation of feature toggle techniques (or feature flags), which involves encapsulating the application's functionalities in *features* and activating its *features* on demand, directly in production, without having to redeploy the code of the application.

Another technique is to use a *blue-green* production infrastructure, which consists of two production environments, one *blue* and one *green*. We first deploy to the *blue* environment, then to the *green*; this will ensure that there is no downtime required:

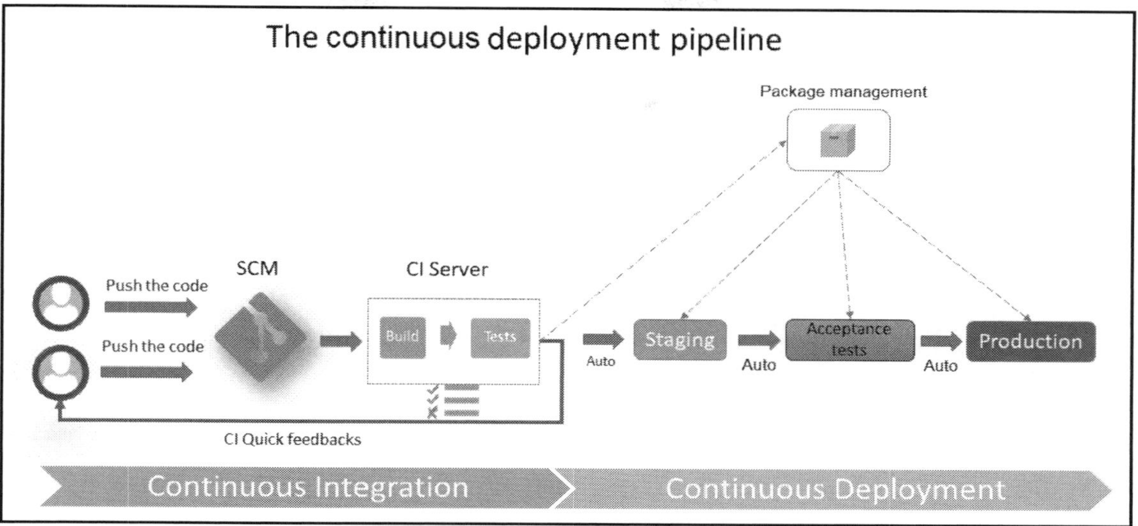

We will look at the feature toggle and blue-green deployment usage in more detail in `Chapter 13`, *Reducing Deployment Downtime*.

The preceding diagram is almost the same as that of CD, but with the difference that it depicts automated end-to-end deployment.

CI/CD processes are therefore an essential part of DevOps culture, with CI allowing teams to integrate and test the coherence of its code and to obtain quick feedback very regularly. CD automatically deploys on one or more staging environments and hence offers the possibility to test the entire application until it is deployed in production.

Finally, continuous deployment automates the deployment of the application from commit to the production environment.

We will see how to implement all of these processes in practice with Jenkins, Azure DevOps, and GitLab CI in `Chapter 6`, *Continuous Integration and Continuous Delivery*.

In this section, we have discussed practices essential to DevOps culture, which are continuous integration, continuous delivery, and continuous deployment.

In the next section, we will go into detail about another DevOps practice, which is IaC.

Understanding IaC practices

IaC is a practice that consists of writing the code of the resources that make up an infrastructure.

This practice began to take effect with the rise of DevOps culture and with the modernization of cloud infrastructure. Indeed, Ops teams that deploy infrastructures manually take time to deliver infrastructure changes due to inconsistent handling and the risk of errors. Also, with the modernization of the cloud and its scalability, the way an infrastructure is built requires a review of provisioning and change practices by adapting a more automated method.

IaC is the process of writing the code of the provisioning and configuration steps of infrastructure components to automate its deployment in a repeatable and consistent manner.

Before we look at the use of IaC, we will see what the benefits of this practice are.

The benefits of IaC

The benefits of IaC are as follows:

- The standardization of infrastructure configuration reduces the risk of error.
- The code that describes the infrastructure is versioned and controlled in a source code manager.
- The code is integrated into CI/CD pipelines.
- Deployments that make infrastructure changes are faster and more efficient.
- There's better management, control, and a reduction in infrastructure costs.

IaC also brings benefits to a DevOps team by allowing Ops to be more efficient on infrastructure improvement tasks rather than spending time on manual configuration and by giving Dev the possibility to upgrade their infrastructures and make changes without having to ask for more Ops resources.

IaC also allows the creation of self-service, ephemeral environments that will give developers and testers more flexibility to test new features in isolation and independently of other environments.

IaC languages and tools

The languages and tools used to code the infrastructure can be of different types; that is, scripting and declarative types.

Scripting types

These are scripts such as Bash, PowerShell, or any other languages that use the different clients (SDKs) provided by the cloud provider; for example, you can script the provisioning of an Azure infrastructure with the Azure CLI or Azure PowerShell.

For example, here is the command that creates a resource group in Azure:

- Using the Azure CLI (the documentation is at `https://bit.ly/2V1OfxJ`), we have the following:

```
az group create -location westeurope -name MyAppResourcegroup
```

- Using Azure PowerShell (the documentation is at `https://bit.ly/2VcASeh`), we have the following:

```
New-AzResourceGroup -Name MyAppResourcegroup -Location westeurope
```

The problem with these languages and tools is that they require a lot of lines of code because we need to manage the different states of the manipulated resources and it is necessary to write all of the steps of the creation or update of the desired infrastructure.

However, these languages and tools can be very useful for tasks that automate repetitive actions to be performed on a list of resources (selection and query) or that require complex processing with a certain logic to be performed on infrastructure resources such as a script that automates the deletion of VMs that carry a certain tag.

Declarative types

These are languages in which it is sufficient to write the state of the desired system or infrastructure in the form of configuration and properties. This is the case, for example, for Terraform and Vagrant from HashiCorp, Ansible, the Azure ARM template, PowerShell DSC, Puppet, and Chef. The user only has to write the final state of the desired infrastructure and the tool takes care of applying it.

For example, the following is the Terraform code that allows you to define the desired configuration of an Azure resource group:

```
resource "azurerm_resource_group" "myrg" {
     name = "MyAppResourceGroup"
     location = "West Europe"

     tags = {
         environment = "Bookdemo"
    }
}
```

In this example, if you want to add or modify a tag, just modify the `tags` property in the preceding code and Terraform will do the update itself.

Here is another example that allows you to install and restart `nginx` on a server using Ansible:

```
---
- hosts: all
 tasks:
 - name: install and check nginx latest version
 apt: name=nginx state=latest
 - name: start nginx
 service:
 name: nginx
 state: started
```

And to ensure that the service is not installed, just change the preceding code, with `service` as an absent value and the `state` property with the `stopped` value:

```
---
- hosts: all
 tasks:
 - name: stop nginx
 service:
 name: nginx
 state: stopped
 - name: check nginx is not installed
 apt: name=nginx state=absent
```

In this example, it was enough to change the `state` property to indicate the desired state of the service.

For details regarding the use of Terraform and Ansible, see `Chapter 2`, *Provisioning Cloud Infrastructure with Terraform*, and `Chapter 3`, *Using Ansible for Configuring IaaS Infrastructure*.

The IaC topology

In a cloud infrastructure, IaC is divided into several typologies:

- The deployment and provisioning of the infrastructure
- The server configuration and templating
- The containerization
- The configuration and deployment in Kubernetes

Let's deep dive into each topology.

The deployment and provisioning of the infrastructure

Provisioning is the act of instantiating the resources that make up the infrastructure. They can be of the **Platform as a Service** (**PaaS**) and serverless resource types, such as a web app, Azure function, or Event Hub but also the entire network part that is managed, such as VNet, subnets, routing tables, or Azure Firewall. For virtual machine resources, the provisioning step only creates or updates the VM cloud resource but not its content.

There are different provisioning tools such as Terraform, the ARM template, AWS Cloud training, the Azure CLI, Azure PowerShell, and also Google Cloud Deployment Manager. Of course, there are many more, but it is difficult to mention them all. In this book, we will look at, in detail, the use of Terraform to provide an infrastructure.

Server configuration

This step concerns the configuration of virtual machines, such as the configuration of hardening, directories, disk mounting, network configuration (firewall, proxy, and so on), and middleware installation.

There are different configuration tools, such as Ansible, PowerShell DSC, Chef, Puppet, and SaltStack. Of course, there are many more, but, in this book, we will look at, in detail, the use of Ansible to configure a virtual machine.

To optimize server provisioning and configuration times, it is also possible to create and use server models, also called images, that contain all of the configuration (hardening, middleware, and so on) of the servers. It will be during the provisioning of the server that we will indicate the template to use, and hence, we will have, in a few minutes, a configured server ready to be used.

There are also many IaC tools for creating server templates, such as aminator (used by Netflix) or HashiCorp Packer.

Here is an example of Packer file code that creates an Ubuntu image with package updates:

```
{
"builders": [{
      "type": "azure-arm",
      "os_type": "Linux",
      "image_publisher": "Canonical",
      "image_offer": "UbuntuServer",
      "image_sku": "16.04-LTS",
      "managed_image_resource_group_name": "demoBook",
      "managed_image_name": "SampleUbuntuImage",
      "location": "West Europe",
      "vm_size": "Standard_DS2_v2"
 }],
 "provisioners": [{
      "execute_command": "chmod +x {{ .Path }}; {{ .Vars }} sudo -E sh '{{
.Path }}'",
      "inline": [
      "apt-get update",
      "apt-get upgrade -y",
      "/usr/sbin/waagent -force -deprovision+user && export HISTSIZE=0 &&
sync"
 ],
 "inline_shebang": "/bin/sh -x",
 "type": "shell"
 }]
}
```

This script creates a template image for the `Standard_DS2_V2` virtual machine based on the Ubuntu OS (the `builders` section). Additionally, Packer will update all packages during the creation of the image with the `apt-get update` command and, after this execution, Packer deprovisions the image to delete all user information (the `provisioners` section).

The Packer part will be discussed in detail in `Chapter 4`, *Optimizing Infrastructure Deployment with Packer*.

Immutable infrastructure with containers

Containerization consists of deploying applications in containers instead of deploying them in VMs.

Today, it is very clear that the container technology to be used is Docker and that the configuration of a Docker image is also done in code in a Dockerfile. This file contains the declaration of the base image, which represents the *bone* to be used, the installation of additional middleware to be installed on the image, only the files and binaries necessary for the application, and the network configuration of the ports. Unlike VMs, containers are said to be immutable; the configuration of a container cannot be modified during its execution.

Here is a simple example of a Dockerfile:

```
FROM ubuntu
RUN apt-get update
RUN apt-get install -y nginx
ENTRYPOINT ["/usr/sbin/nginx","-g","daemon off;"]
EXPOSE 80
```

In this Docker image, we use a basic Ubuntu image, install `nginx`, and expose port `80`.

The Docker part will be discussed in detail in `Chapter 7`, *Containerizing Your Application with Docker*.

Configuration and deployment in Kubernetes

Kubernetes is a container orchestrator—it is the technology that most embodies IaC, in my opinion, because the way it deploys containers, the network architecture (load balancer, ports, and so on), and the volume management, as well as the protection of sensitive information, are described completely in the YAML specification files.

Here is a simple example of a YAML specification file:

```
apiVersion: apps/v1
kind: Deployment
metadata:
   name: nginx-demo
   labels:
       app: nginx
spec:
   replicas: 2
   selector:
     matchLabels:
        app: nginx
   template:
   metadata:
      labels:
        app: nginx
  spec:
    containers:
    - name: nginx
        image: nginx:1.7.9
      ports:
      - containerPort: 80
```

We can see in the preceding specification file, the name of the image to deploy (`ngnix`), the port to open (`80`), and the number of replicas (`2`).

The Kubernetes part will be discussed in detail in `Chapter 8`, *Managing Containers Effectively with Kubernetes*.

IaC best practices

IaC, like software development, requires the implementation of practices and processes that allow the evolution and maintenance of the infrastructure code.

Among these practices are those of software development, as in these examples:

- Have good principles of nomenclature.
- Do not overload the code with unnecessary comments.
- Use small functions.
- Implement error handling.

To learn more about good software development practices, read the excellent book, which is, for my part, a reference on the subject, *Clean Code* by Robert Martin.

But there are more specific practices that I think deserve more attention:

- **Everything must be automated in the code**: When doing IaC, it is indeed necessary to code and automate all of the provisioning steps and not to leave manual steps out of code that *distort* the automation of the infrastructure and that can generate errors. And if necessary, do not hesitate to use several tools such as Terraform and Bash with the Azure CLI scripts.
- **The code must be in a source control manager**: The infrastructure code must also be in an SCM to be versioned, tracked, merged, and restored, and hence have better visibility of the code between Dev and Ops.
- **The infrastructure code must be with the application code**: In some cases, this may be difficult, but if possible, it is much better to place the infrastructure code in the same repository as the application code. This is to have a better work organization between developers and operations, who will share the same workspace.
- **Separation of roles and directories**: It is good to separate the code from the infrastructure according to the role of the code, so you can create one directory for provisioning and for configuring VMs and another directory that will contain the code for testing the integration of the complete infrastructure.
- **Integration into a CI/CD process**: One of the goals of IaC is to be able to automate the deployment of the infrastructure, so from the beginning of its implementation, it is necessary to set up a CI/CD process that will integrate the code, test it, and deploy it in different environments. Some tools, such as Terratest, allow you to write tests on infrastructure code. One of the best practices is to integrate the CI/CD process of the infrastructure into the same pipeline as the application.
- **The code must be idempotent**: The execution of the infrastructure deployment code must be idempotent; that is, automatically executable at will. This means that scripts must take into account the state of the infrastructure when running it and not generate an error if the resource to be created already exists or if a resource to be deleted has already been deleted. We will see that declarative languages, such as Terraform, take on this aspect of idempotence natively. The code of the infrastructure, once fully automated, must allow the construction and complete destruction of the application infrastructure.

- **To be used as documentation**: The code of the infrastructure must be clear and must be able to serve as documentation. Indeed, infrastructure documentation takes a long time to be written and in many cases, it is not updated as the infrastructure evolves.
- **The code must be modular**: In an infrastructure, the components very often have the same code—the only difference is the value of their properties. Also, these components are used several times in the company's applications. It is therefore important to optimize the writing times of code, by factoring it with modules (or roles, for Ansible) that will be called as functions. Another advantage of using modules is the ability to standardize resource nomenclature and compliance on some properties.
- **Having a development environment**: The problem with IaC is that it is difficult to test its infrastructure code under development in environments used for integration and to test the application because changing the infrastructure can have an impact. It is therefore important to have a development environment even for IaC that can be impacted or even destroyed at any time.

For local infrastructure tests, some tools simulate a local environment, such as Vagrant (from HashiCorp), so you should use them to test code scripts as much as possible.

Of course, the full list of good practices is longer than this list; all methods and processes of software engineering practices are also applicable.

IaC is, therefore, like CI/CD processes, a key practice of DevOps culture that allows, by writing code, the deployment and configuration of an infrastructure. However, IaC can only be effective with the use of appropriate tools and the implementation of good practices.

Summary

In this first chapter, we saw that DevOps culture is a story of collaboration, processes, and tools. Then, we detailed the different steps of the CI/CD process and explained the difference between and that continuous deployment.

Finally, the last part explained how to use IaC, with its best practices.

In the next chapter, we will start with the implementation of IaC and how to provision an infrastructure with Terraform.

Questions

1. Of which words is DevOps a contraction?
2. Is DevOps a term that represents: the name of a tool, a culture or a society, or the title of a book?
3. What are the three axes of DevOps culture?
4. What is the objective of continuous integration?
5. What is the difference between continuous delivery and continuous deployment?
6. What is IaC?

Further reading

If you want to know more about DevOps culture, here are some resources:

- The DevOps Resource Center (Microsoft resources): `https://docs.microsoft.com/en-us/azure/devops/learn/`
- 2018 State of DevOps Report (by Puppet): `https://puppet.com/resources/whitepaper/state-of-devops-report`

2

Provisioning Cloud Infrastructure with Terraform

In the previous chapter, we introduced the tools, practices, and benefits of **Infrastructure as Code (IaC)** and its impact on DevOps culture. Out of all of the IaC tools that have been mentioned, one that is particularly powerful is **Terraform**, which is part of the HashiCorp tools suite.

In this chapter, we will look at the basics of using Terraform to provision a cloud infrastructure, using Azure as an example. We will start with an overview of its strengths compared to other IaC tools. We will see how to install it in both manual and automatic mode, and then we will create our first Terraform script to provision an Azure infrastructure with the use of best practices and its automation in a CI/CD process. Finally, we will go a little deeper with the implementation of a remote backend for tfstate.

In this chapter, the following topics will be covered:

- Installing Terraform
- Configuring Terraform for Azure
- Writing a Terraform script to deploy an Azure infrastructure
- Running Terraform for deployment
- Understanding the Terraform's life cycle with the different command-lines options
- Protecting the tfstate file with a remote backend

Technical requirements

This chapter will explain how to use Terraform to provision an Azure infrastructure as an example of a cloud infrastructure, so you will need an Azure subscription, which you can get here for free: `https://azure.microsoft.com/en-us/free/`.

In addition, we will require a code editor to write the Terraform code. There are several editors out there, but I will be using Visual Studio Code, which is free, lightweight, and multiplatform and has several extensions for Terraform. You can download it from `https://code.visualstudio.com/`. The complete source code of this chapter is available here: `https://github.com/PacktPublishing/Learning_DevOps/tree/master/CHAP02`.

Check out the following video to see the Code in Action:
`http://bit.ly/2MJs7TW`

Installing Terraform

Terraform is a command-line tool that, in its basic version, is open source, uses the **HashiCorp Configuration Language** (**HCL**), is declarative, and is relatively easy to read. Its main advantage is the use of the same language to deploy on a multitude of cloud providers such as Azure, AWS, and Google—the complete list is available here: `https://www.terraform.io/docs/providers/`.

Terraform has other advantages:

- It's multiplatform, and it can be installed on Windows, Linux, and Mac.
- It allows a preview of infrastructure changes before they are implemented.
- It allows the parallelization of operations by taking into account resource dependencies.
- It integrates a very large number of providers.

Terraform can be installed in your system in a number of ways. Let's begin by looking at the manual installation method.

Manual installation

To install Terraform manually, perform the following steps:

1. Go to the official download page, `https://www.terraform.io/downloads.html`, and download the package corresponding to your operating system.
2. After downloading, unzip and copy the binary into an execution directory (for example, inside `c:\Terraform`).
3. Then, the `PATH` environment variable must be completed with the path to the `binary` directory. For detailed instructions, view the video here: `https://learn.hashicorp.com/terraform/getting-started/install.html`.

Now that we've learned how to install Terraform manually, let's look at the options available to us to install it using a script.

Installation by script

Script installation automates the installation or update of Terraform on a remote server that will be in charge of executing Terraform code, such as on a Jenkins slave or an Azure Pipelines agent.

Installing Terraform by script on Linux

The Linux Terraform installation script is as follows:

```
TERRAFORM_VERSION="0.12.8" #Update with your desired version

curl -Os
https://releases.hashicorp.com/terraform/${TERRAFORM_VERSION}/terraform_${T
ERRAFORM_VERSION}_linux_amd64.zip \
&& curl -Os
https://releases.hashicorp.com/terraform/${TERRAFORM_VERSION}/terraform_${T
ERRAFORM_VERSION}_SHA256SUMS \
&& curl https://keybase.io/hashicorp/pgp_keys.asc | gpg --import \
&& curl -Os
https://releases.hashicorp.com/terraform/${TERRAFORM_VERSION}/terraform_${T
ERRAFORM_VERSION}_SHA256SUMS.sig \
&& gpg --verify terraform_${TERRAFORM_VERSION}_SHA256SUMS.sig
terraform_${TERRAFORM_VERSION}_SHA256SUMS \
&& shasum -a 256 -c terraform_${TERRAFORM_VERSION}_SHA256SUMS 2>&1 | grep
"${TERRAFORM_VERSION}_linux_amd64.zip:\sOK" \
&& unzip -o terraform_${TERRAFORM_VERSION}_linux_amd64.zip -d
/usr/local/bin
```

This script does the following:

- Set the `TERRAFORM_VERSION` parameter with the version to download.
- Download the Terraform package by checking the checksum.
- Unzip the package in the user local directory.

This script is also available in the GitHub source of this book: `https://github.com/PacktPublishing/Learning_DevOps/blob/master/CHAP02/Terraform_install_Linux.sh`.

For executing this script, follow these steps:

1. Open a command-line Terminal.
2. Copy and paste the preceding script.
3. Execute it by hitting *Enter* in the command-line Terminal.

The following screenshot displays an execution of the script for installing Terraform on Linux:

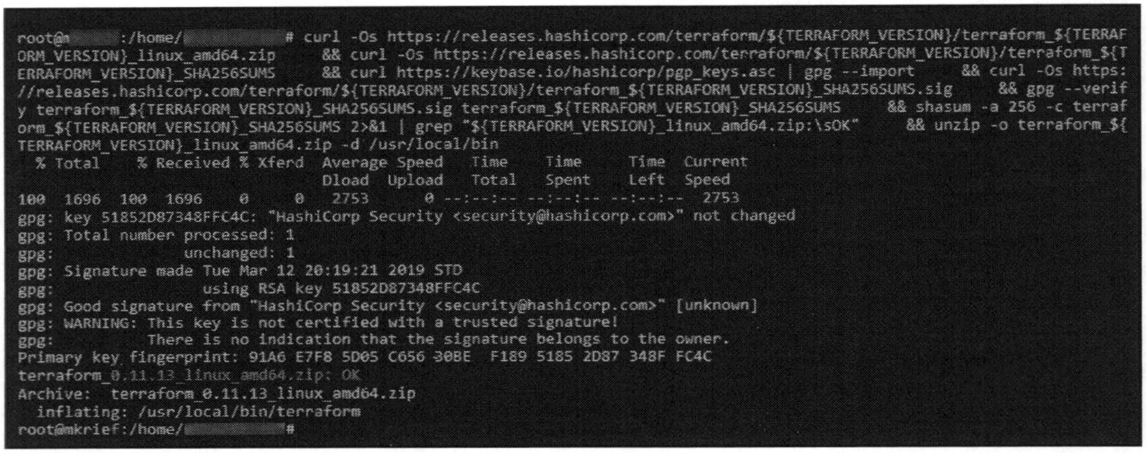

```
root@n     :/home/          # curl -Os https://releases.hashicorp.com/terraform/${TERRAFORM_VERSION}/terraform_${TERRAF
ORM_VERSION}_linux_amd64.zip     && curl -Os https://releases.hashicorp.com/terraform/${TERRAFORM_VERSION}/terraform_${T
ERRAFORM_VERSION}_SHA256SUMS     && curl https://keybase.io/hashicorp/pgp_keys.asc | gpg --import     && curl -Os https:
//releases.hashicorp.com/terraform/${TERRAFORM_VERSION}/terraform_${TERRAFORM_VERSION}_SHA256SUMS.sig     && gpg --verif
y terraform_${TERRAFORM_VERSION}_SHA256SUMS.sig terraform_${TERRAFORM_VERSION}_SHA256SUMS     && shasum -a 256 -c terraf
orm_${TERRAFORM_VERSION}_SHA256SUMS 2>&1 | grep "${TERRAFORM_VERSION}_linux_amd64.zip:\sOK"     && unzip -o terraform_${
TERRAFORM_VERSION}_linux_amd64.zip -d /usr/local/bin
  % Total    % Received % Xferd  Average Speed   Time    Time     Time  Current
                                 Dload  Upload   Total   Spent    Left  Speed
100  1696  100  1696    0     0   2753      0 --:--:-- --:--:-- --:--:--  2753
gpg: key 51852D87348FFC4C: "HashiCorp Security <security@hashicorp.com>" not changed
gpg: Total number processed: 1
gpg:              unchanged: 1
gpg: Signature made Tue Mar 12 20:19:21 2019 STD
gpg:                using RSA key 51852D87348FFC4C
gpg: Good signature from "HashiCorp Security <security@hashicorp.com>" [unknown]
gpg: WARNING: This key is not certified with a trusted signature!
gpg:          There is no indication that the signature belongs to the owner.
Primary key fingerprint: 91A6 E7F8 5D05 C656 30BE  F189 5185 2D87 348F FC4C
terraform_0.11.13_linux_amd64.zip: OK
Archive:  terraform_0.11.13_linux_amd64.zip
  inflating: /usr/local/bin/terraform
root@mkrief:/home/          #
```

In the execution of the preceding script, we can see the download of the Terraform ZIP package (with the `curl` tool) and the unzip operation of this package inside the `/usr/local/bin` folder.

We have just seen the installation of Terraform on Linux; now, let's look at its installation on Windows.

Installing Terraform by script on Windows

If we use Windows, we can use **Chocolatey**, which is a free public package manager, such as NuGet or npm, but dedicated to software; it is widely used for the automation of software on Windows servers or even local machines.

The Chocolatey official website is here: `https://chocolatey.org/`, and its installation documentation is at `https://chocolatey.org/install`.

Once Chocolatey is installed, we just need to run the following command in PowerShell or in the CMD tool:

```
choco install terraform -y
```

The following is a screenshot of the Terraform installation for Windows with Chocolatey:

```
PS C:\WINDOWS\system32> choco install terraform -y
Chocolatey v0.10.7
Installing the following packages:
terraform
By installing you accept licenses for the packages.
Progress: Downloading terraform 0.11.13... 100%

terraform v0.11.13 [Approved]
terraform package files install completed. Performing other installation steps.
Removing old terraform plugins
Downloading terraform 64 bit
  from 'https://releases.hashicorp.com/terraform/0.11.13/terraform_0.11.13_windows_amd64.zip'
Progress: 100% - Completed download of C:\Users\Mikael\AppData\Local\Temp\chocolatey\terraform\0.11.13\terraform_0.11.13
_windows_amd64.zip (20.18 MB).
Download of terraform_0.11.13_windows_amd64.zip (20.18 MB) completed.
Hashes match.
Extracting C:\Users\Mikael\AppData\Local\Temp\chocolatey\terraform\0.11.13\terraform_0.11.13_windows_amd64.zip to C:\Pro
gramData\chocolatey\lib\terraform\tools...
C:\ProgramData\chocolatey\lib\terraform\tools
 ShimGen has successfully created a shim for terraform.exe
 The install of terraform was successful.
  Software installed to 'C:\ProgramData\chocolatey\lib\terraform\tools'

Chocolatey installed 1/1 packages.
 See the log for details (C:\ProgramData\chocolatey\logs\chocolatey.log).
PS C:\WINDOWS\system32> _
```

Executing the `choco install terraform` command installs the latest version of Terraform from Chocolatey.

Once installed, we can check the Terraform version by running the following command:

```
terraform version
```

This command displays the installed Terraform version.

We can also check out the different commands that Terraform offers by running the following command:

```
terraform --help
```

The following screenshot lists the different commands and their functions:

```
:~$ terraform --help
Usage: terraform [-version] [-help] <command> [args]

The available commands for execution are listed below.
The most common, useful commands are shown first, followed by
less common or more advanced commands. If you're just getting
started with Terraform, stick with the common commands. For the
other commands, please read the help and docs before usage.

Common commands:
    apply              Builds or changes infrastructure
    console            Interactive console for Terraform interpolations
    destroy            Destroy Terraform-managed infrastructure
    env                Workspace management
    fmt                Rewrites config files to canonical format
    get                Download and install modules for the configuration
    graph              Create a visual graph of Terraform resources
    import             Import existing infrastructure into Terraform
    init               Initialize a Terraform working directory
    output             Read an output from a state file
    plan               Generate and show an execution plan
    providers          Prints a tree of the providers used in the configuration
    push               Upload this Terraform module to Atlas to run
    refresh            Update local state file against real resources
    show               Inspect Terraform state or plan
    taint              Manually mark a resource for recreation
    untaint            Manually unmark a resource as tainted
    validate           Validates the Terraform files
    version            Prints the Terraform version
    workspace          Workspace management

All other commands:
    debug              Debug output management (experimental)
    force-unlock       Manually unlock the terraform state
    state              Advanced state management
```

Let's now look at the installation of Terraform on macOS.

Installing Terraform by script on macOS

On macOS, we can use **Homebrew**, the macOS package manager (`https://brew.sh/`), for installing Terraform by executing the following command in your Terminal:

```
brew install terraform
```

That's all for the installation of Terraform by script. Let's look at another solution to use Terraform in Azure without having to install it—by using **Azure Cloud Shell**.

Integrating Terraform with Azure Cloud Shell

If we are using Terraform to deploy an infrastructure in Azure, we should also know that the Azure team has integrated Terraform into Azure Cloud Shell.

To learn more about Azure Cloud Shell, refer to its documentation: `https://azure.microsoft.com/en-us/features/cloud-shell/`.

To use it from the Azure Cloud Shell, follow these steps:

1. Connect to the Azure portal by opening the URL, `https://portal.azure.com`, and sign in with your Azure account:

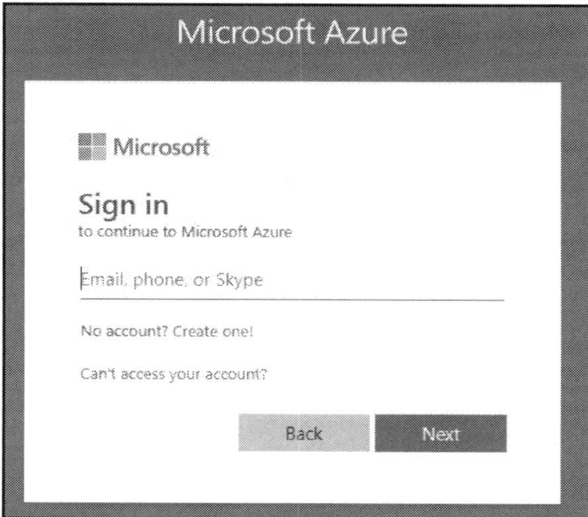

2. Open the Cloud Shell and choose the mode we want, that is, Bash or PowerShell.
3. Then, we can run the Terraform command line in the shell.

The following is a screenshot of the execution of Terraform in Azure Cloud Shell:

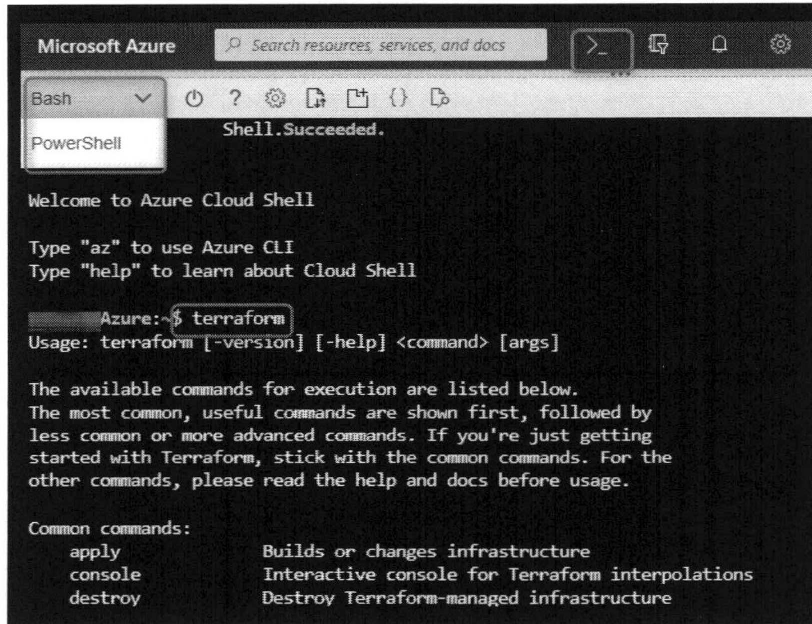

The advantage of using this solution is that we don't need any software to install; we can upload your Terraform files to Cloud Shell and run them in it. Additionally, we are already connected to Azure so no configuration is required (refer to the following section, *Configuring Terraform for Azure*).

However, this solution is only to be used in development mode and not for local or automatic use of Terraform. It is for this reason, in this chapter, we will discuss the configuration of Terraform for Azure.

Now that we have installed Terraform, we can now start using it locally to provision an Azure infrastructure starting with the first step, which is to configure Terraform for Azure.

Configuring Terraform for Azure

Before writing Terraform code to provision a cloud infrastructure such as Azure, we must configure Terraform to allow the manipulation of resources in an Azure subscription.

To do this, we will first create a new Azure **service principal** (**SP**) in Azure **Active Directory** (**AD**), which, in Azure, is an *application user* who has permission to manage Azure resources.

For more details about the Azure SP, read the documentation here: `https://docs.microsoft.com/en-us/azure/active-directory/develop/app-objects-and-service-principals`.

For this Azure SP, we have to assign to it the contributing permissions on the subscription in which we will create resources.

Creating the Azure SP

This operation can be done either via the Azure portal (all steps are detailed on the official documentation here: `https://docs.microsoft.com/en-us/azure/active-directory/develop/howto-create-service-principal-portal`) or via a script by executing the following `az cli` command (which we can launch in Azure Cloud Shell).

The following is a template `az cli` script that you have to run for creating an service principal, where you have to enter your SP name, role, and scope:

```
az ad sp create-for-rbac --name="<ServicePrincipal name>" --
role="Contributor" --scopes="/subscriptions/<subscription Id>"
```

See the following example:

```
az ad sp create-for-rbac --name="SPForTerraform" --role="Contributor" --
scopes="/subscriptions/8921-1444-..."
```

This sample script creates a new service principal named `SPForTerraform` and gives it the contributor permission on the subscription ID, `8921....`

For more details about the Azure CLI command to create an Azure SP, see the documentation: `https://docs.microsoft.com/en-us/cli/azure/create-an-azure-service-principal-azure-cli?view=azure-cli-latest`.

The following screenshot shows the execution of the script that creates an Azure SP:

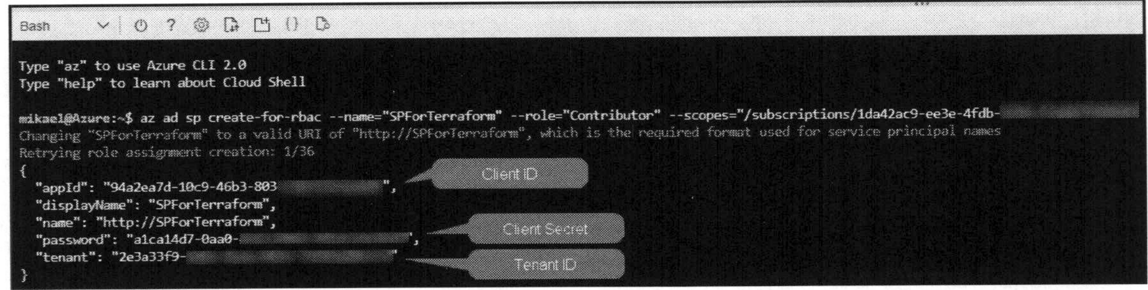

The creation of this service principal returns three pieces of identification information:

- The application ID, also called the client ID
- The client secret
- The tenant ID

And the SP is created in Azure AD. The following screenshot shows the Azure AD SP:

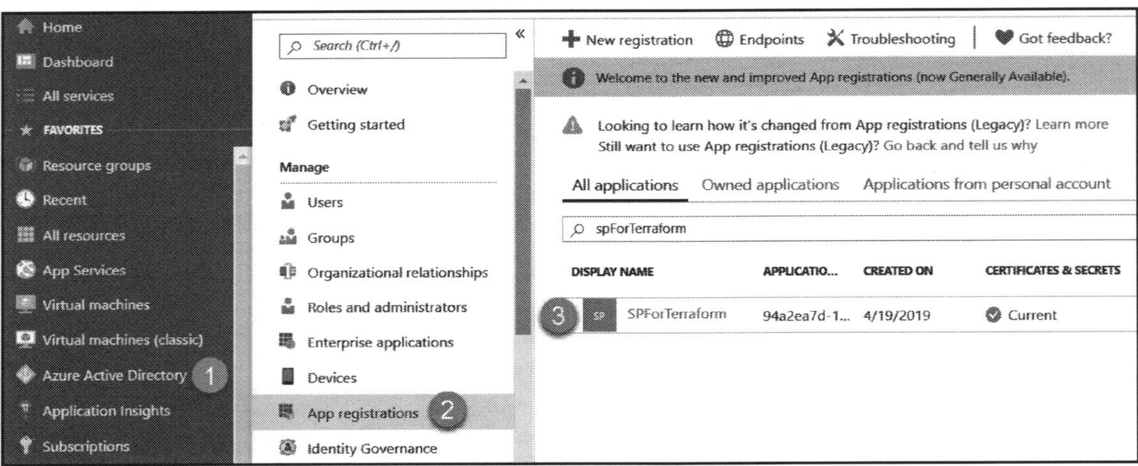

Here, we have just seen how to create a service principal in the Azure AD and we have given it the permission to manipulate the resources of our Azure subscriptions.

Now, let's see how to configure Terraform to use our Azure SP.

Configuring the Terraform provider

Once the Azure SP has been created, we will configure our Terraform code to connect to Azure with this SP. For this, follow these steps:

1. In a directory of your choice, create a new filename, `provider.tf` (`extension.tf` corresponds to Terraform files), which contains the following code:

```
provider "azurerm" {
        subscription_id = "<subscription ID>"
        client_id = "<Client ID>"
        client_secret = "<Client Secret>"
        tenant_id = "<Tenant Id>"
}
```

   In this code, we indicate that the provider we are using is `azurerm` and that the authentication information to Azure is the service principal created. However, for security reasons, it is not advisable to put identification information in code, knowing that this code may be accessible by other people.

2. We will, therefore, improve the previous code by replacing it with this one:

```
provider "azurerm" {}
```

3. So, we delete the credentials in the Terraform code and we will pass the identification values to specific Terraform environment variables:
    - `ARM_SUBSCRIPTION_ID`
    - `ARM_CLIENT_ID`
    - `ARM_CLIENT_SECRET`
    - `ARM_TENANT_ID`

We will see how to set these environment variables later in this chapter, in the *Deploy the infrastructure* section.

As a result, the Terraform code no longer contains any identification information.

We have just seen how to configure Terraform for Azure authentication. We will now explain how to quickly configure Terraform to perform local development and testing.

Terraform configuration for local development and testing

When you work locally and want to test the Terraform code quickly, in a sandbox environment, for example, it may be more convenient and faster to use your own Azure account instead of using an SP.

To do this, it is possible to connect to Azure beforehand with the `az login` command, and then enter your identification information in the window that opens.

The following is a screenshot of the Azure login window:

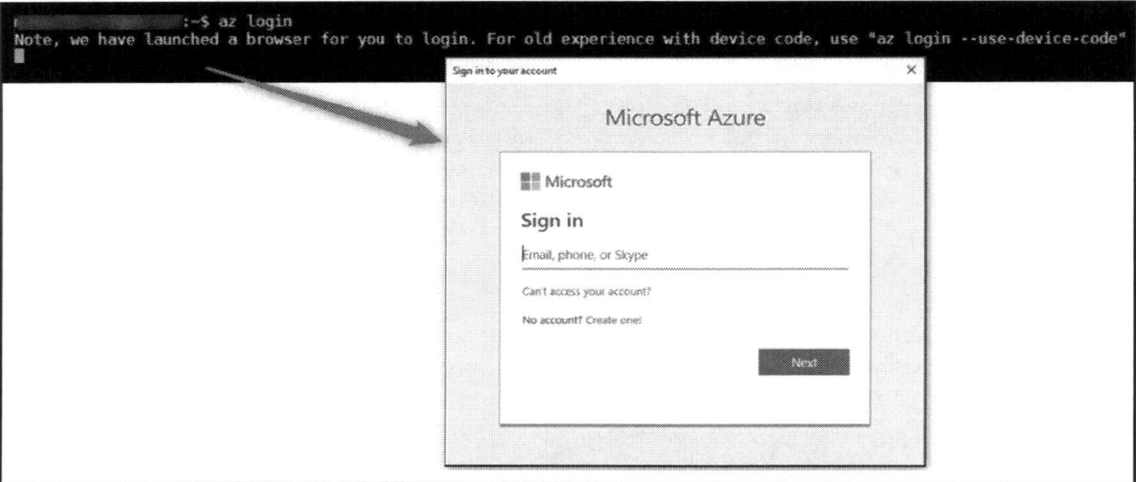

If several subscriptions are accessed, the desired one can be selected with the following command:

```
az account set --subscription="<Subscription ID>"
```

Then, we configure the Terraform provider as before with the *provider*, `"azurerm" { }`.

Of course, this authentication method is not to be done in the case of execution on a remote server.

For more information on the provider configuration, refer to the documentation: `https://www.terraform.io/docs/providers/azurerm/index.html`.

The Terraform configuration for Azure is, therefore, defined by the configuration of the provider that uses the information from an Azure SP.

Once this configuration is complete, we can start writing Terraform code to manage and provision Azure resources.

Writing a Terraform script to deploy Azure infrastructure

To illustrate the use of Terraform to deploy resources in Azure, we will provision a simple Azure architecture with Terraform that is composed of the following:

- There's a group resource.
- There's also a network part composed of a virtual network and a subnet.
- In this subnet, we will create a virtual machine that has a public IP address in order to be publicly available.

For this, in the same directory where we previously created the `provider.tf` file, we will create a `main.tf` file with the following code:

1. Let's start with the code that provides the `resource` group:

```
resource "azurerm_resource_group" "rg" {
    name = "bookRg"
    location = "West Europe"
    tags {
        environment = "Terraform Azure"
    }
}
```

   Any Terraform code is composed of the same syntax model, and the syntax of a Terraform object consists of four parts:

   - A type of `resource` or `data` block
   - A name of the resource to be managed (here, it's `azurerm_resource_group`)
   - An internal Terraform ID (here, it's `rg`)
   - A list of properties that correspond to the real properties of the resource (that is, `name` and `location`)

More documentation on the Terraform syntax is available at `https://www.terraform.io/docs/configuration-0-11/resources.html`.

This code uses the `azurerm_resource_group` Terraform resource and will provision a resource group named `bookRg` that will be stored in the West Europe location.

2. Then, we will write the code for the network part:

```
resource "azurerm_virtual_network" "vnet" {
   name = "book-vnet"
   location = "West Europe"
   address_space = ["10.0.0.0/16"]
   resource_group_name = azurerm_resource_group.rg.name
}

resource "azurerm_subnet" "subnet" {
   name = "book-subnet"
   virtual_network_name = azurerm_virtual_network.vnet.name
   resource_group_name = azurerm_resource_group.rg.name
   address_prefix = "10.0.10.0/24"
}
```

In this Terraform code for the network part, we create the code for a VNet, `book-vnet`, and in it we create a subnet, `book-subnet`.

If we look at this code carefully, we can see that, for dependencies between resources, we do not put in clear IDs, but we use pointers on Terraform resources.

The VNet and subnet are the property of the resource group with `${azurerm_resource_group.rg.name}`, which tells Terraform that the VNet and subnet will be created just after the resource group. As for the subnet, it is dependent on its VNet with the use of the `${azurerm_virtual_network.vnet.name}` value; it's the explicit dependence concept.

Let's now write the Terraform provisioning code of the virtual machine, which is composed of the following:

- A network interface
- A public IP address
- A storage for the diagnostic boot (boot information logs)
- A virtual machine

The sample code for the **network interface** with the IP configuration is as follows:

```
resource "azurerm_network_interface" "nic" {
     name = "book-nic"
     location = "West Europe"
     resource_group_name = azurerm_resource_group.rg.name

     ip_configuration {
         name = "bookipconfig"
         subnet_id = azurerm_subnet.subnet.id
         private_ip_address_allocation = "Dynamic"
         public_ip_address_id = azurerm_public_ip.pip.id
    }
}
```

In this Terraform code, we use an `azurerm_network_interface` block (`https://www.terraform.io/docs/providers/azurerm/r/network_interface.html`), in which we configure the name, region, resource group, and IP configuration with the dynamic IP address of the network interface.

The code for `public ip address`, which has an IP address in the subnet we just created, is as follows:

```
resource "azurerm_public_ip" "pip" {
  name = "book-ip"
  location = "West Europe"
  resource_group_name = "${azurerm_resource_group.rg.name}"
  public_ip_address_allocation = "Dynamic"
  domain_name_label = "bookdevops"
}
```

In this Terraform code, we use an `azurerm_public_ip` block at `https://www.terraform.io/docs/providers/azurerm/r/public_ip.html`, in which we configure the dynamic allocation of the IP address and the DNS label.

The code for `storage account`, which we use for the boot diagnostic logs, is as follows:

```
resource "azurerm_storage_account" "stor" {
  name = "bookstor"
  location = "West Europe"
  resource_group_name = azurerm_resource_group.rg.name
  account_tier = "Standard"
  account_replication_type = "LRS"
}
```

In this Terraform code, we use an `azurerm_storage_account` block at https://www.terraform.io/docs/providers/azurerm/r/storage_account.html, in which we configure the name, region, resource group, and type of storage, which, in our case, is Standard LRS.

The documentation for the storage account is here: https://docs.microsoft.com/en-us/azure/storage/common/storage-account-overview.

And the code for the Ubuntu virtual machine which contains the ID of the network interface created previously, is as follows:

```
resource "azurerm_virtual_machine" "vm" {
  name = "bookvm"
  location = "West Europe"
  resource_group_name = azurerm_resource_group.rg.name
  vm_size = "Standard_DS1_v2"
  network_interface_ids = ["${azurerm_network_interface.nic.id}"]

 storage_image_reference {
     publisher = "Canonical"
     offer = "UbuntuServer"
     sku = "16.04-LTS"
     version = "latest"
 }
....
}
```

In this Terraform code, we use an `azurerm_virtual_machine` block at https://www.terraform.io/docs/providers/azurerm/r/virtual_machine.html, in which we configure the name, size (`Standard_DS1_V2`), reference to the `network_interface` Terraform object, and type of virtual machine OS system (Ubuntu).

All of these code sections are exactly like the previous ones with the use of explicit dependency to specify the relationships between resources.

This complete source code is available here: https://github.com/PacktPublishing/Learning_DevOps/tree/master/CHAP02/01-terraform_simple_script.

We have just created a complete Terraform script that allows us to provision a small Azure infrastructure, but as in any language, there are good practices regarding file separation, applying a clear and readable code, and, finally, the use of built-in functions.

Following some Terraform good practices

We have just seen an example of Terraform code to provision an Azure infrastructure, but it is also useful to look at some good practices for writing Terraform code.

Better visibility with the separation of files

When executing Terraform code, all files in the `execution` directory that have the `.tf` extension are automatically executed; in our example, we have `provider.tf` and `main.tf`. It is good to separate the code into several files in order to improve the readability of the code and its evolution.

Using our example script, we can do better and separate it with the following:

- `Rg.tf`, which contains the code for the resource group
- `Network.tf`, which contains the code for the VNet and subnet
- `Compute.tf`, which contains the code for the network interface, public IP, storage, and virtual machine

The complete code with separate files is here: `https://github.com/PacktPublishing/Learning_DevOps/tree/master/CHAP02/terraform_separate_files`.

Protection of sensitive data

Care must be taken with sensitive data in the Terraform code, such as passwords and accesses permissions. We have already seen that, for access authentication to Azure, it is not necessary to leave them in the code. Also, in our example concerning the administrator account of the VM, we can notice that the password of the admin account of the virtual machine is specified clearly in this Terraform code. To remedy this, we can use a strong password chest to store passwords, such as Azure Key Vault or HashiCorp Vault, and recover them via Terraform.

Dynamizing the code with variables and interpolation functions

When writing Terraform code, it is important to take into account from the beginning that the infrastructure that will host an application is very often the same for all stages, but only some information will vary from one stage to another, such as the name of the resources and the number of instances.

To give more flexibility to the code, we must use variables in the code with the following steps:

1. Declare the variables by adding the following sample code in the global Terraform code, or we can add it in another file (such as `variables.tf`) for better readability of the code:

```
variable "resource_group_name" {
    description ="Name of the resource group"
}

variable "location" {
    description ="Location of the resource"
    default ="West Europe"
}

variable "application_name" {
    description ="Name of the application"
}
```

2. Instantiate their values in another file, `.tfvars`, named `terraform.tfvars` with the syntax, `variable_name=value`, like the following:

```
resource_group_name ="bookRg"
application_name ="book"
```

3. Use these variables in code with `var.<name of the variable>`; for example, in the resource group Terraform code, we can write the following:

```
resource "azurerm_resource_group" "rg" {
     name = var.resoure_group_name
     location = var.location
     tags {
         environment = "Terraform Azure"
     }
}
```

In addition, Terraform has a large list of built-in functions that can be used to manipulate data or variables. To learn more about these functions, consult the documentation here: `https://www.terraform.io/docs/configuration/functions.html`.

Of course, there are many other good practices, but these are to be applied from the first lines of code to make sure that your code is well maintained.

The complete final code of this sample is available here: `https://github.com/PacktPublishing/Learning_DevOps/tree/master/CHAP02/03-terraform_vars_interp`.

We have written a Terraform script, with best practices, which allows us to create a simple cloud infrastructure in Azure that provides a network and a virtual machine. Let's now look at how to run Terraform with our code for provisioning this infrastructure.

Deploying the infrastructure with Terraform

With the Terraform code written, we now need to run Terraform to deploy our infrastructure.

However, before any execution, it is necessary to first provide authentication with the Azure SP to ensure that Terraform can manage Azure resources.

For this, we can either set the environment variables specific to Terraform to contain the information of the SP created earlier in the *Configuring Terraform for Azure* section or use the `az cli` script.

The following script exports the four Terraform environment variables in the Linux OS:

```
export ARM_SUBSCRIPTION_ID=xxxxx-xxxxx-xxxx-xxxx
export ARM_CLIENT_ID=xxxxx-xxxxx-xxxx-xxxx
export ARM_CLIENT_SECRET=xxxxxxxxxxxxxxxxx
export ARM_TENANT_ID=xxxxx-xxxxx-xxxx-xxxx
```

We can also use the `az cli` script with the `login` command:

```
az login
```

Once authenticated, we can run Terraform.

In our scenario, for example, we start with an empty Azure subscription without any group resources, but in the real world, our subscription may already contain group resources.

Before running Terraform, in the Azure portal, check that we do not have a resource group in our subscription:

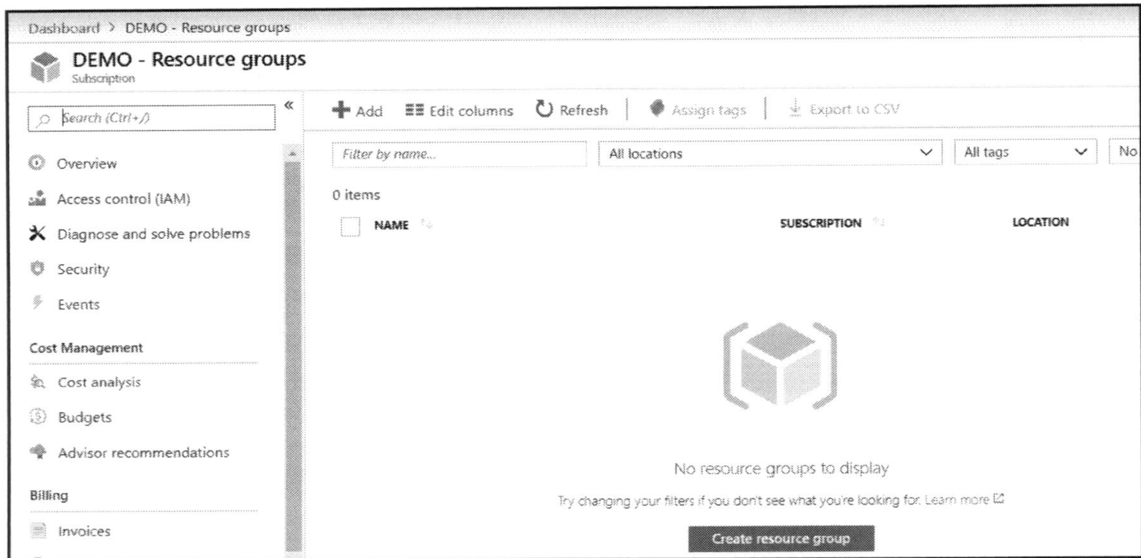

To run Terraform, we must open a command-line Terminal such as CMD, PowerShell, or Bash and go to the directory where the Terraform code files we wrote are located.

The Terraform code is executed in several steps such as initialization, the preview of changes, and the application of those changes.

Let's look in detail at the execution of these steps next, starting with the initialization step.

Initialization

The initialization step allows Terraform to do the following:

- Initialize the Terraform context to check and make the connection between the Terraform provider and remote service—in our case, with Azure.
- Download the plugin(s) of the provider(s)—in our case, it will be the `azurerm` provider.
- Check the code variables.

To execute the initialization, run the `init` command:

```
terraform init
```

The following is a screenshot of `terraform init`:

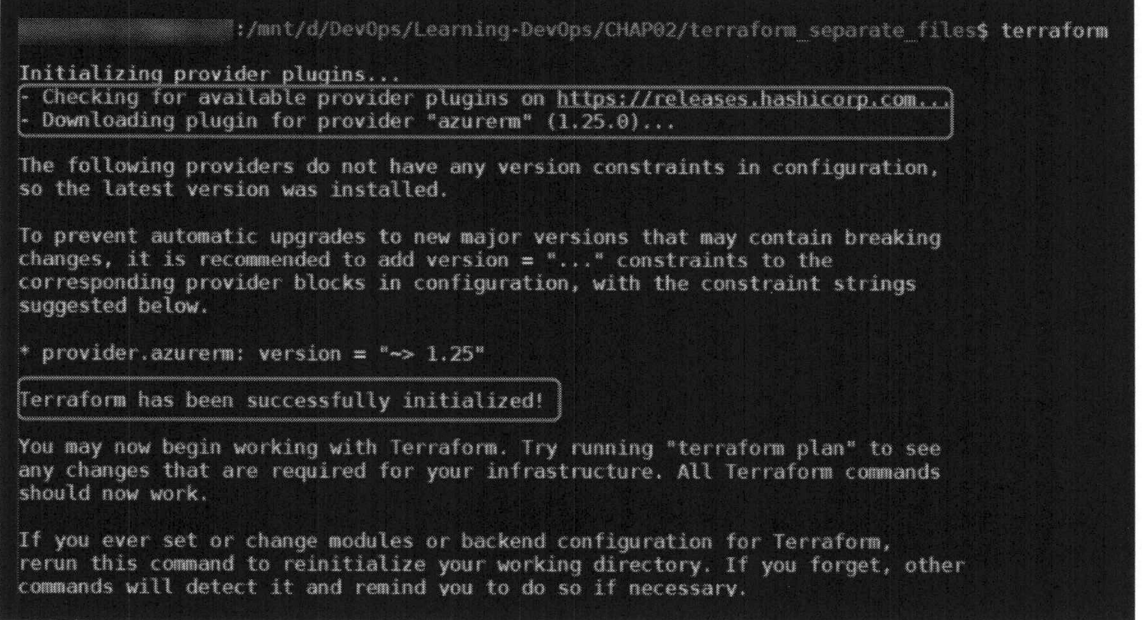

As can be seen during its execution of the preceding command, Terraform does the following:

- It downloads the latest version of the `azurerm` plugin.
- It creates a working `.terraform` directory.

The following is a screenshot of the `.terraform` directory:

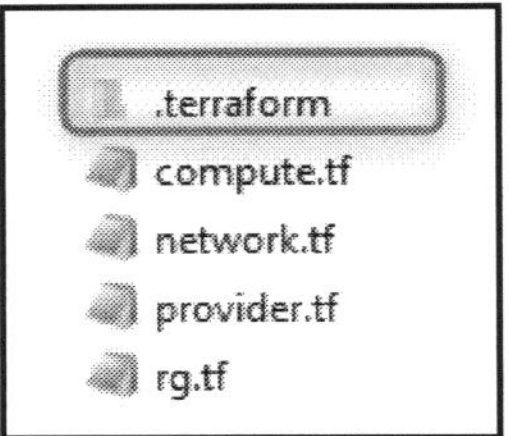

For more information about the `init` command line, see the documentation: `https://www.terraform.io/docs/commands/init.html`.

Once the initialization step is done, we can proceed to the next step, which is previewing the changes.

Previewing changes

The next step is the preview of the changes made to the infrastructure before applying them.

For this, run Terraform with the `plan` command and, when executed, the plan automatically uses the `terraform.tfvars` file to set the variables.

To execute it, launch the `plan` command:

```
terraform plan
```

The following output shows the execution of the `terraform plan` command:

```
:/mnt/d/DevOps/Learning-DevOps/CHAP02/terraform_separate_files$ terraform plan
Refreshing Terraform state in-memory prior to plan...
The refreshed state will be used to calculate this plan, but will not be
persisted to local or remote state storage.


------------------------------------------------------------------------

An execution plan has been generated and is shown below.
Resource actions are indicated with the following symbols:
  + create

Terraform will perform the following actions:

  + azurerm_network_interface.nic
      id:                                                              <computed>
      applied_dns_servers.#:                                           <computed>
      dns_servers.#:                                                   <computed>
      enable_accelerated_networking:                                   "false"
      enable_ip_forwarding:                                            "false"
      internal_dns_name_label:                                         <computed>
      internal_fqdn:                                                   <computed>
      ip_configuration.#:                                              "1"
      ip_configuration.0.application_gateway_backend_address_pools_ids.#: <computed>
      ip_configuration.0.application_security_group_ids.#:             <computed>
      ip_configuration.0.load_balancer_backend_address_pools_ids.#:    <computed>
      ip_configuration.0.load_balancer_inbound_nat_rules_ids.#:        <computed>
      ip_configuration.0.name:                                         "bookipconfig"
      ip_configuration.0.primary:                                      <computed>
      ip_configuration.0.private_ip_address_allocation:                "dynamic"
      ip_configuration.0.private_ip_address_version:                   "IPv4"
      ip_configuration.0.public_ip_address_id:                         "${azurerm_public_ip.pip.id}"
      ip_configuration.0.subnet_id:                                    "${azurerm_subnet.subnet.id}"
      location:                                                        "westeurope"

  + azurerm_virtual_network.vnet
      id:                                                              <computed>
      address_space.#:                                                 "1"
      address_space.0:                                                 "10.0.0.0/16"
      location:                                                        "westeurope"
      name:                                                            "book-vnet"
      resource_group_name:                                             "bookRg"
      subnet.#:                                                        <computed>
      tags.%:                                                          <computed>


Plan: 7 to add, 0 to change, 0 to destroy.

------------------------------------------------------------------------

Note: You didn't specify an "-out" parameter to save this plan, so Terraform
can't guarantee that exactly these actions will be performed if
```

During the execution of the `plan` command, the command displays the name and properties of the resources that will be impacted by the change, with also the number of new resources and the number of resources that will be modified, as well as the number of resources that will be deleted.

For more information about the `plan` command line, see the documentation: `https://www.terraform.io/docs/commands/plan.html`.

We have, therefore, just predicted the changes that will be applied to our infrastructure; we will now see how to apply them.

Applying the changes

After having validated that the result of the `plan` command corresponds to our expectations, the last step is the application of the Terraform code in real time to provision and apply the changes to our infrastructure.

For this, we execute the `apply` command:

```
terraform apply
```

This command does the same operation as the `plan` command and interactively asks the user for confirmation that we want to implement the changes.

The following is a screenshot of the `terraform apply` confirmation:

```
:/mnt/d/DevOps/Learning-DevOps/CHAP02/terraform_separate_files$ terraform apply

An execution plan has been generated and is shown below.
Resource actions are indicated with the following symbols:
  + create

Terraform will perform the following actions:

  + azurerm_network_interface.nic
      id:                                                   <computed>
      applied_dns_servers.#:                                <computed>


      tags.%:                                               <computed>

Plan: 7 to add, 0 to change, 0 to destroy.

Do you want to perform these actions?
  Terraform will perform the actions described above.
  Only 'yes' will be accepted to approve.

  Enter a value:
```

The confirmation is given by inputting `yes` (or `no` to cancel), then Terraform applies the changes to the infrastructure.

The following is a screenshot of the `terraform apply` execution:

```
  Only 'yes' will be accepted to approve.

  Enter a value: yes

azurerm_resource_group.rg: Creating...
  location:          "" => "westeurope"
  name:              "" => "bookRg"
  tags.%:            "" => "1"
  tags.environment:  "" => "Terraform Azure"

azurerm_virtual_machine.vm: Still creating... (2m30s elapsed)
azurerm_virtual_machine.vm: Creation complete after 2m32s (ID: /subscriptions/1da42a

apply complete! Resources: 7 added, 0 changed, 0 destroyed.
```

The output of the `apply` command displays all actions executed by Terraform, with all changes and the impacted resources. It ends with a summary line that displays the sum of all added, changed, or destroyed resources.

For more information about the `apply` command line, see the documentation: `https://www.terraform.io/docs/commands/apply.html`.

Since the Terraform `apply` command has been executed correctly, we can check in the Azure portal that the resources described in the Terraform code are present.

The following is a screenshot of Azure resources by Terraform:

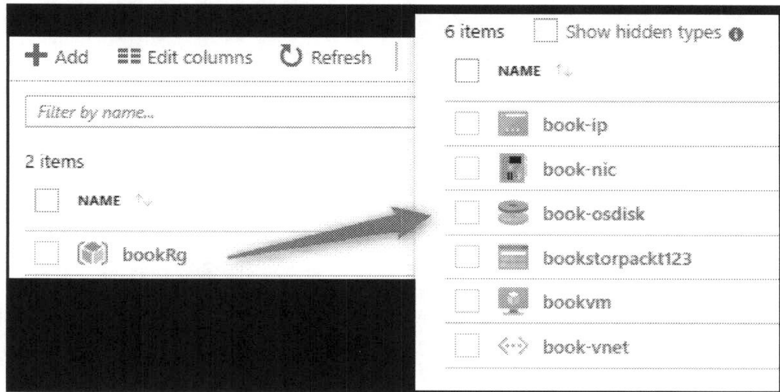

We can see from the portal that the resources specified in the Terraform code have been provisioned successfully.

We have just seen how Terraform is useful for provisioning infrastructure with three main commands:

- The `init` command that initializes the context
- The `plan` command to preview the changes
- The `apply` command to apply the changes

In the next section, we will explore other Terraform commands and the Terraform's life cycle.

Terraform command lines and life cycle

We have just seen that applying changes to an infrastructure with Terraform is mainly done with three commands: the `init`, `plan`, and `apply` commands. But Terraform has other very practical and important commands in order to manage the best life cycle of our infrastructure, and the question of how to execute Terraform in an automation context such as in a CI/CD pipeline must also be considered.

Among the other operations that can be done on an infrastructure is cleaning up resources by removing them, and this is done either to better rebuild or to remove temporary infrastructure.

Using destroy to better rebuild

One of the steps in the life cycle of an infrastructure that is maintained by IaC is the removal of the infrastructure because, let's not forget that one of the objectives and benefits of IaC, is to be able to make rapid changes to an infrastructure but also create environments on demand. That is to say, we create and keep environments as long as we need them, and we will destroy them when they are no longer used and hence allow the company to make financial savings.

To accomplish this, it is necessary to automate the removal of an infrastructure in order to be able to rebuild it quickly.

To destroy an infrastructure previously provisioned with Terraform, execute the following command:

```
terraform destroy
```

The execution of this command gives the following output:

```
:/mnt/d/DevOps/Learning-DevOps/CHAP02/terraform_separate_files$ terraform destroy
azurerm_resource_group.rg: Refreshing state... (ID: /subscriptions/1da42ac9-ee3e-4fdb-b294-f7a607f589d5/resourceGroups/bookRg)
azurerm_storage_account.stor: Refreshing state... (ID: /subscriptions/1da42ac9-ee3e-4fdb-b294-...orage/storageAccounts/bookstorpackt123)
azurerm_public_ip.pip: Refreshing state... (ID: /subscriptions/1da42ac9-ee3e-4fdb-b294-...soft.Network/publicIPAddresses/book-ip)
azurerm_virtual_network.vnet: Refreshing state... (ID: /subscriptions/1da42ac9-ee3e-4fdb-b294-...soft.Network/virtualNetworks/book-vnet)
azurerm_subnet.subnet: Refreshing state... (ID: /subscriptions/1da42ac9-ee3e-4fdb-b294-...Networks/book-vnet/subnets/book-subnet)
azurerm_network_interface.nic: Refreshing state... (ID: /subscriptions/1da42ac9-ee3e-4fdb-b294-...oft.Network/networkInterfaces/book-nic)
azurerm_virtual_machine.vm: Refreshing state... (ID: /subscriptions/1da42ac9-ee3e-4fdb-b294-...crosoft.Compute/virtualMachines/bookvm)

An execution plan has been generated and is shown below.
Resource actions are indicated with the following symbols:
  - destroy

Terraform will perform the following actions:

  - azurerm_network_interface.nic

  - azurerm_public_ip.pip

  - azurerm_resource_group.rg

  - azurerm_storage_account.stor

  - azurerm_subnet.subnet

  - azurerm_virtual_machine.vm

  - azurerm_virtual_network.vnet


Plan: 0 to add, 0 to change, 7 to destroy.

Do you really want to destroy all resources?
  Terraform will destroy all your managed infrastructure, as shown above.
  There is no undo. Only 'yes' will be accepted to confirm.

  Enter a value:
```

This command, just as for `apply`, requires confirmation from the user before applying the destruction:

```
azurerm_resource_group.rg: Still destroying... (ID: /subscriptions/1da42ac9-ee3e-4fdb-b294-f7a607f589d5/resourceGroups/bookRg, 2m40s elapsed)
azurerm_resource_group.rg: Destruction complete after 2m47s

Destroy complete! Resources: 7 destroyed.
```

Once validated, wait for the confirmation message that the infrastructure has been destroyed.

The `destroy` command only destroys the resources configured in the current Terraform code. Other resources (created manually or by another Terraform code) are not affected. However, if our Terraform code provides a resource group, it will destroy all of its content.

For more information about the `destroy` command line, see the documentation: https://www.terraform.io/docs/commands/destroy.html.

We have just seen that Terraform also allows us to destroy resources on the command line. Let's now look at how to format and validate your Terraform code.

Formatting and validating the code

After seeing how to destroy resources with Terraform, it is also important to emphasize the importance of having well-formatted code that meets Terraform's style rules and to validate that the code does not contain syntax or variable errors.

Formatting the code

Terraform has a command that allows the code to be properly aligned with Terraform's styles and conventions.

The following command automatically formats the code:

```
terraform fmt
```

The following is a screenshot of a Terraform-arranged file:

```
/d/Repos/Learning-DevOps/CHAP02/terraform_separate_files# terraform fmt
compute.tf
network.tf
provider.tf
rg.tf
```

The command re-formats the code and indicates the list of arranged files.

For more information on the Terraform style guide, refer to https://www.terraform.io/docs/configuration/style.html, and for information about the `terraform fmt` command line, read https://www.terraform.io/docs/commands/fmt.html.

Validating the code

Along the same lines, Terraform has a command that validates the code and allows us to detect possible errors before executing the `plan` or `apply` command.

Let's take the example of this code extract:

```
resource "azurerm_public_ip" "pip" {
    name = var.ip-name
    location = var.location
    resource_group_name = "${azurerm_resource_group.rg.name}"
    allocation_method = "Dynamic"
    domain_name_label = "bookdevops"
}
```

In the `name` property, we use an `ip-name` variable that has not been declared or instantiated with any value.

Executing the `terraform plan` command would return an error:

```
:/d/Repos/Learning-DevOps/CHAP02/terraform_vars_interp# terraform plan
Error: resource 'azurerm_public_ip.pip' config: unknown variable referenced: 'ip-name'; define it with a 'variable' block
```

And because of this error, in a CI/CD process, it could delay the deployment of the infrastructure.

In order to detect errors in the Terraform code as early as possible in the development cycle, execute the following command, which validates all Terraform files in the directory:

```
terraform validate
```

The following screenshot shows the execution of this command:

```
:/d/Repos/Learning-DevOps/CHAP02/terraform_vars_interp# terraform validate
Error: resource 'azurerm_public_ip.pip' config: unknown variable referenced: 'ip-name'; define it with a 'variable' block
```

We observe the same error as the one returned by the `plan` command.

We have just seen Terraform's main command lines. Let's go a little deeper with the integration of Terraform into a CI/CD process.

Terraform's life cycle in a CI/CD process

So far, we have seen and executed, *on the local machine*, the various Terraform commands that allow us to initialize, preview, apply, and destroy an infrastructure and to format and validate Terraform code. When using Terraform locally, in a development context, its execution life cycle is as follows:

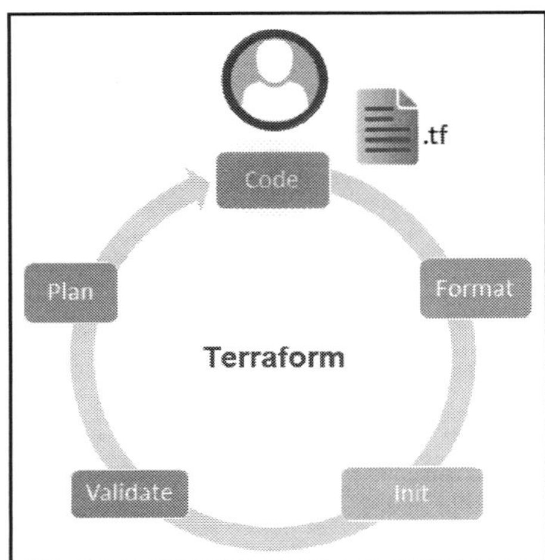

The steps shown in the diagram follow this sequence:

1. Code development
2. Code formatting with `terraform fmt`
3. Initialization with `terraform init`
4. Code validation with `terraform validate`
5. Planning with `terraform plan`
6. Manual verification of Terraform changes on infrastructure

But IaC, like an application, must be deployed or executed in an **automatic CI/CD process**. It starts with the archiving of the Terraform code of the team members, triggers the CI, and executes the Terraform commands that we have studied in this chapter.

The following is a screenshot of the Terraform life cycle in CI/CD automation:

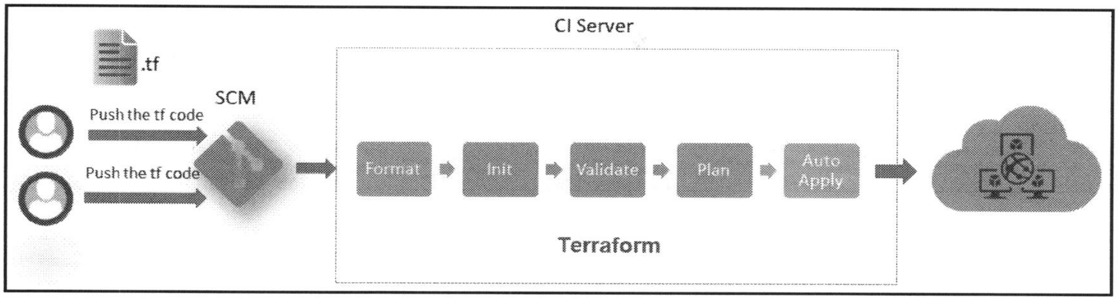

The steps of CI/CD by the CI server (in which Terraform is installed) for Terraform are as follows:

1. Retrieving the code from the SCM
2. Code formatting with `terraform fmt`
3. Initialization with `terraform init`
4. Code validation with `terraform validate`
5. Display a preview of the infrastructure changes with `terraform plan -out=out.tfplan`
6. Application of changes in automatic mode with `terraform apply --auto-approve out.tfplan`

By adding the `--auto-approve` option to the `apply` and `destroy` commands, Terraform can also be executed in automatic mode, so as not to ask for confirmation from the user to validate the changes that need to be applied. With this automation, Terraform can be integrated with CI/CD tools.

In the `plan` command, an `out` option is added to specify a file with the `.tfplan` format that corresponds to a file that contains the output of the `plan` command. This `out.tfplan` file is then used by the `apply` command. The advantage of this procedure is that it is possible to execute the application on a later plan, which can be used in a rollback case.

We have seen, in this section, that, apart from the usual Terraform commands of `init`, `plan`, `apply`, and `destroy`, Terraform also has options that will allow us to improve the readability of the code and validate the code syntax. We also explained that Terraform also allows a perfect integration into a CI/CD pipeline with a life cycle and automation options.

In the next section, we will see what the tfstate file is and how to protect it in a remote backend.

Protecting tfstate in a remote backend

When Terraform handles resources, it writes the state of these resources in a tfstate file. This file is in JSON format and preserves the resources and their properties throughout the execution of Terraform.

By default, this file, called `terraform.tfstate`, is created locally when the first execution of the `apply` command is executed. It will then be used by Terraform each time the `plan` command is executed in order to compare its state (written in this tfstate) with that of the target infrastructure, and hence return the preview of what will be applied.

When using Terraform in an enterprise, this locally stored tfstate file poses many problems:

- Knowing that this file contains the status of the infrastructure, it should not be deleted. If deleted, Terraform may not behave as expected when it is executed.
- It must be accessible at the same time by all members of the team handling resources on the same infrastructure.
- This file can contain sensitive data, so it must be secure.
- When provisioning multiple environments, it is necessary to be able to use multiple tfstate files.

With all of these points, it is not possible to keep this tfstate file locally or even to archive it in an SCM.

To solve this problem, Terraform allows this tfstate file to be stored in a shared and secure storage called the **remote backend**.

Terraform supports several types of remote backends; the list is available here: `https://www.terraform.io/docs/backends/types/remote.html`.

In our case, we will use an **azurerm remote backend** to store our tfstates files with a storage account and a blob for the tfstate file.

We will, therefore, implement and use a remote backend in three steps:

1. The creation of the storage account
2. Terraform configuration for the remote backend
3. The execution of Terraform with the use of this remote backend

Let's look in detail at the execution of these steps:

1. To create an Azure Storage Account and a blob container, we can use either the Azure portal (`https://docs.microsoft.com/en-gb/azure/storage/common/storage-quickstart-create-account?tabs=azure-portal`) or an `az cli` script:

   ```
   # 1-Create resource group
   az group create --name MyRgRemoteBackend --location westeurope

   # 2-Create storage account
   az storage account create --resource-group MyRgRemoteBackend --name storageremotetf --sku Standard_LRS --encryption-services blob

   # 3-Get storage account key
   ACCOUNT_KEY=$(az storage account keys list --resource-group MyRgRemoteBackend --account-name storageremotetf --query [0].value -o tsv)

   # 4-Create blob container
   az storage container create --name tfbackends --account-name storageremotetf --account-key $ACCOUNT_KEY
   ```

   This script creates a `MyRgRemoteBackend` resource group and a storage account, `storageremotetf`.

   Then, the script retrieves the key account from the storage account and creates a blob container, `tfbackends`, in this storage account.

   This script can be run in Azure Cloud Shell, and the advantage of using a script rather than using the Azure portal is that this script can be integrated into a CI/CD process.

2. Then, to configure Terraform to use the previously created remote backend, we must add the configuration section in the `Terraform.tf` file:

   ```
   terraform {
       backend "azurerm" {
           storage_account_name    = "storageremotetfdemo"
           container_name          = "tfbackends"
           key                     = "myappli.tfstate"
       }
   }
   ```

   The `storage_account_name` property contains the name of the storage account, the `container_name` property contains the container name, and the `key` property contains the name of the blob tfstate object.

However, there is still one more configuration information to be provided to Terraform so that it can connect and have permissions on the storage account. This information is the access key, which is a private authentication and authorization key on the storage account. To provide the storage key to Terraform, as with the Azure SP information, set an `ARM_STORAGE_KEY` environment variable with its value.

The following is a screenshot of the Azure storage access key:

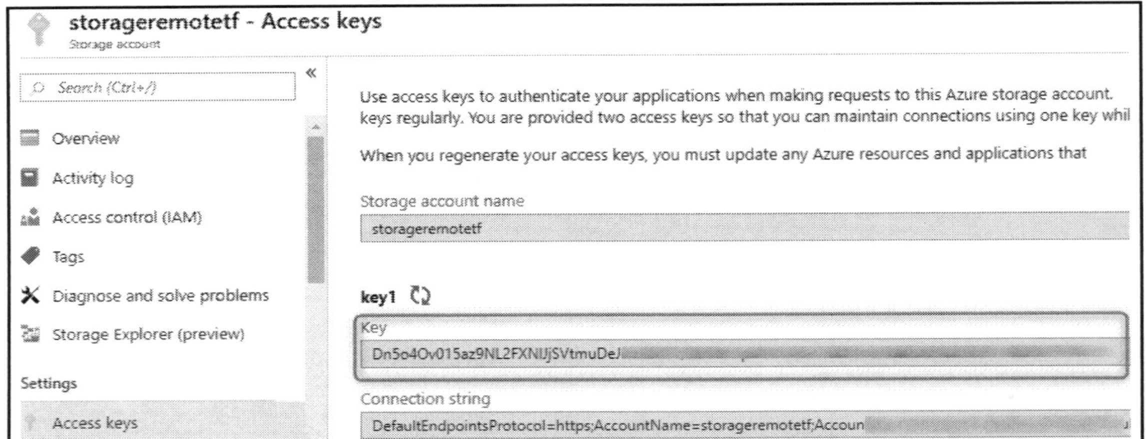

Terraform supports other types of authentication on the storage account such as the use of a SAS token or by using an service principal. For more information on configuring Terraform for an `azurerm` remote backend, refer to the documentation: `https://www.terraform.io/docs/backends/types/azurerm.html`.

3. Finally, once the Terraform configuration is completed, Terraform can be run with this new remote backend. It is during `init` that Terraform initializes the context of the tfstate file and, by default, the `init` command remains unchanged with `terraform init`.

However, if several tfstates in several environments are used, it is possible to create several remote backend configurations with the simplified code in the `.tf` file:

```
terraform {
    backend "azurerm" {}
}
```

Then, create several `backend.tfvars` files that only contain the properties of the backends.

These `backend` properties are the storage account name, the name of the blob container, and the blob name of the tfstate:

```
storage_account_name    = "storageremotetf"
container_name          = "tfbackends"
key                     = "myappli.tfstate"
```

In this case, when executing the `init` command, we can specify the `backend.tfvars` file to use with the following command:

```
terraform init -backend-config="backend.tfvars"
```

The `-backend-config` argument is the path to the backend configuration file.

Personally, I prefer this way of doing things as it allows me to decouple the code by externalizing the values of the backend properties and for better readability of the code. So, here is the execution of Terraform:

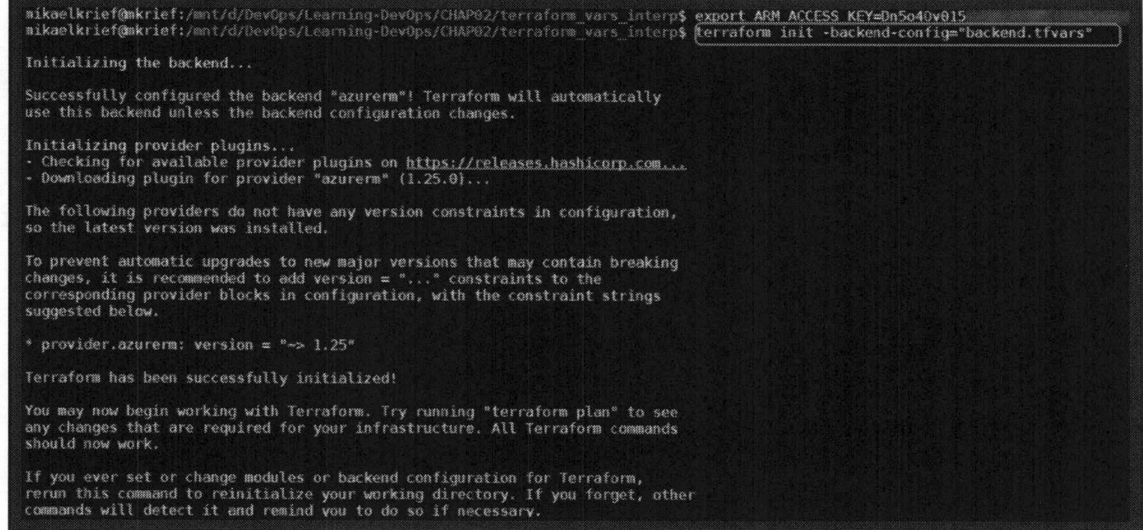

In this execution, we can see the export of the `ARM_ACCESS_KEY` environment variable, as well as the Terraform `init` command that determines the backend configuration with the `-backend-config` option.

With this remote backend, the tfstate file will no longer be stored locally, but on a storage account, which is a shared space. Therefore, it can be used at the same time by several users. This storage account offers, at the same time, security to protect the sensitive data of the tfstate but also the possibilities of backups and restoration of the tfstate files, which are an essential and critical element of Terraform as well.

The entire source code of this chapter is available here: `https://github.com/PacktPublishing/Learning_DevOps/tree/master/CHAP02`, and the final Terraform code is in the `terraform_vars_interp` folder. This provided Terraform code is used for **Terraform 0.12**; the code for Terraform 0.11 is available here: `https://github.com/PacktPublishing/Learning_DevOps/tree/master/CHAP02/terraform_0.11`.

Summary

In this chapter dedicated to Terraform, we have seen that its installation can be done either manually or with scripts.

To apply it, we have detailed the different steps of its configuration to provision an Azure infrastructure using an Azure service principal.

We also explained step by step its local execution with its main command lines, which are `init`, `plan`, `apply`, and `destroy` as well as its life cycle in a CI/CD process. This chapter ends with the protection of the tfstate in an Azure remote backend.

Terraform is, therefore, a tool that is in line with the principles of IaC. The Terraform code is readable and understandable by users, and its execution integrates very well into a CI/CD pipeline that will allow you to provision a cloud infrastructure automatically.

Throughout this book, we will continue to talk about Terraform, with additions about its use with Packer, Azure Kubernetes Services, and downtime reduction.

In the next chapter, we will see the next step of the IaC, which is configuration management by using Ansible. It will cover its installation, its usage for configuring our provisioned virtual machine, and how to protect secrets with Ansible Vault.

Questions

1. What is the language used by Terraform?
2. What is Terraform's role?
3. Is Terraform a scripting tool?
4. Which command allows you to display the installed version?
5. When using Terraform for Azure, what is the name of the Azure object that connects Terraform to Azure?
6. What are the three main orders of Terraform?
7. Which Terraform command allows you to destroy resources?
8. What is the option added to the `apply` command to automate the application of infrastructure changes?
9. What is the purpose of the tfstate file?
10. Is it a good practice to leave the tfstate locally? If not, what should be done?

Further reading

If you want to know more about Terraform, here are some resources:

- Terraform documentation: `https://www.terraform.io/`
- Terraform download and installation: `https://www.terraform.io/downloads.html`
- Terraform Azure provider: `https://www.terraform.io/docs/providers/azurerm/index.html`
- Azure documentation for Terraform: `https://docs.microsoft.com/en-us/azure/terraform/terraform-overview`
- Getting started with Terraform, second edition: `https://www.packtpub.com/networking-and-servers/getting-started-terraform-second-edition`
- Online learning for Terraform: `https://learn.hashicorp.com/terraform`

3

Using Ansible for Configuring IaaS Infrastructure

In the previous chapter, we talked about the provisioning of an Azure cloud infrastructure with Terraform. If this infrastructure contains **virtual machines** (**VMs**), after their provisioning, it is necessary to configure their systems and install all middleware. This configuration will be necessary for the proper functioning of the applications that will be hosted on the VM.

There are several **Infrastructure as Code** (**IaC**) tools available for configuring VMs and the best known are Ansible, Puppet, Chef, SaltStack, and PowerShell DSC. Among them, Ansible from Red Hat (`https://www.ansible.com/overview/it-automation`) stands out for its many assets:

- It is declarative and uses the easy-to-read YAML language.
- Ansible only works with one executable.
- It does not require agents installed on the VMs to be configured.
- A simple SSL/WinRM connection is required for Ansible to connect to remote VMs.
- It has a template engine and a vault to encrypt/decrypt sensitive data.
- It is idempotent.

It should also be noted that Ansible does not only configure VMs, but it can also do infrastructure provisioning and security compliance.

In this chapter, we will see how to install Ansible, and then use it to configure a VM with an inventory and a playbook. We will also see how to protect sensitive data with Ansible Vault and finally, this chapter will discuss the use of a dynamic inventory in Azure.

The following topics are covered in this chapter:

- Installing Ansible
- Creating an Ansible inventory
- Executing the first playbook
- Protecting data with Ansible Vault
- Using a dynamic inventory for an Azure infrastructure

Technical requirements

- To install Ansible, we need an OS: Red Hat, Debian, CentOS, macOS, or any of the BSDs. For those who have Windows, we can install the **Windows Subsystem for Linux** (**WSL**); refer to the documentation here: `https://docs.microsoft.com/en-us/windows/wsl/install-win10`.
- An Ansible playbook uses YAML configuration files, so any code editor would work; however, we will be using Visual Studio Code as it is very suitable. You can download it here: `https://code.visualstudio.com/`.
- Most of this chapter will not focus on a particular cloud provider, except for the last section on Azure. We will need an Azure subscription that we can get for free from here: `https://azure.microsoft.com/en-us/free/`.
- To run the Ansible dynamic inventory for Azure, we need to install the Azure Python SDK: `https://docs.microsoft.com/en-us/azure/python/python-sdk-azure-install?view=azure-python`.
- The complete source code of this chapter is available here: `https://github.com/PacktPublishing/Learning_DevOps/tree/master/CHAP03`.

Check out the following video to see the Code in Action:
`http://bit.ly/2BCZ51B`

Installing Ansible

Before we start using Ansible, we must first know on which OS we can use it and how to install and configure it. Then, we must learn about some of the concepts of the artifacts that it needs to operate.

In this section, we will look at the installation of Ansible in a local or server machine and the integration of Ansible in the Azure Cloud Shell. Then, we will talk about the different elements or artifacts that make up Ansible. Finally, we will end this section with the configuration of Ansible.

To get started, we will see how to download and install Ansible with an automatic script.

Installing Ansible with a script

Unlike Terraform, Ansible is not multiplatform and can only be installed on the following OSes: Red Hat, Debian, CentOS, macOS, or any of the BSDs, and its installation is done by a script that differs according to your OS.

For example, to install its latest version on Ubuntu, we must run the following script in a Bash Terminal:

```
sudo apt-get update
sudo apt-get install software-properties-common
sudo apt-add-repository --yes --update ppa:ansible/ansible
sudo apt-get install ansible
```

This script is also available here: `https://github.com/PacktPublishing/Learning_DevOps/blob/master/CHAP03/install_ansible_ubuntu.sh`.

This script updates the packages, installs the `software-properties-common` dependency, adds the Ansible repository, and finally, installs the latest version of Ansible.

Ansible installation scripts for all distribution types are available here: `https://docs.ansible.com/ansible/latest/installation_guide/intro_installation.html#installing-the-control-machine`.

To install Ansible locally on a Windows OS machine, there is no native solution, but it can be installed on the WSL. WSL allows developers who are on a Windows OS to test their scripts and applications directly on their workstation without having to install a virtual machine.

For more details about WSL, read the documentation here: `https://docs.microsoft.com/en-us/windows/wsl/about`.

To test whether it has been successfully installed, we can run the following command to check its installed version:

```
ansible --version
```

The result of the execution of this command provides some information on the installed version of Ansible, like this:

```
mikael@vmAnsible:~$ ansible --version
ansible 2.8.3
  config file = /etc/ansible/ansible.cfg
  configured module search path = [u'/home/mikael/.ansible/plugins/modules', u'/usr/share/ansible/plugins/modules']
  ansible python module location = /usr/lib/python2.7/dist-packages/ansible
  executable location = /usr/bin/ansible
  python version = 2.7.15+ (default, Oct  7 2019, 17:39:04) [GCC 7.4.0]
```

To display a list of all Ansible commands and options, execute the `ansible` command with the `--help` argument:

```
ansible --help
```

The following screenshot shows the execution of this command:

```
:~$ ansible --help
Usage: ansible <host-pattern> [options]

Define and run a single task 'playbook' against a set of hosts

Options:
  -a MODULE_ARGS, --args=MODULE_ARGS
                        module arguments
  --ask-vault-pass      ask for vault password
  -B SECONDS, --background=SECONDS
                        run asynchronously, failing after X seconds
                        (default=N/A)
  -C, --check           don't make any changes; instead, try to predict some
                        of the changes that may occur
  -D, --diff            when changing (small) files and templates, show the
                        differences in those files; works great with --check
  -e EXTRA_VARS, --extra-vars=EXTRA_VARS
                        set additional variables as key=value or YAML/JSON, if
```

The installation of Ansible on a local machine or a remote machine is therefore quite simple and can be automated by a script. If we deploy an infrastructure in Azure, we can also use Ansible as it is integrated into Azure Cloud Shell.

Let's now look at how Ansible is integrated into Azure Cloud Shell.

Integrating Ansible into Azure Cloud Shell

If we have an Azure subscription, we learned in `Chapter 2`, *Provisioning Cloud Infrastructure with Terraform*, that **Azure Cloud Shell** integrates third-party tools that can be used in Azure without having to install them on a VM. Among these tools is Terraform, which we saw in detail in the previous chapter, but there is also Ansible, which Microsoft has integrated to facilitate the automatic configuration of the VMs that are hosted in Azure.

To use Ansible in Azure Cloud, we must do the following:

1. Connect to the Azure portal at `https://portal.azure.com`.
2. Open Cloud Shell.
3. Choose Bash mode.
4. In the Terminal that opens, we now have access to all Ansible commands.

The following screenshot shows the Ansible command in Azure Cloud Shell:

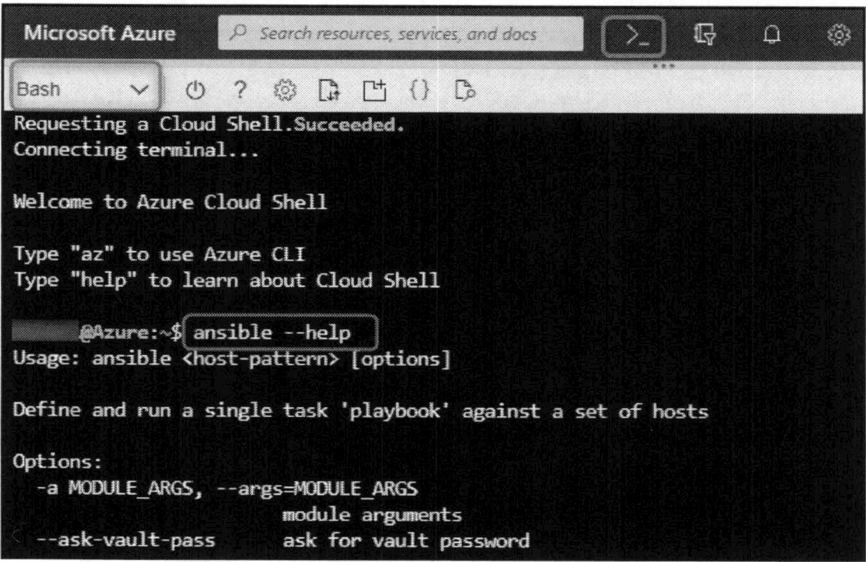

This way, it will be possible to use Ansible for development and testing without installing any software.

Also, Ansible has modules that allow the provisioning of an Azure infrastructure (such as Terraform, but this aspect of Ansible will not be covered in this book), and therefore its integration into Azure Cloud Shell allows a simplified authentication.

Detailed documentation on integrating Ansible into Azure Cloud Shell is available here: `https://docs.microsoft.com/en-us/azure/ansible/ansible-run-playbook-in-cloudshell`.

Before we start using Ansible, we will review the important concepts (or artifacts) of Ansible, which will serve us throughout this chapter.

Ansible artifacts

To configure a system, Ansible needs several main artifacts:

- **The hosts**: These are target systems that Ansible will configure; the host can also be a local system.
- **The inventory**: This is a file in INI or YAML format that contains the list of target hosts on which Ansible will perform configuration actions. This inventory can also be a script, which is the case with a dynamic inventory.

We will see the implementation of the inventory next, in the *Creating an inventory for targeting Ansible host* section, and a dynamic inventory in the *Using a dynamic inventory for Azure infrastructure* section.

- **The playbook**: This is the Ansible configuration script that will be executed to configure hosts.

Writing playbooks will be seen in the *Writing the first playbook* section later in this chapter.

After seeing how to install Ansible, we have just gone through the essential elements of Ansible, which are hosts, inventory, and playbooks. We will now see how to configure Ansible.

Configuring Ansible

By default, the Ansible configuration is in the `/etc/ansible/ansible.cfg` file that is created during the installation of Ansible. This file contains several configuration keys, such as an SSL connection, a user, a protocol, transport, and many others.

This file is created by default during the installation of Ansible and, to guide the user to get started, initial content is placed in it. This content contains a multitude of configuration keys that are commented out so that they are not applied by Ansible but can be activated at any time by the user.

The following screenshot shows an extract from this `/etc/ansible/ansible.cfg` configuration file with some keys in comment with the `#` symbol:

```
# config file for ansible -- https://ansible.com/
# ===============================================

# nearly all parameters can be overridden in ansible-playbook
# or with command line flags. ansible will read ANSIBLE_CONFIG,
# ansible.cfg in the current working directory, .ansible.cfg in
# the home directory or /etc/ansible/ansible.cfg, whichever it
# finds first

[defaults]

# some basic default values...

#inventory      = /etc/ansible/hosts
#library        = /usr/share/my_modules/
#module_utils   = /usr/share/my_module_utils/
#remote_tmp     = ~/.ansible/tmp
#local_tmp      = ~/.ansible/tmp
#plugin_filters_cfg = /etc/ansible/plugin_filters.yml
#forks          = 5
#poll_interval  = 15
#sudo_user      = root
#ask_sudo_pass = True
#ask_pass      = True
#transport      = smart
#remote_port    = 22
#module_lang    = C
#module_set_locale = False
```

If we want to change the default Ansible configuration, we can modify this file.

For more details about all Ansible configuration keys, see the official documentation: https://docs.ansible.com/ansible/latest/reference_appendices/config.html#ansible-configuration-settings.

We can also view and modify this configuration using the `ansible-config` command. For example, for displaying the Ansible configuration file, execute the following command:

```
ansible-config view
```

The following screenshot shows the execution of this command:

```
:/home/m                  # ansible-config view
# config file for ansible -- https://ansible.com/
# ===============================================

# nearly all parameters can be overridden in ansible-playbook
# or with command line flags. ansible will read ANSIBLE_CONFIG,
# ansible.cfg in the current working directory, .ansible.cfg in
# the home directory or /etc/ansible/ansible.cfg, whichever it
# finds first

[defaults]

# some basic default values...

#inventory      = /etc/ansible/hosts
#library        = /usr/share/my_modules/
#module_utils   = /usr/share/my_module_utils/
#remote_tmp     = ~/.ansible/tmp
#local_tmp      = ~/.ansible/tmp
#plugin_filters_cfg = /etc/ansible/plugin_filters.yml
#forks          = 5
#poll_interval  = 15
#sudo_user      = root
#ask_sudo_pass = True
#ask_pass      = True
#transport      = smart
#remote_port    = 22
#module_lang    = C
#module_set_locale = False
```

In this section, we have seen how to install Ansible and then we have explored some of Ansible's artifacts. Finally, we saw different ways to configure Ansible.

In the next section, we will detail a static Ansible inventory and how to create it for targeting hosts.

Creating an inventory for targeting Ansible hosts

The inventory contains the list of hosts on which Ansible will perform administration and configuration actions.

There are two types of inventories:

- **Static inventory**: Hosts are listed in a text file in INI (or YAML) format; this is the basic mode of Ansible inventory. The static inventory is used in cases where we know the host addresses (IP or FQDN).
- **Dynamic inventory**: The list of hosts is dynamically generated by an external script (for example, with a Python script). The dynamic inventory is used in case we do not have the addresses of the hosts, for example, as with an infrastructure that is composed of on-demand environments.

In this section, we will see how to create a static inventory in `init` format, starting with a basic example, and then we will look at the groups and host configuration.

Let's start by seeing how to create a static inventory file.

The inventory file

For Ansible to configure hosts when running the playbook, it needs to have a file that contains the list of hosts, that is, the list of IP or **Fully Qualified Domain Name** (**FQDN**) addresses of the target machines. This list of hosts is noted in a static file called the **inventory file**.

By default, Ansible contains an inventory file created during its installation; this file is `/etc/ansible/hosts`, and it contains several inventory configuration examples. In our case, we will manually create and fill this file in a directory of our choice, such as `devopsansible`.

Let's do this step by step:

1. The first step is the creation of the directory with the following basic command:

```
mkdir devopsansible
cd devopsansible
```

2. Then, let's create a file named `myinventory` (without an extension), in which we write the IP addresses or the FQDN of the targets hosts, as in this example:

```
192.10.14.10
mywebserver.entreprise.com
localhost
```

When Ansible is executed based on this inventory, it will execute all of the requested actions (playbook) on all hosts mentioned in this inventory.

For more information about the inventory file, read the documentation: `https://docs.ansible.com/ansible/latest/user_guide/intro_inventory.html`.

However, in the real usage of Ansible in enterprise, the same Ansible code (or playbook) contains the configuration actions performed for all of the VMs of an application. Since these VMs have different roles within the application, such as an application that consists of one (or more) web server and one database server, we must divide our inventory to properly separate the VMs by functional roles.

To group VMs by role, in the inventory, we will organize our VMs into groups that will be noted between `[]`, which gives us the following inventory:

```
[webserver]
192.10.20.31
mywebserver.exemple.com

[database]
192.20.34.20
```

In this example, we have defined two groups, `webserver` and `database`, and all hosts are distributed in each of their groups.

For another example, we can also group the hosts by environments with this sample inventory:

```
[dev]
192.10.20.31
192.10.20.32

[qa]
192.20.34.20
192.20.34.21

[prod]
192.10.12.10
192.10.12.11
```

We will see later in this chapter how these groups will be used in playbook writing.

Now, let's look at how to complete our inventory with the configuration of hosts.

Configuring hosts in the inventory

As we have seen, the entire Ansible configuration is in the `ansible.cfg` file. However, this configuration is generic and applies to all Ansible executions as well as connectivity to hosts.

However, when using Ansible to configure VMs from different environments or roles with different permissions, it is important to have different connectivity configurations, such as different admin users and SSL keys per environment. For this reason, it is possible to override the default Ansible configuration in the inventory file by configuring specific parameters per host defined in this inventory.

The main configuration parameters that can be overridden are as follows:

- `ansible_user`: This is the user who connects to the remote host.
- `ansible_port`: It is possible to change the default value of the SSH port.
- `ansible_host`: This is an alias for the host.

- `ansible_connection`: This is the type of connection to the remote host and can be Paramiko, SSH, or local.
- `ansible_private_key_file`: This is the private key used to connect to the remote host.

The complete list of parameters is available in the following documentation: https://docs.ansible.com/ansible/latest/user_guide/intro_inventory.html#list-of-behavioral-inventory-parameters.

Here is an example of an inventory in which we configure the connection of hosts:

```
[webserver]
webserver1 ansible_host=192.10.20.31 ansible_port=2222
webserver2 ansible_host=192.10.20.31 ansible_port=2222

[database]
database1 ansible_host=192.20.34.20 ansible_user=databaseuser
database2 ansible_host=192.20.34.21 ansible_user=databaseuser

[dev]
webserver1
database1

[qa]
webserver2
database2
```

The following can be seen in this inventory example:

- The connection information has been specified beside each host.
- The alias implementation (such as `webserver1` and `webserver2`) is used in another group (such as the `qa` group in this example).

Having completed the implementation of an Ansible inventory, we will now look at how to test this inventory.

Testing the inventory

Once the inventory is written, it is possible to test whether all of the hosts mentioned are accessible from Ansible. To do this, we can execute the following command:

```
ansible -i inventory all -u demobook -m ping
```

The `-i` argument is the path of the inventory file, the `-u` argument corresponds to the remote user name used to connect to the remote machine, and `-m` is the command to execute—here, we execute the `ping` command on all machines in the inventory.

The following screenshot shows the execution of this command:

```
/devopsansible# ansible -i inventory  all -u demobook -m ping
webserver2 | SUCCESS => {
    "changed": false,
    "ping": "pong"
}
database1 | SUCCESS => {
    "changed": false,
    "ping": "pong"
}
webserver1 | SUCCESS => {
    "changed": false,
    "ping": "pong"
}
```

We can also test connectivity only on hosts of a particular group by calling this command with the group name instead of `all`. For example, in our case, we will execute this command:

```
ansible -i inventory webserver -u demobook -m ping
```

The following screenshot shows the execution of this command:

```
/devopsansible# ansible -i inventory  webserver -u demobook -m ping
webserver1 | SUCCESS => {
    "changed": false,
    "ping": "pong"
}
webserver2 | SUCCESS => {
    "changed": false,
    "ping": "pong"
}
```

In this section, we learned that Ansible needs an inventory file to configure hosts. Then, we created and tested our first inventory file and finally, we saw how to go further in the configuration of this file.

In the next section, we will see how to set up and write the configuration action code in the Ansible playbooks.

Writing the first playbook

One of the essential elements of Ansible is the playbook because, as stipulated in the introduction, it contains the code of the actions or tasks that need to be performed to configure or administer a VM.

Indeed, once the VM is provisioned, it must be configured, with the installation of all of the middleware needed to run the applications that will be hosted on this VM. Also, it is necessary to perform administrative tasks concerning the configuration of directories and their access.

In this section, we will see what a playbook is made up of, its modules, and how to improve our playbook with roles.

Now, let's start studying how to write a basic playbook.

Writing a basic playbook

The code of a playbook is written in YAML, a declarative language that allows us to easily visualize the configuration steps.

To understand what a playbook looks like, let's look at a simple and classic example that is the installation of an nginx server on an Ubuntu VM. Previously, we created a working `devopsansible` directory, inside which we will create a `playbook.yml` file and insert the following content code:

```
---
- hosts: all
  tasks:
  - name: install and check nginx latest version
      apt: name=nginx state=latest
  - name: start nginx
     service:
     name: nginx
     state: started
---
```

Let's take a look at this in detail:

- First of all, the YAML file starts and ends with the optional `---` characters.
- The `- hosts` property contains the list of hosts to configure. Here, we have written the value to this property as `all` to install nginx on all of the VMs listed in our inventory. If we want to install it only on a particular group, for example, on the `webserver` group, we will note it as follows:

```
---
- hosts: webserver
```

- Then, we indicate the list of tasks or actions to be performed on these VMs, with the property of the list of tasks.
- Under the `tasks` element, we describe the list of tasks and, for each of them, a name that serves as a label, in the `name` property. Under the name, we call the function to be executed using the *Ansible modules* and their properties. Here, in our example, we have used two modules:
    - `apt`: This allows us to retrieve a package (the `apt-get` command) to get the latest version of the `nginx` package.
    - `service`: This allows us to start or stop a service—in this example, to start the nginx service.

What we notice is that it does not require any knowledge of development or IT scripting to use Ansible; the important thing is to know the list of actions to perform on VMs to configure them. The Ansible playbook is, therefore, a sequence of actions that are encoded in Ansible modules.

We have just seen that the tasks used in playbooks use modules, so I now propose a brief overview of the modules and their use.

Understanding Ansible modules

In the previous section, we learned that, in Ansible playbooks, we use modules. This has made Ansible so popular today that there is a huge list of public modules provided by Ansible natively (+200). The complete list is available here: `https://docs.ansible.com/ansible/latest/modules/list_of_all_modules.html`.

These modules allow us to perform all of the tasks and operations to be performed on a VM for its configuration and administration without having to write any lines of code or scripts.

Within an enterprise, we can also create our custom modules and publish them in a private registry internally. More information can be found here: `https://docs.ansible.com/ansible/latest/dev_guide/developing_modules_general.html`.

After seeing the writing of a simple playbook and the use of modules, we will now improve it with roles.

Improving your playbooks with roles

Within an enterprise, when configuring a VM, we notice a certain repetition of tasks for each application, for example, several applications require the identical installation of nginx, which must be performed in the same way.

With Ansible, this repetition will require duplicating the playbook code, seen in our playbook example in the *Writing a basic playbook* section, between several playbooks (because each application contains its playbook). To avoid this duplication and, hence, save time, avoid errors, and homogenize installation and configuration actions, we can encapsulate the playbook code in a directory called `role` that can be used by several playbooks.

To create the `nginx` role corresponding to our example, we will create the following directory and file tree within our `devopsansible` directory:

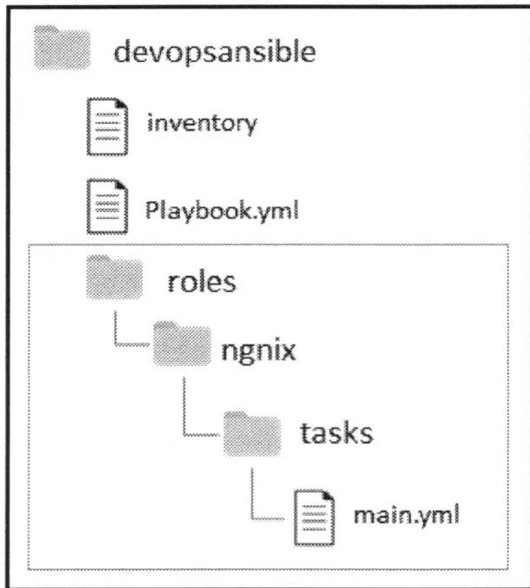

Then, in the `main.yml` file, which is located in `tasks`, we will copy and paste the following code from our playbook in the file that is created:

```
- name: install and check nginx latest version
    apt: name=nginx state=latest
- name: start nginx
     service:
     name: nginx
     state: started
```

Then, we will modify our playbook to use this role with the following content:

```
---
- hosts: webserver
  roles:
    - nginx
```

Following the node roles, we provide a list of roles (names of the `role` directories) to be used. So, this `nginx` role is now centralized and can be used in several playbooks simply without having to rewrite its code.

The following is the code of a playbook that configures a VM web server with Apache and another VM that contains a MySQL database:

```
---
- hosts: webserver
  roles:
    - php
    - apache
- hosts: database
  roles:
    - mysql
```

For more information on role creation, read the official documentation at `https://docs.ansible.com/ansible/latest/user_guide/playbooks_reuse_roles.html`.

However, before we start creating a role, we can use Ansible Galaxy (`https://galaxy.ansible.com/`), which contains a large number of roles provided by the community and covers a high number of configuration and administration needs.

Within an enterprise, we can also create custom roles and publish them in a private galaxy within the company. More information can be found here: `https://docs.ansible.com/ansible/latest/dev_guide/developing_modules_general.html`.

In this section, we have detailed the writing of a playbook as well as its improvement with the use of roles. All of our artifacts are finally ready, so we will now be able to execute Ansible.

Executing Ansible

We have so far seen the installation of Ansible, listed the hosts in the inventory, and set up our Ansible playbook; now, we can run Ansible to configure our VMs.

For this, we will run the Ansible tool with the `ansible-playbook` command, like this:

```
ansible-playbook -i inventory playbook.yml
```

The basic options of this command are as follows:

- The `-i` argument with the inventory file path
- The path of the playbook file

The following is the execution of this command:

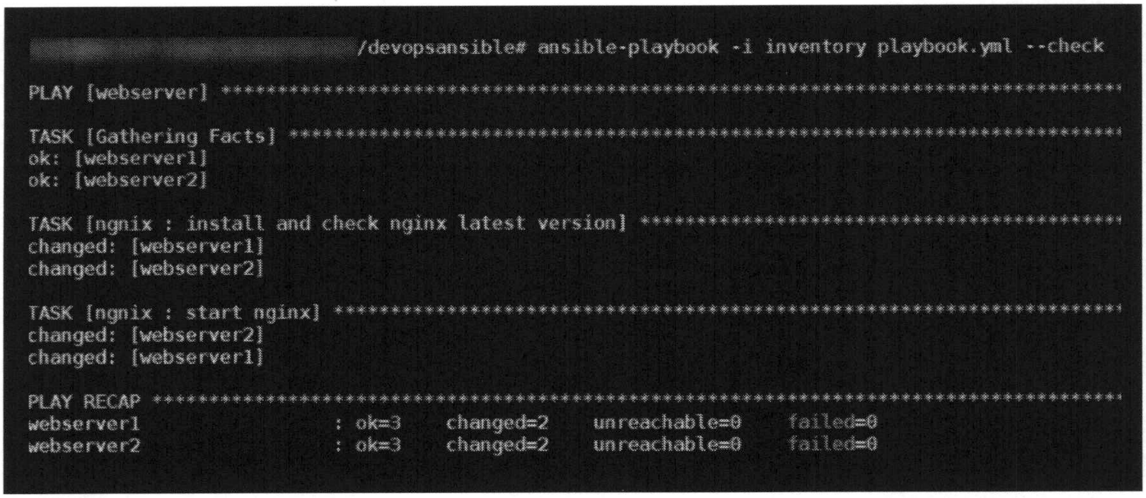

```
/devopsansible# ansible-playbook -i inventory playbook.yml --check

PLAY [webserver] ****************************************************************************

TASK [Gathering Facts] **********************************************************************
ok: [webserver1]
ok: [webserver2]

TASK [ngnix : install and check nginx latest version] ***************************************
changed: [webserver1]
changed: [webserver2]

TASK [ngnix : start nginx] ******************************************************************
changed: [webserver2]
changed: [webserver1]

PLAY RECAP **********************************************************************************
webserver1                 : ok=3    changed=2    unreachable=0    failed=0
webserver2                 : ok=3    changed=2    unreachable=0    failed=0
```

The execution of this command applies the playbook to the hosts in the inventory in several steps:

1. Gathering facts: Ansible checks that the hosts are reachable.
2. The tasks' playbook is executed on hosts.
3. `PLAY Recap`: This is the status of the changes that were executed on each host; the value of this status can be as follows:

`ok`	This is the number of playbook tasks that have been correctly applied to the host.
`changed`	This is the number of changes applied.
`unreachable`	The host is unreachable.
`failed`	Execution failed on this host.

If we need to upgrade our playbook to add or modify middleware on our VMs, during the second execution of Ansible with this upgraded playbook, it can be noticed that Ansible did not reapply the complete configuration of the VMs, but only applied the differences.

The following screenshot shows the second execution of Ansible with no changes made to our playbook:

```
/devopsansible# ansible-playbook -i inventory playbook.yml

PLAY [webserver] ****************************************************************

TASK [Gathering Facts] **********************************************************
ok: [webserver1]
ok: [webserver2]

TASK [ngnix : install and check nginx latest version] ***************************
ok: [webserver2]
ok: [webserver1]

TASK [ngnix : start nginx] ******************************************************
ok: [webserver1]
ok: [webserver2]

PLAY RECAP **********************************************************************
webserver1                 : ok=3    changed=0    unreachable=0    failed=0
webserver2                 : ok=3    changed=0    unreachable=0    failed=0
```

We can see in the recap step that Ansible didn't change anything on the hosts (`changed=0`).

We can also add some useful options to this command to provide the following:

- A preview of Ansible changes before applying the changes
- More logs in the execution output

These options are not only important for the playbook development phase, but also for debugging them in case of errors during their execution.

Now, let's look at how to use these preview options.

Using the preview or dry run option

When coding an Ansible playbook, we often need to test different steps without applying them directly to infrastructure. Hence, it is very useful, especially in the automation of VM configuration with Ansible, to have a preview of its execution, to check that the syntax of the playbook is maintaining good consistency with the system configuration that already exists on the host.

With Ansible, it's possible to check the execution of a playbook on hosts by adding the `--check` option to the command:

```
ansible-playbook -i inventory playbook.yml --check
```

Here is an example of this dry run execution:

```
/devopsansible# ansible-playbook -i inventory playbook.yml --check

PLAY [webserver] ****************************************************************************

TASK [Gathering Facts] **********************************************************************
ok: [webserver1]
ok: [webserver2]

TASK [ngnix : install and check nginx latest version] ***************************************
changed: [webserver1]
changed: [webserver2]

TASK [ngnix : start nginx] ******************************************************************
changed: [webserver2]
changed: [webserver1]

PLAY RECAP **********************************************************************************
webserver1                 : ok=3    changed=2    unreachable=0    failed=0
webserver2                 : ok=3    changed=2    unreachable=0    failed=0
```

With this option, Ansible does not apply configuration changes on the host; it only checks and previews the changes made to the hosts.

For more information on the `--check` option, refer to the following documentation: `https://docs.ansible.com/ansible/latest/user_guide/playbooks_checkmode.html`.

We have just seen that Ansible allows us to check a playbook before applying it only on a host; it is also necessary to know that there are other tools to test the good functioning of a playbook (without having to simulate its execution), such as Vagrant by HashiCorp.

Vagrant allows us to locally create a test environment composed of VMs very quickly, on which we can really run our playbooks and see the real results of their executions. For more information on the use of Ansible and Vagrant, refer to the documentation here: `https://docs.ansible.com/ansible/latest/scenario_guides/guide_vagrant.html`.

We have just seen how to preview the changes that will be applied by Ansible. Now, let's look at how to increase the log level output of Ansible's execution.

Increasing the log level output

In case of error, it is possible to add more logs during the output by adding the `-v`, `-vvv`, or `-vvvv` option to the Ansible command.

The `-v` option enables the basic verbose mode, the `-vvv` option enables the verbose mode with more outputs, and the `-vvvv` option adds the verbose mode and the connection debugging information.

Executing the following command applies a playbook and will display more log information using the `-v` option that has been added:

```
ansible-playbook -i inventory playbook.yml -v
```

This can be useful for debugging in case of Ansible errors.

The complete documentation on the `ansible-playbook` command is available here: `https://docs.ansible.com/ansible/2.4/ansible-playbook.html`.

We have just studied the execution of Ansible with its inventory and playbook by exploring some options that allow the following:

- Previewing the changes that will be made by Ansible
- Increasing the level of logs to make debugging easier

In the next section, we will talk about data security with the use of Ansible Vault.

Protecting data with Ansible Vault

We have seen so far how to use Ansible with an inventory file that contains the list of hosts to configure, and with a playbook that contains the code of the host's configuration actions. But in all IaC tools, it will be necessary to extract some data that is specific to a context or environment inside variables.

In this section, we will look at how to use variables in Ansible and how to protect sensitive data with Ansible Vault.

To illustrate this use and protection of variables, we will complete our example with the installation of a MySQL server on the database server.

Let's begin by looking at the use and utility of variables in Ansible.

Using variables in Ansible for better configuration

When deploying infrastructure with IaC, the code used is often composed of two parts:

- A part that describes the elements or resources that make up the infrastructure
- Another part that differentiates the properties of this infrastructure from one environment to another

This second part of differentiation for each environment is done with the use of variables, and Ansible has a whole system that allows us to inject variables into playbooks.

To see the use of variables in Ansible, we will complete our code and add a role called `mysql` in the `roles` directory with the following tree structure:

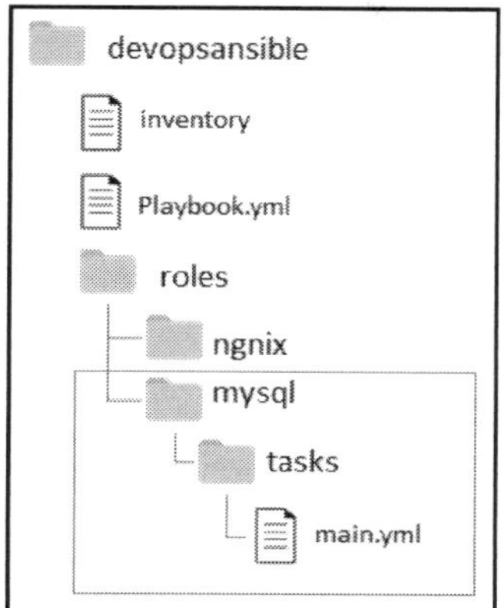

In the `main.yml` file of this role, we write the following code:

```
---
- name: Update apt cache
  apt: update_cache=yes cache_valid_time=3600
- name: Install required software
  apt: name="{{ packages }}" state=present
  vars:
  packages:
  - python-mysqldb
  - mysql-server
- name: Create mysql user
  mysql_user:
    name={{ mysql_user }}
    password={{ mysql_password }}
    priv=*.*:ALL
    state=present
```

In this code, some static information has been replaced by variables. They are as follows:

- `packages`: This contains a list of packages to install and this list is defined in the following code.
- `mysql_user`: This contains the user admin of the MySQL database.
- `mysql_password`: This contains the admin password of the MySQL database.

The different tasks of this role are as follows:

1. Updating packages
2. The installation of the MySQL server and Python MySQL packages
3. The creation of a MySQL user

The complete source code of this role is available at `https://github.com/PacktPublishing/Learning_DevOps/tree/master/CHAP03/devopsansible/roles/mysql`.

As we can see, in the user creation task, we have put in the `mysql_user` and `mysql_password` variables for the name and password. Hence, this information may be different depending on the environment, or may be instantiated dynamically when running Ansible.

To define the values of these variables, we will create a `group_vars` directory, which will contain all of the values of variables for each group defined in our inventory.

Then, in this `group_vars` folder, we create a `database` subdirectory corresponding to the database group defined in the inventory and a `main.yml` subfile.

In this `main.yml` file, we put the desired values of these variables as follows:

```
---
mysql_user: mydbuserdef
mysql_password: mydbpassworddef
```

Finally, we complete our playbook with a call to the `mysql` role by adding the following code:

```
- hosts: database
become: true
roles:
  - mysql
```

We execute Ansible with the same command as the previous one, `ansible-playbook -i inventory playbook.yml`, and this is the output generated:

```
root@DESKTOP-9Q2U73J:/d/devopsansible# ansible-playbook -i inventory playbook.yml
PLAY [webserver] ***************************************************************************

TASK [Gathering Facts] *********************************************************************
ok: [webserver1]
ok: [webserver2]

TASK [ngnix : install and check nginx latest version] **************************************
ok: [webserver1]
ok: [webserver2]

TASK [ngnix : start nginx] *****************************************************************
ok: [webserver1]
ok: [webserver2]

PLAY [database] ****************************************************************************

TASK [Gathering Facts] *********************************************************************
ok: [database1]

TASK [mysql : Update apt cache] ************************************************************
ok: [database1]

TASK [mysql : Install required software] ***************************************************
changed: [database1]

TASK [mysql : Create mysql user] ***********************************************************
changed: [database1]

PLAY RECAP *********************************************************************************
database1                  : ok=4    changed=2    unreachable=0    failed=0
webserver1                 : ok=3    changed=0    unreachable=0    failed=0
webserver2                 : ok=3    changed=0    unreachable=0    failed=0
```

Ansible has therefore updated the database server with two changes that are the list of packages to be installed and the MySQL admin user. We have just seen how to use variables in Ansible, but this one is clear in the code, which raises security issues.

For more information about Ansible variables, read the complete documentation here: `https://docs.ansible.com/ansible/latest/user_guide/playbooks_variables.html`.

Let's now look at how to use Ansible Vault to protect playbook variables.

Protecting sensitive data with Ansible Vault

The configuration of a system often requires sensitive information that should not be in the wrong hands. In the Ansible tool, there is a sub-tool called **Ansible Vault** that protects the data transmitted to Ansible through playbooks.

We will see in our example how to manipulate Ansible Vault to encrypt and decrypt the information of the MySQL user.

The first step is to encrypt the `group_vars/database/main.yml` file that contains the values of the variables by executing the following command:

```
ansible-vault encrypt group_vars/database/main.yml
```

Ansible Vault requests the inclusion of a password that will be required to decrypt the file and then shows the execution of this command to encrypt the content of a file:

```
/devopsansible# ansible-vault encrypt group_vars/database/main.yml
New Vault password:
Confirm New Vault password:
Encryption successful
```

After the execution of this command, the content of the file is encrypted, so the values are no longer clear. The following is a sample from it:

```
$ANSIBLE_VAULT;1.1;AES256
6231313630336362323534353031313939396232646461666137313363335
3236313832323736373064353465353266653336383231660a6134653265
3434346639633066343438386133383930383038663934336633323465
3336633033643735310a6261343131643239333631393234373737303161
356331633035623831626462666534303831383634313934663266303137
3335633630373039376666656635383639366563383663396432613630333
626263303233626330323063363136653962
```

To decrypt the file to modify it, it will be necessary to execute the `decrypt` command:

```
ansible-vault decrypt group_vars/database/main.yml
```

Ansible Vault requests the password that was used to encrypt the file, and the file becomes readable again.

In an Ansible usage automation process, it is preferable to store the password in a file in a protected location, for example, in the `~/.vault_pass.txt` file.

Then, to encrypt the variable file with this file, we execute the `ansible-vault` command and add the `--vault-password-file` option:

```
ansible-vault encrypt group_vars/database/main.yml --vault-password-file
~/.vault_pass.txt
```

Now that the file is encrypted and the data is protected, we will run Ansible with the following commands:

In **interactive mode**, we will run the following:

```
ansible-playbook -i inventory playbook.yml --ask-vault-pass
```

Ansible asks the user to enter the password shown in the following screenshot:

```
/devopsansible# ansible-playbook -i inventory playbook.yml --ask-vault-pass
Vault password:
```

In **automatic mode**, that is, in a CI/CD pipeline, we can add the `--vault-password-file` parameter with the path of the file that contains the password to decrypt the data:

```
ansible-playbook -i inventory playbook.yml --vault-password-file
~/.vault_pass.txt
```

That's all right. We just executed Ansible with data that is no longer clear in the code and with the use of the `ansible-vault` command.

The entire source code of the inventory, playbook, and roles is available here: `https://github.com/PacktPublishing/Learning_DevOps/tree/master/CHAP03/devopsansible`.

In this section, we have seen how to protect sensitive data in your playbooks using the `ansible-vault` utility. We encrypted and decrypted variable files to protect them, and then re-ran Ansible with these encrypted files.

In the following section, we will see how to use Ansible with a dynamic inventory.

Using a dynamic inventory for Azure infrastructure

When configuring an infrastructure that is composed of several VMs, along with ephemeral environments that are built on demand, the observation often made is that maintaining a static inventory, as we saw in the *Creating an inventory for targeting Ansible hosts* section, can quickly become complicated and its maintenance takes a lot of time.

To overcome this problem, Ansible allows inventories to be obtained dynamically by calling a script (for example, in Python) that is either provided by cloud providers or a script that we can develop and that aims to return the contents of the inventory.

In this section, we will see the different steps to use Ansible to configure VMs in Azure using a dynamic inventory:

1. The first step is to configure Ansible to be able to access Azure resources. For this, we will create an Azure Service Principal in Azure AD, exactly in the same way as we did for Terraform (see the *Configuring Terraform for Azure* section in `Chapter 2`, *Provisioning Cloud Infrastructure with Terraform*). Then, export the information of four service principal IDs to the following environment variables:

```
export AZURE_SUBSCRIPTION_ID=<subscription_id>
export AZURE_CLIENT_ID=<client ID>
export AZURE_SECRET=<client Secret>
export AZURE_TENANT=<tenant ID>
```

For more information on Azure environment variables for Ansible, refer to the Azure documentation here: `https://docs.microsoft.com/fr-fr/azure/virtual-machines/linux/ansible-install-configure#create-azure-credentials`.

2. Then, to be able to generate an inventory with groups and to filter VMs, it is necessary to add `Tags` to the VMs. Tags can be added using Terraform, an `az cli` command line, or an Azure PowerShell script.

Here are some examples of scripts with `az cli`:

```
az resource tag --tags role=webserver -n VM01 -g demoAnsible --
resource-type "Microsoft.Compute/virtualMachines"
```

The preceding script adds a `role` tag of the `webserver` value to the VM, `VM01`. Then, we do the same operation with the `VM02` VM (just change the value of the `-n` parameter to `VM02` in the preceding script).

The following screenshot shows the VM tag in the Azure portal:

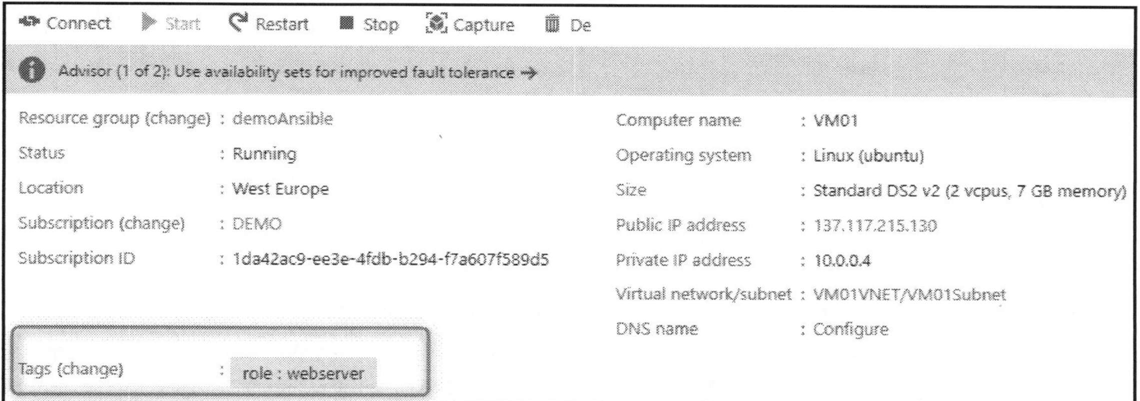

And then, we add the tag on our VM that contains the database with this script:

```
az resource tag --tags role=database -n VM04 -g demoAnsible --
resource-type "Microsoft.Compute/virtualMachines"
```

This script adds a `role` tag to `VM04`, which has the value `database`.

The `az cli` documentation for the management of Azure tags can be found here: `https://docs.microsoft.com/fr-fr/cli/azure/resource?view=azure-cli-latestviewFallbackFrom=azure-cli-latest.md#az-resource-tag`.

3. To use a dynamic inventory in Azure, we need to do the following:
    - Install the Python Azure SDK on the machine that runs Ansible with the `pip install azure` command.
    - At the root of the `devopsansible` folder, create a new folder named `inventories`.

- Download the Python inventory script, `https://raw.githubusercontent.com/ansible/ansible/devel/contrib/inventory/azure_rm.py`, and the second file, the `azure_rm.ini` configuration file: `https://raw.githubusercontent.com/ansible/ansible/devel/contrib/inventory/azure_rm.ini`.
- Copy these two scripts to the created folder, `inventories`.
- Give the user execution permission with the `chmod +x azure_rm.py` command.
- Update the configuration of the `azure_rm.ini` file by allowing the grouping of the returned VM list in the inventory only by their tag.

The following screenshot shows the content of the `azure_rm.ini` file in which the value of the `group_by_tag` property has been changed to `yes`:

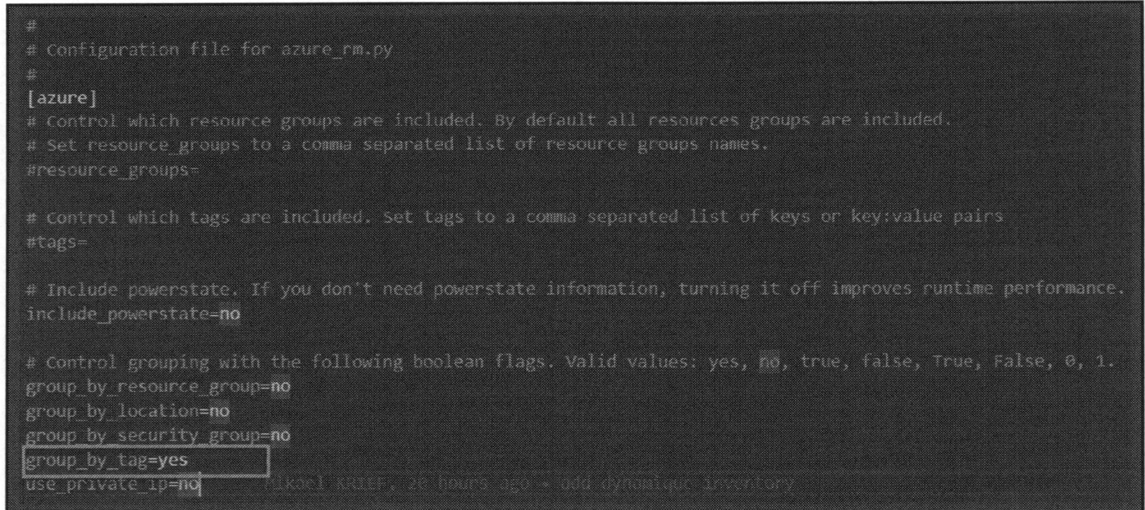

A complete list of all dynamic inventory scripts for all cloud providers provided by Ansible can be found here: `https://github.com/ansible/ansible/tree/devel/contrib/inventory`.

4. As regards the inventory, we could very well stop there, but the problem is that we have to reconstitute an inventory that contains the same group of hosts as our original static inventory. To ensure that our dynamic inventory can be used by our playbook, we need to install nginx on all hosts in the `webserver` group and install MySQL on all hosts in the `database` group.

   To have a dynamic inventory that contains our web server and database groups, we will, in our `inventories` folder, create a file called `demo` that will have the following content:

   ```
   ;init azure group to remove warning
   [webserver:children]
   role_webserver

   [database:children]
   role_database

   [all:vars]
   ansible_user=demobook11
   ansible_password=xxxxxxx #if login to vm need password
   ```

   The `[webserver:children]` code indicates that we are creating a `webserver` group that will be composed (as children) of other groups, and here it is the `role_webserver` group. This group is returned by the dynamic inventory script that returns all VMs that have the `role` tag with the `webserver` value.

   This is how the dynamic inventory creates its groups with the `tagname_value` notation. It is exactly the same for the `[database:children]` group, which is composed of the `role_database` group; it will contain all of the VMs that have the `role` tag with the `database` value.

   Finally, in the last part, we define the user name, which is the same for all VMs.

To summarize, here is the final content of our `devopsansible` working folder:

The complete source code of the `inventories` folder is available here: `https://github.com/PacktPublishing/Learning_DevOps/tree/master/CHAP03/devopsansible/inventories`.

5. After having set up all of the artifacts for our Ansible dynamic inventory in Azure, it is good to test its proper functioning, which includes the following:
    - That there are no execution errors.
    - The connection and authentication to our Azure environment are done correctly.
    - Its execution returns the Azure VMs from our infrastructure.

As mentioned in the *Technical requirements* section, before running the following commands, we need to have the Azure Python SDK installed on the machine. Read the following link for more information: `https://azure.github.io/azure-sdk-for-python/`.

To perform this test, the following command is executed:

```
ansible-inventory -i inventories/azure_rm.py --list
```

This command allows us to display as output the return of the inventory script in list format. The Python script returns the complete list of all VMs and their Azure properties in JSON format. Here is a small sample screen from this execution:

```
/devopsansible# ansible-inventory -i inventories/azure_rm.py --list


{
    "_meta": {
        "hostvars": {
            "VM01": {
                "ansible_host": "137.117.215.130",
                "computer_name": "VM01",
                "fqdn": null,
                "id": "/subscriptions/1da42ac9-ee3e-4                    /resourceGroups/DEMOANSIBLE/providers/Microsoft
                "image": {
                    "offer": "UbuntuServer",
                    "publisher": "Canonical",
                    "sku": "18.04-LTS",
                    "version": "latest"
                },
                "location": "westeurope",
                "mac_address": "00-0D-3A-39-73-B0",
                "name": "VM01",
                "network_interface": "VM01VMNic",
                "network_interface_id": "/subscriptions/1da42a                    /resourceGroups/demoAnsible/provide
rfaces/VM01VMNic",
                "os_disk": {
                    "name": "VM01_disk1_5dcc564907a1494ba78c587a3e836723",
                    "operating_system_type": "linux"
                },
                "plan": null,
                "private_ip": "10.0.0.4",
                "private_ip_alloc_method": "Dynamic",
                "provisioning_state": "Succeeded",
                "public_ip": "137.117.215.130",
                "public_ip_alloc_method": "Dynamic",
                "public_ip_id": "/subscriptions/1da42ac9-ee3                    /resourceGroups/demoAnsible/providers
M01PublicIP",
                "public_ip_name": "VM01PublicIP",
```

We can also display this inventory in graph mode by running the same command, but with the `--graph` option, as follows:

```
/devopsansible# ansible-inventory -i inventories/azure_rm.py --graph

@all:
  |--@azure:
  |  |--VM01
  |  |--VM02
  |  |--VM03
  |  |--VM04
  |--@linux:
  |  |--VM01
  |  |--VM02
  |  |--VM03
  |  |--VM04
  |--@role:
  |  |--VM01
  |  |--VM02
  |  |--VM04
  |--@role_database:
  |  |--VM04
  |--@role_webserver:
  |  |--VM01
  |  |--VM02
  |--@ungrouped:
```

With the `--graph` option, we have a better visualization of the VMs according to their tags.

With the test being conclusive, we can proceed to the final step, which is the execution of Ansible with a dynamic inventory.

6. Once we have tested our dynamic inventory in Azure, we just have to run Ansible on it, using the tags applied on the VMs. For this, we execute the first Ansible command with the `--check` option, which allows us to check that the playbook execution is based on the right VM groups:

```
ansible-playbook -i inventories/ playbook.yml --check --vault-password-file ~/.vault_pass.txt
```

As we see in this command, we used the `inventories` folder as an argument. The Ansible command will use all of its content (the Python script and its configuration as well as the inventory file) to obtain the list of hosts.

The following screenshot shows the execution of the `check` command:

```
/devopsansible# ansible-playbook -i inventories/  playbook.yml --check --vault-password-file ~/.vault_pass.txt

PLAY [webserver] ****************************************************************************************************

TASK [Gathering Facts] **********************************************************************************************
ok: [VM01]
ok: [VM02]

TASK [ngnix : install and check nginx latest version] ***************************************************************
ok: [VM02]
ok: [VM01]

TASK [ngnix : start nginx] ******************************************************************************************
ok: [VM01]
ok: [VM02]

PLAY [database] *****************************************************************************************************

TASK [Gathering Facts] **********************************************************************************************
ok: [VM04]

TASK [mysql : Update apt cache] *************************************************************************************
changed: [VM04]

TASK [mysql : Install required software] ****************************************************************************
ok: [VM04]

TASK [mysql : Create mysql user] ************************************************************************************
ok: [VM04]

PLAY RECAP **********************************************************************************************************
VM01                       : ok=3    changed=0    unreachable=0    failed=0
VM02                       : ok=3    changed=0    unreachable=0    failed=0
VM04                       : ok=4    changed=1    unreachable=0    failed=0
```

This check shows us that the playbook will be executed on the right groups of VMs.

Let's now run our playbook with the same command, minus the `--check` option:

```
ansible-playbook -i inventories/ playbook.yml --vault-password-file
~/.vault_pass.txt
```

This screenshot shows the execution of the Ansible playbook with the dynamic inventory:

```
/devopsansible# ansible-playbook -i inventories/  playbook.yml  --vault-password-file ~/.vault_pass.txt

PLAY [webserver] ************************************************************************************

TASK [Gathering Facts] ******************************************************************************
ok: [VM01]
ok: [VM02]

TASK [ngnix : install and check nginx latest version] ***********************************************
ok: [VM02]
ok: [VM01]

TASK [ngnix : start nginx] **************************************************************************
ok: [VM01]
ok: [VM02]

PLAY [database] *************************************************************************************

TASK [Gathering Facts] ******************************************************************************
ok: [VM04]

TASK [mysql : Update apt cache] *********************************************************************
ok: [VM04]

TASK [mysql : Install required software] ************************************************************
ok: [VM04]

TASK [mysql : Create mysql user] ********************************************************************
ok: [VM04]

PLAY RECAP ******************************************************************************************
VM01                       : ok=3    changed=0    unreachable=0    failed=0
VM02                       : ok=3    changed=0    unreachable=0    failed=0
VM04                       : ok=4    changed=0    unreachable=0    failed=0
```

From now on, each time a VM of our Azure infrastructure has a `role=webserver` or `role=database` tag, it will be automatically taken into account by the dynamic inventory, and therefore, no code modification will be necessary.

In terms of another method to use a dynamic inventory on Azure, you can also consult the Azure documentation on the use of a dynamic inventory at `https://docs.microsoft.com/fr-fr/azure/ansible/ansible-manage-azure-dynamic-inventories`.

The use of a dynamic inventory, therefore, allows full advantage to be taken of the scalability of the cloud with an automatic VM configuration and without code change.

In this section, we have seen how to use a dynamic inventory for Azure with the implementation of its configuration and the necessary script recovery and finally, we have executed Ansible with this dynamic inventory.

Summary

In this chapter, we saw that Ansible is a very powerful and complete tool that allows the automation of server configuration and administration. To work, it uses an inventory that contains the list of hosts to be configured and a playbook in which the list of configuration actions is coded.

Roles, modules, and variables also allow for better management and centralization of playbook code. Ansible also has a Vault that protects sensitive playbook data. Finally, for dynamic environments, inventory writing is simplified with the implementation of dynamic inventories.

In the next chapter, we will see how to optimize infrastructure deployment with the use of Packer to create server templates.

Questions

1. What is the role of Ansible that is detailed in this chapter?
2. Can we install Ansible on a Windows OS?
3. What are the two artifacts studied in this chapter that Ansible needs to run?
4. What is the name of the option added to the `ansible-playbook` command that is used to preview the changes that will be applied?
5. What is the name of the utility used to encrypt and decrypt Ansible data?
6. When using a dynamic inventory in Azure, on which properties of the VMs is the inventory script used to return the list of VMs?

Further reading

If you want to know more about Ansible, here are some resources:

- The Ansible documentation: `https://docs.ansible.com/ansible/latest/index.html`
- Quick Start video: `https://www.ansible.com/resources/videos/quick-start-video?extIdCarryOver=truesc_cid=701f2000001OH6uAAG`
- Ansible on Azure documentation: `https://docs.microsoft.com/en-us/azure/ansible/`

- **Visual Studio Code Ansible extension:** `https://marketplace.visualstudio.com/items?itemName=vscoss.vscode-ansible`
- **Mastering Ansible:** `https://www.packtpub.com/virtualization-and-cloud/mastering-ansible-third-edition`
- **Ansible Webinars Training:** `https://www.ansible.com/resources/webinars-training`

4
Optimizing Infrastructure Deployment with Packer

In the previous chapters, we learned how to provision a cloud infrastructure using Terraform and then we continued with the automated configuration of VMs with Ansible. This automation allows us to benefit from a real improvement in productivity and very visible time-saving.

However, despite this automation, we notice the following:

- Configuring a VM can be very time-consuming because it depends on its hardening as well as the middleware that will be installed and configured on this VM.
- Between each environment or application, the middleware versions are not identical because their automation script is not necessarily identical or maintained over time. Hence, for example, the production environment, being more critical, will be more likely to have the latest version of packages, which is not the case in ante production environments. And with this situation, we often encounter issues with the behavior of applications in production.
- Configuration and security compliance is not often applied or updated.

To address these issues, all cloud providers have integrated a service that allows them to create or generate custom VM images. These images contain all of the configurations of the VMs with their security administration and middleware configurations and can then be used as a basis to create VMs for applications.

The benefits of using these images are as follows:

- The provisioning of a VM from an image is very fast.
- Each VM is uniform in configuration and, above all, is safety compliant.

Among the **Infrastructure as Code** (**IaC**) tools, there is Packer from the HashiCorp tools, which allows us to create VM images from a file (or template).

In this chapter, we will learn how to install Packer in different modes. We will discuss the syntax of Packer templates to create custom VM images in Azure that use scripts or Ansible playbooks.

We will detail the execution of Packer with these templates. Finally, we will see how Terraform uses the images created by Packer. Through this chapter, we'll understand that Packer is a simple tool that simplifies the creation of VMs in a DevOps process and integrates very well with Terraform.

In this chapter, we will cover the following:

- An overview of Packer
- Creating Packer templates using scripts
- Creating Packer templates using Ansible
- Executing Packer
- Using images created by Packer with Terraform

Technical requirements

This chapter will explain how to use Packer to create a VM image in an Azure infrastructure as an example of cloud infrastructure. So, you will need an Azure subscription, which you can get here for free: `https://azure.microsoft.com/en-us/free/`.

In the *Creating Packer templates using Ansible* section, we will learn how to write Packer templates that use Ansible, so you will need to install Ansible on your machine and understand how it works, which is detailed in `Chapter 3`, *Using Ansible for Configuring IaaS Infrastructure*.

The last section of this chapter will give an example of using Terraform with a Packer image; for its application, it will be necessary to install Terraform and understand its operation, which is detailed in `Chapter 2`, *Provisioning Cloud Infrastructure with Terraform*.

The entire source code of this chapter is available here: `https://github.com/PacktPublishing/Learning_DevOps/blob/master/CHAP04/`.

Check out the following video to see the Code in Action:
`http://bit.ly/32MVTMQ`

An overview of Packer

Packer is part of the HashiCorp open source suite of tools, and this is the official Packer page: `https://www.packer.io/`. It's an open source command-line tool that allows us to create custom VM images of any OS (these images are also called **templates**) on several platforms from a JSON file.

Packer's operation is simple; it is based on the basic OS provided by the different cloud providers and configures a temporary VM by executing the scripts described in the JSON template. Then, from this temporary VM, Packer generates a custom image ready to be used to provision VMs.

Apart from VM images, Packer also provides other types of images such as Docker images or Vagrant images. After this brief overview of Packer, let's look at the different installation modes.

Installing Packer

Packer, like Terraform, is a cross-platform tool and can be installed on Windows, Linux, or macOS. The installation of Packer is almost identical to the Terraform installation (see `Chapter 2`, *Provisioning Cloud Infrastructure with Terraform*) and can be done in two ways: either manually or via a script.

Installing manually

To install Packer manually, use the followings steps:

1. Go to the official download page (`https://www.packer.io/downloads.html`) and download the package corresponding to your operating system.
2. After downloading, unzip and copy the binary into an execution directory (for example, inside `c:\Packer`).
3. Then, the `PATH` environment variable must be set with the path to the binary directory.

For detailed instructions on how to update the `PATH` environment variable on Windows, refer to this article: `https://www.architectryan.com/2018/03/17/add-to-the-path-on-windows-10/`, and for Linux, refer to this one: `https://www.techrepublic.com/article/how-to-add-directories-to-your-path-in-linux/`.

Now that we've learned how to install Packer manually, let's look at the options available to us to install it using a script.

Installing by script

It is also possible to install Packer with an automatic script that can be installed on a remote server and be used on a CI/CD process. Indeed, Packer can be used locally, as we will see in this chapter, but its real goal is to be integrated into a CI/CD pipeline. This automatic DevOps pipeline will allow the construction and publication of uniform VM images that will guarantee the integrity of middleware and VM security based on these images.

Let's see the structure of these scripts for the different OSes, that is, Linux, Windows, and macOS.

Installing Packer by script on Linux

The installation script for a **Linux** machine is as follows:

```
PACKER_VERSION="1.4.3" #Update with your desired version

curl -Os
https://releases.hashicorp.com/packer/${PACKER_VERSION}/packer_${PACKER_VER
SION}_linux_amd64.zip \
&& curl -Os
https://releases.hashicorp.com/packer/${PACKER_VERSION}/packer_${PACKER_VER
SION}_SHA256SUMS \
&& curl https://keybase.io/hashicorp/pgp_keys.asc | gpg --import \
&& curl -Os
https://releases.hashicorp.com/packer/${PACKER_VERSION}/packer_${PACKER_VER
SION}_SHA256SUMS.sig \
&& gpg --verify packer_${PACKER_VERSION}_SHA256SUMS.sig
packer_${PACKER_VERSION}_SHA256SUMS \
&& shasum -a 256 -c packer_${PACKER_VERSION}_SHA256SUMS 2>&1 | grep
"${PACKER_VERSION}_linux_amd64.zip:\sOK" \
&& unzip -o packer_${PACKER_VERSION}_linux_amd64.zip -d /usr/local/bin
```

The code of this script is also available here: `https://github.com/PacktPublishing/Learning_DevOps/blob/master/CHAP04/install_packer.sh`.

This script performs the following actions:

1. Download the Packer version 1.4.0 package and check the checksum.
2. Unzip and copy the package into a local directory, `/usr/local/bin` (by default, this folder is in the `PATH` environment variable).

The following is a screenshot of the execution of the script for installing Packer on Linux:

```
/CHAP04# sh install_packer.sh
                                   Speed   Time    Time    Time  Current
                           Dload  Upload   Total   Spent   Left  Speed
100  1696  100  1696    0     0   3981      0 --:--:-- --:--:-- --:--:--  3971
gpg: key 51852D87348FFC4C: "HashiCorp Security <security@hashicorp.com>" not changed
gpg: Total number processed: 1
gpg:              unchanged: 1
gpg: Signature made Thu Apr 11 20:30:03 2019 DST
gpg:                using RSA key 91A6E7F85D05C65630BEF18951852D87348FFC4C
gpg: Good signature from "HashiCorp Security <security@hashicorp.com>" [unknown]
gpg: WARNING: This key is not certified with a trusted signature!
gpg:          There is no indication that the signature belongs to the owner.
Primary key fingerprint: 91A6 E7F8 5D05 C656 30BE  F189 5185 2D87 348F FC4C
packer_1.4.0_linux_amd64.zip: OK
Archive:  packer_1.4.0_linux_amd64.zip
  inflating: /usr/local/bin/packer
```

Installing Packer by script on Windows

On Windows, we can use **Chocolatey**, which is a software package manager. Chocolatey is a free public package manager, like NuGet or npm, but dedicated to software. It is widely used for the automation of software on Windows servers or even local machines. Chocolatey's official website is here: `https://chocolatey.org/`, and its installation documentation is here: `https://chocolatey.org/install`.

Once Chocolatey is installed, we just need to run the following command in PowerShell or in the CMD tool:

```
choco install packer -y
```

The following is a screenshot of the Packer installation for Windows with Chocolatey:

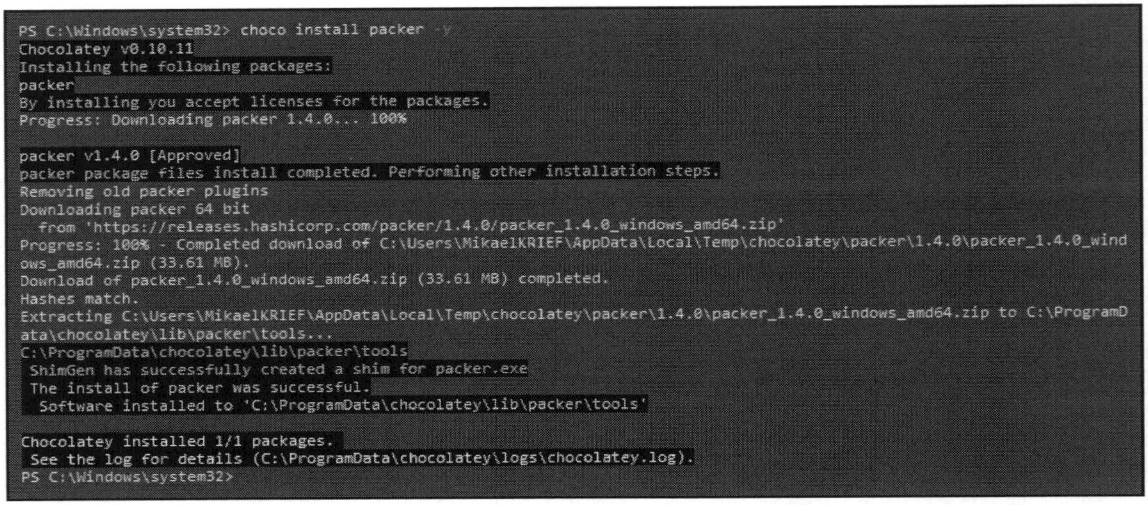

```
PS C:\Windows\system32> choco install packer -y
Chocolatey v0.10.11
Installing the following packages:
packer
By installing you accept licenses for the packages.
Progress: Downloading packer 1.4.0... 100%

packer v1.4.0 [Approved]
packer package files install completed. Performing other installation steps.
Removing old packer plugins
Downloading packer 64 bit
  from 'https://releases.hashicorp.com/packer/1.4.0/packer_1.4.0_windows_amd64.zip'
Progress: 100% - Completed download of C:\Users\MikaelKRIEF\AppData\Local\Temp\chocolatey\packer\1.4.0\packer_1.4.0_wind
ows_amd64.zip (33.61 MB).
Download of packer_1.4.0_windows_amd64.zip (33.61 MB) completed.
Hashes match.
Extracting C:\Users\MikaelKRIEF\AppData\Local\Temp\chocolatey\packer\1.4.0\packer_1.4.0_windows_amd64.zip to C:\ProgramD
ata\chocolatey\lib\packer\tools...
C:\ProgramData\chocolatey\lib\packer\tools
 ShimGen has successfully created a shim for packer.exe
 The install of packer was successful.
  Software installed to 'C:\ProgramData\chocolatey\lib\packer\tools'

Chocolatey installed 1/1 packages.
 See the log for details (C:\ProgramData\chocolatey\logs\chocolatey.log).
PS C:\Windows\system32>
```

The execution of `choco install packer -y` installs the latest version of Packer from Chocolatey.

Installing Packer by script on macOS

On macOS, we can use **Homebrew**, the macOS package manager (`https://brew.sh/`), for installing Packer by executing the following command in our Terminal:

```
brew install packer
```

Integrating Packer with Azure Cloud Shell

Just as we learned in detail for Terraform in the *Integrating Terraform with Azure Cloud Shell* section in `Chapter 2`, *Provisioning Cloud Infrastructure with Terraform*, Packer is also integrated with Azure Cloud Shell, as shown in the following screenshot:

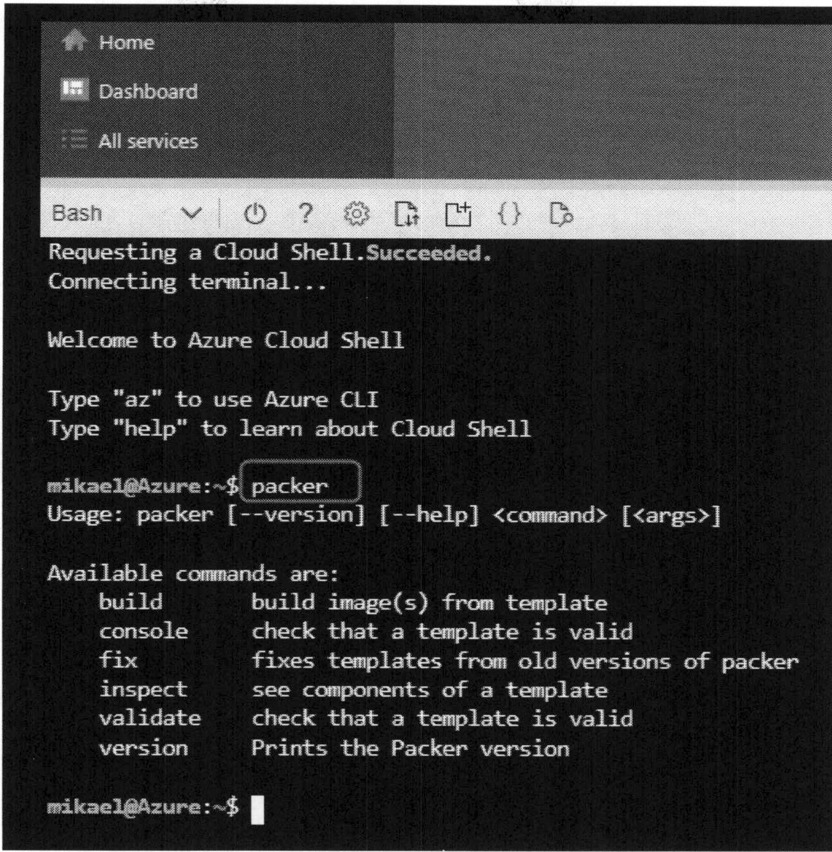

Now that we have seen the installation of Packer on different operating systems and its integration with Azure Cloud Shell, we will next check its installed version.

Checking the Packer installation

Once installed, we can check the installed version of Packer by running the following command:

```
packer --version
```

This command displays the installed Packer version:

```
:~$ packer --version
1.4.2
```

To see all of the Packer command-line options, we can execute the following command:

```
packer --help
```

After executing, we will see a list of available commands, as shown in the following screenshot:

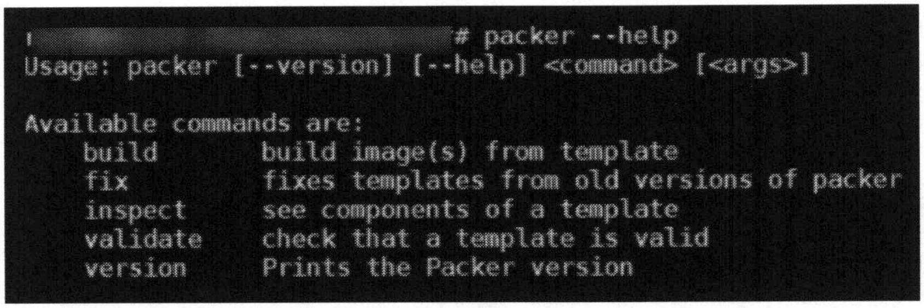

```
# packer --help
Usage: packer [--version] [--help] <command> [<args>]

Available commands are:
    build       build image(s) from template
    fix         fixes templates from old versions of packer
    inspect     see components of a template
    validate    check that a template is valid
    version     Prints the Packer version
```

We have just seen the manual installation procedure for Packer and installation with a script on different OSes, as well as its integration with Azure Cloud Shell.

We will now write a template to create a VM image in Azure with Packer using scripts.

Creating Packer templates for Azure VMs with scripts

As mentioned in the introduction, to create a VM image, Packer is based on a file (template) that is in JSON format. We will first see the structure and composition of this JSON template, and then we will put into practice how to create a template that will create a VM image in Azure.

The structure of the Packer template

The Packer JSON template is composed of several main sections such as `builders`, `provisioners`, and `variables`. The structure of a template is as follows:

```
{
    "variables": {
    // list of variables
    ...
    },
    "builders": [
```

```
            {
            //builders properties
            ...
            }
        ],
        "provisioners": [
            {
            // list of scripts to execute for image provisionning
            ...
            }
        ]
}
```

Let's look at the details of each section.

The builders section

The `builders` section is mandatory and contains all of the properties that define the image and its location, such as its name, the type of image, the cloud provider on which the image will be generated, connection information to the cloud, the base image to use, and other properties that are specific to the image type.

Here is an example code of a `builders` section:

```
{
"builders": [{
  "type": "azure-rm",
  "client_id": "xxxxxxxx",
  "client_secret": "xxxxxxxx",
  "subscription_id": "xxxxxxxxxx",
  "tenant_id": "xxxxxx",
  "os_type": "Linux",
  "image_publisher": "Canonical",
  "image_offer": "UbuntuServer",
  "location": "westus"
.......
 }]
}
```

In this sample, the `builders` section defines an image that will be stored in the Azure cloud and is based on the Linux Ubuntu OS. We also configure the authentication keys for the cloud.

The documentation on the `builders` section is here: `https://www.packer.io/docs/templates/builders.html`.

If we want to create the same image but on several providers, we can indicate in the same template file multiple block `builders` which will contains the provider properties. For example, see the following code sample:

```
{
"builders": [
    {
     "type": "azure-rm",
     "location": "westus",
    ....
     },
    {
     "type": "docker",
     "image": "alpine:latest",
    ...
    }
]
}
```

In this code sample, we define in the Packer template, the information for an image of an AWS VM, and the information for a Docker image based on Alpine. The advantage of this is to standardize the scripts that will be detailed in the provisioning section of these two images.

Just after the details of each section, we will see a concrete example with a `builders` section to create an image in Azure.

Let's move on to explaining the `provisioners` section.

The provisioners section

The `provisioners` section, which is optional, contains a list of scripts that will be executed by Packer on a temporary VM base image in order to build our custom VM image according to our needs.

If the Packer template does not contain a `provisioners` section, no configuration will be made on the base images.

The actions defined in this section are available for Windows as well as Linux images, and the actions be of several types such as executing a local or remote script, executing a command, or copying a file.

The `provisioners` type proposed natively by Packer is detailed in the documentation: `https://www.packer.io/docs/provisioners/index.html`.

It is also possible to extend Packer by creating custom provisioning types. To learn more about custom `provisioners`, refer to the documentation here: `https://www.packer.io/docs/extending/custom-provisioners.html`.

The following is a sample of a `provisioners` section:

```
{
...
    "provisioners": [
        {
         "type": "shell",
         "script": "hardening-config.sh"
        },
        {
         "type": "file",
         "source": "scripts/installers",
         "destination": "/tmp/scripts"
        }
    ]
...
}
```

In this `provisioners` section, Packer will upload and execute the local script, `hardening-config.sh`, to apply the hardening configuration on the remote temporary VM base image, and copy the content of the `scripts/installers` local folder to the remote folder, `/tmp/scripts`, to configure the image.

So, in this section, we list all of the configuration actions for the image to be created.

However, when creating an image of a VM, it's necessary to generalize it—in other words, delete all of the personal user information that was used to create this image.

For example, for a Windows VM image, we will use the Sysprep tool as the last step of `provisioners` with this following code:

```
"provisioners": [
...
{
 "type": "powershell",
 "inline": ["& C:\\windows\\System32\\Sysprep\\Sysprep.exe /oobe
/generalize /shutdown /quiet"]}
]
```

Another example of Sysprep usage in Packer templates is available here: `https://www.packer.io/docs/builders/azure.html`.

And for deleting the personal user information on a Linux image, we will use the following code:

```
"provisioners": [
.....
{
  "type": "shell",
  "execute_command": "sudo sh -c '{{ .Vars }} {{ .Path }}'",
  "inline": [
     "/usr/sbin/waagent -force -deprovision+user && export HISTSIZE=0 &&
sync"
  ]
 }
]
```

For more information about the `provisioners` section, refer to the documentation here: `https://www.packer.io/docs/templates/provisioners.html`, and the list of actions can be found here: `https://www.packer.io/docs/provisioners/index.html`.

After the `provisioners` section, let's talk about variables.

The variables section

In the Packer template, we may often need to use values that are not static in the code. This optional `variables` section is used to define variables that will be filled either as command-line arguments or as environment variables. These variables will then be used in the `builders` or `provisioners` sections.

Here is an example of a `variables` section:

```
{
 "variables": {
     "access_key": "{{env `ACCESS_KEY`}}",
     "image_folder": "/image",
     "vm_size": "Standard_DS2_v2"
 },
....
}
```

In this example, we initialize the following:

- The `access_key` variable with the `ACCESS_KEY` environment variable
- The `image_folder` variable with the `/image` value
- The value of the VM image size, which is the `vm_size` variable

To use these so-called user variables, we use the `{{user 'variablename' }}` notation, and here is an example of using these variables in the `builders` section:

```
"builders": [
     {
         "type": "azure-arm",
         "access_key": "{{user `access_key`}}",
         "vm_size": "{{user `vm_size`}}",
      ...
     }
 ],
```

And in the `provisioners` section, we use the variables defined in the `variables` section, as shown here:

```
"provisioners": [
     {
         "type": "shell",
         "inline": [
         "mkdir {{user `image_folder`}}",
         "chmod 777 {{user `image_folder`}}",
         ...
     ],
         "execute_command": "sudo sh -c '{{ .Vars }} {{ .Path }}'"
     },
 ...
 ]
```

We, therefore, define the properties of the image with variables that will be provided when executing the Packer template. We can also use these variables in the `provisioners` section for centralizing these properties and not have to redefine them, here, the path of the images (`/image`) that will be repeated several times in the templates.

Apart from the variables provided by the user, it is also possible to retrieve other variable sources such as secrets stored in HashiCorp Vault or Consul. For more information about variables, refer to the documentation: `https://www.packer.io/docs/templates/user-variables.html`.

We have just seen the structure of the Packer template with the principal sections builders, provisioners, and variables that compose it, which are `builders`, `provisioners`, and `variables`. Now, let's look at a concrete example with the writing of a Packer template to create an image in Azure.

Building an Azure image with the Packer template

With all of the elements we saw earlier, we will now be able to create a Packer template that will create a VM image in Azure.

For this, we will need to first create an Azure AD **Service Principal** (**SP**) that will have the permissions to create resources in our subscription. The creation is exactly the same as we did for Terraform; for more details, see the *Configuring Terraform for Azure* section in `Chapter 2`, *Provisioning Cloud Infrastructure with Terraform*. Then, on the local disk, we will create an `azure_linux.json` file, which will be our Packer template. We will start writing to this file with the `builders` section, as follows:

```
..."builders": [{
     "type": "azure-arm",
     "client_id": "{{user `clientid`}}",
     "client_secret": "{{user `clientsecret`}}",
     "subscription_id": "{{user `subscriptionid`}}",
     "tenant_id": "{{user `tenantid`}}",

    "os_type": "Linux",
     "image_publisher": "Canonical",
     "image_offer": "UbuntuServer",
     "image_sku": "18.04-LTS",
     "location": "West Europe",
     "vm_size": "Standard_DS2_v3",

    "managed_image_resource_group_name": "{{user `resource_group`}}",
```

```
    "managed_image_name": "{{user `image_name`}}-{{user `image_version`}}",

   "azure_tags": {
   "version": "{{user `image_version`}}",
   "role": "WebServer"
     }
}],....
```

This section describes the following:

- It describes the `azure_rm` type, which indicates the provider.
- Also, it describes the `client_id`, `secret_client`, `subscription_id`, and `tenant_id` properties, which contain information from the previously created SP. For security reasons, these values are not written in plain text in the JSON template; they will be placed in variables (which we will see right after the details of the `builders` section).
- The `managed_image_resource_group_name` and `managed_image_name` properties indicate the resource group as well as the name of the image to be created. The name of the image is also placed into a variable with a name and a version number.
- The other properties correspond to the information of the OS type (Ubuntu 18), size (`Standard_DS2_v3`), region, and tag.

Now, we will write the `variables` section that defines the elements that are not fixed:

```
..."variables": {
 "subscriptionid": "{{env `AZURE_SUBSCRIPTION_ID`}}",
 "clientid": "{{env `AZURE_CLIENT_ID`}}",
 "clientsecret": "{{env `AZURE_CLIENT_SECRET`}}",
 "tenantid": "{{env `AZURE_TENANT_ID`}}",

 "resource_group": "rg_images",
 "image_name": "linuxWeb",
 "image_version": "0.0.1"
 },...
```

We have defined the variables and their default values with the following:

- The four pieces of authentication information from the SP will be passed either in the Packer command line or as an environment variable.
- The resource group, name, size, and region of the image to be generated are also in `variables`.
- The `image_version` variable that contains the version of the image (used in the name of the image) is defined.

So, with these variables, we will be able to use the same JSON template file to generate several images with different names and sizes (we will see it when Packer is executed).

Finally, the last action is to write the steps of the `provisioners` image with the following code:

```
"provisioners": [
{
    "type": "shell",
    "execute_command": "sudo sh -c '{{ .Vars }} {{ .Path }}'",
     "inline": [
        "apt-get update",
        "apt-get -y install nginx"
     ]
 },
 {
     "type": "shell",
     "execute_command": "sudo sh -c '{{ .Vars }} {{ .Path }}'",
     "inline": [
       "/usr/sbin/waagent -force -deprovision+user && export HISTSIZE=0 &&
sync"
     ]
   }
]
```

Here is what the previous code block is doing:

- It updates packages with `apt-get update` and `upgrade`.
- It installs nginx.

Then, in the last step before the image is created, the VM is deprovisioned to delete the user information that was used to install everything on the temporary VM using the following command:

```
/usr/sbin/waagent -force -deprovision+user && export HISTSIZE=0 && sync
```

The complete source code of this Packer template is available here: `https://github.com/PacktPublishing/Learning_DevOps/blob/master/CHAP04/templates/azure_linux.json`.

We have just seen the structure of a Packer template, which is mainly composed of three sections, which are `variables`, `builders`, and `provisioners`, and from there we saw a concrete example with the writing of a Packer template to generate a custom VM image in Azure that uses scripts or provisioning commands.

We have our Packer template finished and ready to be run, but first, we will see another type of `provisioners` using Ansible.

Using Ansible in a Packer template

We have just seen how to write a Packer template that uses command scripts (for example, `apt-get`), but it is also possible to use Ansible playbooks to create an image. Indeed, when we use IaC to configure VMs, we are often used to configuring the VMs directly using Ansible before thinking about making them into VM images.

What is interesting about Packer is that we can reuse the same playbook scripts that we used to configure VMs to create our VM images. So, it's a huge time saver because we don't have to rewrite the scripts.

To put this into practice, we will write the following:

- An Ansible playbook that installs nginx
- A Packer template that uses Ansible with our playbook

Let's start with the writing of the Ansible playbook.

Writing the Ansible playbook

The playbook we are going to write is almost identical to the one we set up in `Chapter 3`, *Using Ansible for Configuring IaaS Infrastructure*, but with some changes.

Follow the sample playbook code:

```
---
- hosts: 127.0.0.1
 become: true
 connection: local
 tasks:
 - name: installing Ngnix latest version
   apt:
    name: nginx
```

```
      state: latest
  - name: starting Nginx service
    service:
      name: nginx
      state: started
```

The changes made are as follows:

- There is no inventory because it is Packer that manages the remote host, which is the temporary VM that will be used to create the image.
- The value of `hosts` is, therefore, the local IP address.
- We only keep the installation of nginx in this playbook and we deleted the task that installed the MySQL database.

The code of this playbook is available here: `https://github.com/PacktPublishing/Learning_DevOps/blob/master/CHAP04/templates/ansible/playbookdemo.yml`.

Now that we have written our Ansible playbook, we will see how to integrate its execution into the Packer template.

Integrating an Ansible playbook in a Packer template

In terms of the Packer template, the JSON `builders` and `variables` sections are identical to one of the templates that uses scripts that we detailed earlier in the *Using Ansible in a Packer template* section. What is different is the JSON `provisioners` section, which we will write as follows:

```
provisioners": [
 {
     "type": "shell",
     "execute_command": "sudo sh -c '{{ .Vars }} {{ .Path }}'",
     "inline": [
       "add-apt-repository ppa:ansible/ansible", "apt-get update", "apt-get
install ansible -y"
     ]
 },
 {
    "type": "ansible-local",
    "playbook_file": "ansible/playbookdemo.yml"
 },
```

```
{
   "type": "shell",
   "execute_command": "sudo sh -c '{{ .Vars }} {{ .Path }}'",
   "script": "clean.sh"
},
.....//Deprovision the VM
]
```

The actions described in this `provisioners` section, which Packer will execute using this template are as follows:

1. Install Ansible on the temporary VM.
2. On this temporary VM, the `ansible-local` provisioner runs the playbook `playbookdemo.yaml` that installs and starts nginx. The documentation of this provisioner is here: https://www.packer.io/docs/provisioners/ansible-local.html.
3. The `clean.sh` script deletes Ansible and its dependent packages that are no longer used.
4. Deprovision the VM to delete the local user information.

The complete Packer template is available here: https://github.com/PacktPublishing/Learning_DevOps/blob/master/CHAP04/templates/azure_linux_ansible.json, and the source of the clean script, `clean.sh`, is available here: https://github.com/PacktPublishing/Learning_DevOps/blob/master/CHAP04/templates/clean.sh.

As we can see here, Packer will execute Ansible on the temporary VM that will be used to create the image, but it is also possible to use Ansible remotely by using Packer's Ansible provisioner, the documentation of which is located here: https://www.packer.io/docs/provisioners/ansible.html.

We have seen up to this point how a Packer template is composed of the `builders`, `variables`, and `provisioners` sections, and we have seen that it is possible to use Ansible within a Packer template.

We will now run Packer with these JSON templates to create a VM image in Azure.

Executing Packer

Now that we have created the Packer templates, the next step is to run Packer to generate a custom VM image, which will be used to quickly provision VMs that are already configured and ready to use for your applications.

As a reminder, to generate this image, Packer will, from our JSON template, *create a temporary VM*, on which it will *perform all of the configuration actions* described in this template, and then it will *generate an image* from this image. Finally, at the end of its execution, it *removes the temporary VM* and all of its dependencies.

To generate our VM image in Azure, follow these steps:

1. Configure Packer to authenticate to Azure.
2. Check our Packer template.
3. Run Packer to generate our image.

Let's look in detail at the execution of each of its steps.

Configuring Packer to authenticate to Azure

To allow Packer to create resources in Azure, we will use the Azure AD SP that we created earlier in this chapter in the *Building Azure image with the Packer template* section. To execute Packer in Azure, we will use the four pieces of authentication information (`subscription_id`, `client_id`, `client_secret`, and `tenant_id`) of this SP in the environment variables provided in our Packer template in the `variables` section.

In our following template, we have four variables (`client_id`, `client_secret`, `subscription_id`, and `tenant_id`), which take as their values four environment variables (`ARM_CLIENT_ID`, `ARM_CLIENT_SECRET`, `ARM_SUBSCRIPTION_ID`, and `ARM_TENANT_ID`):

```
"variables": {
    "client_id": "{{env `ARM_CLIENT_ID`}}",
    "client_secret": "{{env `ARM_CLIENT_SECRET`}}",
    "subscription_id": "{{env `ARM_SUBSCRIPTION_ID`}}",
    "tenant_id": "{{env `ARM_TENANT_ID`}}",
    "resource_group": "rg_images",
    "image_name": "linuxWeb",
    "image_version": "0.0.1",
    "location": "West Europe",
    "vm_size": "Standard_DS2_v2"
},
```

So, we can set these environment variables as follows (this is a Linux example):

```
export ARM_SUBSCRIPTION_ID=<subscription_id>
export ARM_CLIENT_ID=<client ID>
export ARM_SECRET_SECRET=<client Secret>
export ARM_TENANT_ID=<tenant ID>
```

For Windows, we can use the PowerShell `$env` command to set an environment variable.

The first step of authentication is done, and we will now check the Packer template we wrote.

Checking the validity of the Packer template

Before executing Packer to generate the image, we will execute the `packer validate` command to check that our template is correct.

So, inside the folder that contains the Packer template, we execute the following command on the template:

```
packer validate azure_linux.json
```

The output of the execution of this command returns the status of the check for whether the template is valid, as shown in the following screenshot:

```
/templates# packer validate azure_linux.json
Template validated successfully.
```

Our Packer template is correct in its syntax, so we can launch Packer to generate our image.

Running Packer to generate our VM image

To generate our image with Packer, we will execute Packer with the `build` command on the template file as follows:

```
packer build azure_linux.json
```

In the output of the Packer execution, we can see the different actions being performed by Packer:

1. First is the creation of the temporary VM, as shown in the following screenshot:

```
/CHAP04/templates# packer build azure_linux.json
azure-arm output will be in this color.

==> azure-arm: Running builder ...
==> azure-arm: Getting tokens using client secret
    azure-arm: Creating Azure Resource Manager (ARM) client ...
==> azure-arm: WARNING: Zone resiliency may not be supported in West Europe, checkout the docs at https:/
==> azure-arm: Creating resource group ...
==> azure-arm:  -> ResourceGroupName : 'packer-Resource-Group-xwzj9da4y4'
==> azure-arm:  -> Location          : 'West Europe'
==> azure-arm:  -> Tags              :
==> azure-arm:  ->> version : 0.0.1
==> azure-arm:  ->> role : WebServer
==> azure-arm: Validating deployment template ...
==> azure-arm:  -> ResourceGroupName : 'packer-Resource-Group-xwzj9da4y4'
==> azure-arm:  -> DeploymentName    : 'pkrdpxwzj9da4y4'
==> azure-arm: Deploying deployment template ...
==> azure-arm:  -> ResourceGroupName : 'packer-Resource-Group-xwzj9da4y4'
==> azure-arm:  -> DeploymentName    : 'pkrdpxwzj9da4y4'
```

In the Azure portal, we see a temporary resource group and its resources created by Packer, as shown in the following screenshot:

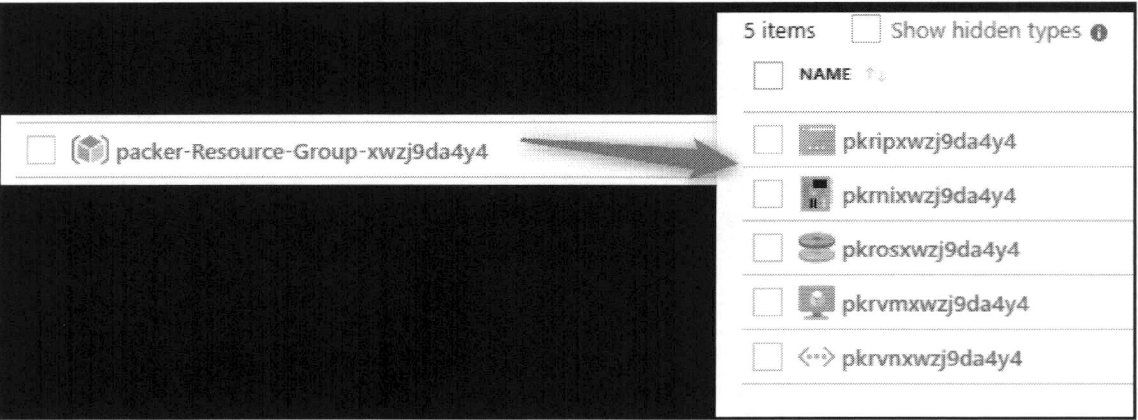

2. The execution time of Packer depends on the actions to be performed on the temporary VM. At the end of its execution, Packer indicates that it has generated the image and deletes the temporary resources.

3. The following screenshot is the end of the output of the Packer execution, which displays the deletion of the temporary resource group and the generation of the image:

```
==> azure-arm: Querying the machine's properties ...
==> azure-arm:  -> ResourceGroupName : 'packer-Resource-Group-uht810tdrw'
==> azure-arm:  -> ComputeName       : 'pkrvmuht810tdrw'
==> azure-arm:  -> Managed OS Disk   : '/subscriptions/8a7aace5-                    /resourceGroups/packer-Resource-Group-uht810tdrw/providers/Microsoft
.Compute/disks/pkrosuht810tdrw'
==> azure-arm: Querying the machine's additional disks properties ...
==> azure-arm:  -> ResourceGroupName : 'packer-Resource-Group-uht810tdrw'
==> azure-arm:  -> ComputeName       : 'pkrvmuht810tdrw'
==> azure-arm: Powering off machine ...
==> azure-arm:  -> ResourceGroupName : 'packer-Resource-Group-uht810tdrw'
==> azure-arm:  -> ComputeName       : 'pkrvmuht810tdrw'
==> azure-arm: Capturing image ...
==> azure-arm:  -> Compute ResourceGroupName : 'packer-Resource-Group-uht810tdrw'
==> azure-arm:  -> Compute Name              : 'pkrvmuht810tdrw'
==> azure-arm:  -> Compute Location          : 'West Europe'
==> azure-arm:  -> Image ResourceGroupName   : 'rg_images'
==> azure-arm:  -> Image Name                : 'linuxWeb-0.0.2'
==> azure-arm:  -> Image Location            : 'westeurope'
==> azure-arm: Deleting resource group ...
==> azure-arm:  -> ResourceGroupName : 'packer-Resource-Group-uht810tdrw'
==> azure-arm:
==> azure-arm: The resource group was created by Packer, deleting ...
==> azure-arm: Deleting the temporary OS disk ...
==> azure-arm:  -> OS Disk : skipping, managed disk was used...
==> azure-arm: Deleting the temporary Additional disk ...
==> azure-arm:  -> Additional Disk : skipping, managed disk was used...
Build 'azure-arm' finished.

==> Builds finished. The artifacts of successful builds are:
--> azure-arm: Azure.ResourceManagement.VMImage:

OSType: Linux
ManagedImageResourceGroupName: rg_images
ManagedImageName: linuxWeb-0.0.2
ManagedImageId: /subscriptions/8a7aace5-                    /resourceGroups/rg_images/providers/Microsoft.Compute/images/linuxWeb-0.0.2
ManagedImageLocation: westeurope
```

4. After the Packer execution, in the Azure portal, we check that the image is present. The following screenshot shows our generated image:

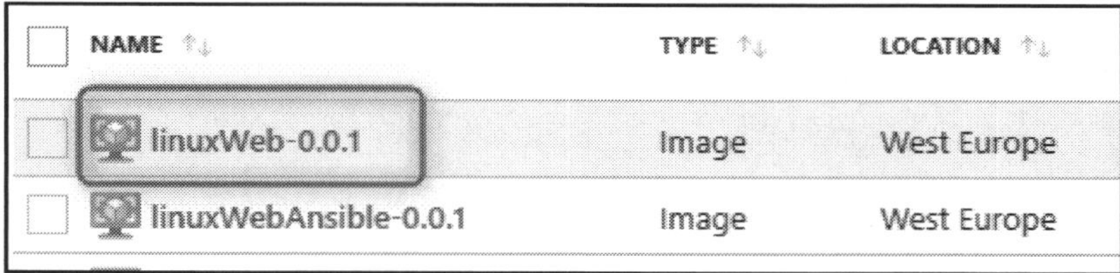

In this screen, we can see our image and the images that we generated with the Packer template, which uses Ansible.

It is also interesting to know that we can override the variables of our template when executing the `packer build` command, as in the following example:

```
packer build -var 'image_version=0.0.2' azure_linux.json
```

We can pass all variables with the `-var` options to the `build` command.

So, with this option, we can change the name of the image without changing the content of the template, and we can do this for all of the variables that are defined in our template.

The complete documentation of the Packer `build` command is available here: `https://www.packer.io/docs/commands/build.html`.

We have just seen the Packer command lines to check the syntax of the JSON Packer template and then to run Packer on a template that generates a VM image in Azure.

We will now learn how to provision, with Terraform, a VM based on this image that we have just generated.

Using a Packer image with Terraform

Now that we have generated a custom VM image, we will provision a new VM based on this new image. For the provisioning of this VM, we will continue to use IaC practices using Terraform from HashiCorp.

For the entire implementation and use of Terraform, read `Chapter 2`, *Provisioning Cloud Infrastructure with Terraform.*

To do this, we will take the Terraform script created in `Chapter 2`, *Provisioning Cloud Infrastructure with Terraform,* and modify it to use the custom image.

In the `compute.tf` script, add the following block of data, which will point to the VM image that we generated with Packer in the last section:

```
## GET THE CUSTOM IMAGE CREATED BY PACKER
data "azurerm_image" "customngnix" {
 name = "linuxWeb-0.0.1"
 resource_group_name = "rg_images"
}
```

In this code, we add a block of `azurerm_image` Terraform data that allows us to retrieve the properties of a VM image in Azure, in which we specify the `name` property with name of the custom image, and the `resource_group_name` property with the resource group of the image.

For more information about this azurerm_image data block and its properties, refer to the documentation: https://www.terraform.io/docs/providers/azurerm/d/image.html.

Then, in the VM Terraform code in the azurerm_virtual_machine resource code (still in the compute.tf file), the storage_image_reference section is modified with this code:

```
resource "azurerm_virtual_machine" "vm" {
...
 ## USE THE CUSTOM IMAGE
 storage_image_reference {
 id = "${data.azurerm_image.customngnix.id}"
 }
...
}
```

In this code, the ID property uses id of the image from the block data that we added earlier.

The entire code for the compute.tf script is available here: https://github.com/PacktPublishing/Learning_DevOps/blob/master/CHAP04/terraform/compute.tf, and the full Terraform code is here: https://github.com/PacktPublishing/Learning_DevOps/tree/master/CHAP04/terraform.

When executing this Terraform code, which is identical to a classic Terraform execution, as seen in Chapter 2, *Provisioning Cloud Infrastructure with Terraform*, the provisioned VM will be based on the custom image generated by Packer.

We have seen that by changing a little bit of our previous Terraform code, adding a data block that retrieves information from a VM image, and using the ID of that image, we can in Terraform use custom VM images generated by Packer.

Summary

In this chapter, we have seen how to install Packer and use it to create custom VM images. The VM image was made from two Packer templates: the first one using scripts and the second one using Ansible. Finally, we modified our Terraform code to use our VM image.

This chapter ends the implementation of IaC practices, starting with **Terraform** to provision a cloud infrastructure, then with **Ansible** for server configuration, and, finally, we finished with **Packer** for VM image creation.

With these VM images created by Packer, we will be able to improve infrastructure provisioning times with faster deployment, ready-to-use VMs, and, therefore, a reduction in downtime.

It is obvious that these are not the only IaC tools; there are many others on the marketplace, and we will have to do technology monitoring to find the ones that best suit your needs.

In the next chapter, we will start a new part, which is the implementation of CI/CD, and we will learn how to use Git for sourcing your code.

Questions

1. What are the two ways to install Packer?
2. What are the mandatory sections of a Packer template that are used to create a VM image in Azure?
3. Which command is used to validate a Packer template?
4. Which command is used to generate a Packer image?

Further reading

If you want to know more about Packer, here are some resources:

- Packer documentation: `https://www.packer.io/docs/`
- Packer Learning: `https://learn.hashicorp.com/packer`
- Using Packer in Azure: `https://docs.microsoft.com/en-us/azure/virtual-machines/linux/build-image-with-packer`
- Designing Immutable Infrastructure with Packer (Pluralsight Video): `https://www.pluralsight.com/courses/packer-designing-immutable-infrastructure`

2

Section 2: DevOps CI/CD Pipeline

This section covers the DevOps pipeline process starting with continuous integration and continuous deployment principles. It will be illustrated with the usage of different tools such as Jenkins, Azure Pipeline, and GitLab.

We will have the following chapters in this section:

- `Chapter 5`, *Managing Your Source Code with Git*
- `Chapter 6`, *Continuous Integration and Continuous Deployment*

5
Managing Your Source Code with Git

A few years ago, when we were developers and writing code as part of a team, we encountered recurring problems that were for the most part as follows:

- How to share my code with my team members
- How to version the update of my code
- How to track changes to my code
- How to retrieve an old state of my code or part of it

Over time, these issues have been solved with the emergence of source code managers, also called a **Version Control System** (**VCS**) or noted more commonly as a **Version Control Manager** (**VCM**).

The goals of these VCSes are mainly to do the following:

- Allow collaboration of developers' code.
- Retrieve the code.
- Version the code.
- Track code changes.

With the advent of agile methods and DevOps culture, the use of a VCS in processes has become mandatory. Indeed, as mentioned in `Chapter 1`, *DevOps Culture and Practices*, the implementation of a CI/CD process can only be done with a VCS as a prerequisite.

In this chapter, we will see how to use one of the best-known VCSes, which is Git. We will start with an overview of Git and see how to install it. Then, we will see its main command lines to familiarize any developer with their uses. Finally, we will see the current process of using Git workflows and usage of GitFlow. The purpose of this chapter is to show the usage of Git daily with simple processes.

This chapter covers the following:

- A Git overview and principal Git command lines
- The Git process and GitFlow pattern

This chapter does not cover the installation of a Git server, so if you want to know more about that, you can refer to the documentation: `https://git-scm.com/book/en/v2/Git-on-the-Server-The-Protocols`.

Technical requirements

The use of Git does not have any technical prerequisites; we just need a command-line Terminal.

To illustrate its usage in this chapter, we will use Azure Repos from the Azure DevOps platform (formally VSTS), which is a cloud platform that has a Git repository manager. You can register here, `https://visualstudio.microsoft.com/team-services/`, for free.

Check out the following video to see the Code in Action:
`http://bit.ly/2MK9Ky0`

Overviewing Git and its command lines

To understand the origin of Git, it is necessary to know that there are two types of VCS: centralized and distributed systems.

The first type to emerge is the **centralized systems**, such as SVN, CVS, Subversion, and TFVC (or SourceSafe). These systems consist of a remote server that centralizes the code of all developers.

We can represent a source centralized system like this:

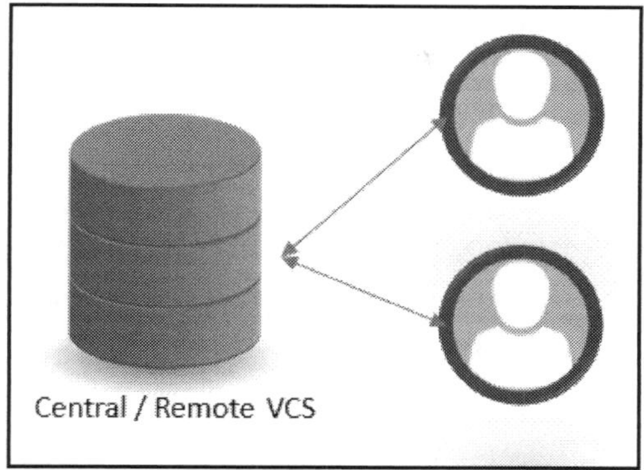

All developers can archive and retrieve their code on the remote server. The system allows better collaboration between teams and a guarantee of code backup. However, it has its drawbacks:

- In case of no connection (for a network problem or internet disconnection) between the developers and the remote server, no more archiving or code recovery actions can be performed.
- If the remote server no longer works, the code, as well as the history, will be lost.

The second type of CVS, which appeared later, is a **distributed system**, such as Mercurial or Git. These systems consist of a remote repository and a local copy of this repository on each developer's local machine, as shown in the following schema:

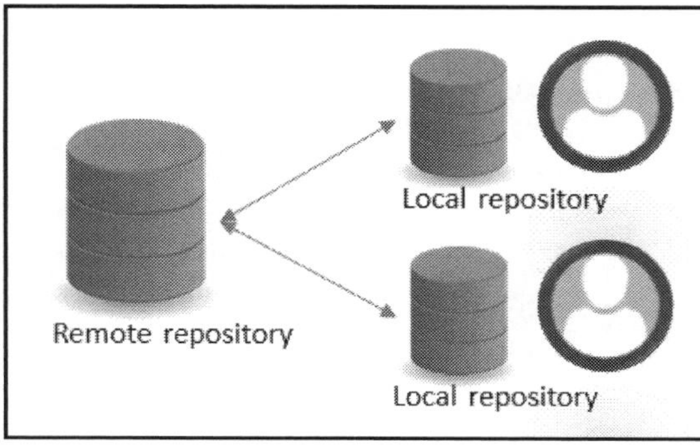

So, with this distributed system, even in the event of disconnection from the remote repository, the developer can continue to work with the local repository, and synchronization will be done when the remote repository is accessible again. And on the other hand, a copy of the code and its history is also present in the local repository.

Git is, therefore, a distributed CVS that was created in 2005 by Linus Torvalds and the Linux development community.

To learn a little more about Git's history, read this page: `https://git-scm.com/book/en/v2/Getting-Started-A-Short-History-of-Git`.

Since its creation, Git has become a very powerful and mature tool that can be used by anyone for coding.

Git is a free, cross-platform tool, and it can be installed on a local machine for people who manipulate code, that is, in client mode, but can also be installed on servers to host and manage remote repositories.

Git is a command-line tool with a multitude of options. Nevertheless, today there are many graphical tools, such as Git GUI, Git Kraken, GitHub desktop, or SourceTree, that allow you to interact with Git operations more easily and graphically without having to use command lines yourself. However, these graphical tools do not contain all of the operations and options available on the command line. Fortunately, many code editors such as Visual Studio Code, Visual Studio, JetBrains, and SublimeText allow direct code integration with Git and remote repositories.

For remote repositories, there are several clouds and free solutions such as GitHub, GitLab, Azure DevOps, or Bitbucket cloud. Also, there are other solutions called on-premises, which can be installed in an enterprise, such as Azure DevOps Server, Bitbucket, or GitHub Enterprise.

In this chapter, we will see the usage of Git, for example, with Azure DevOps as a remote repository, and we will see the use of GitLab and GitHub in future chapters.

In this section, we have introduced Git, and we will now see how to install it on a local machine to develop and version our sources.

Git installation

We will now detail the steps of installing and configuring Git on Windows, Linux, and macOS systems.

To install on a Windows machine manually, we must download the **Git for Windows tool** executable from `https://gitforwindows.org/` and, once downloaded, we click on the executable file and follow the following different configuration steps during the installation:

1. Choose the integration of Git in Windows Explorer by marking the **Windows Explorer integration** checkbox:

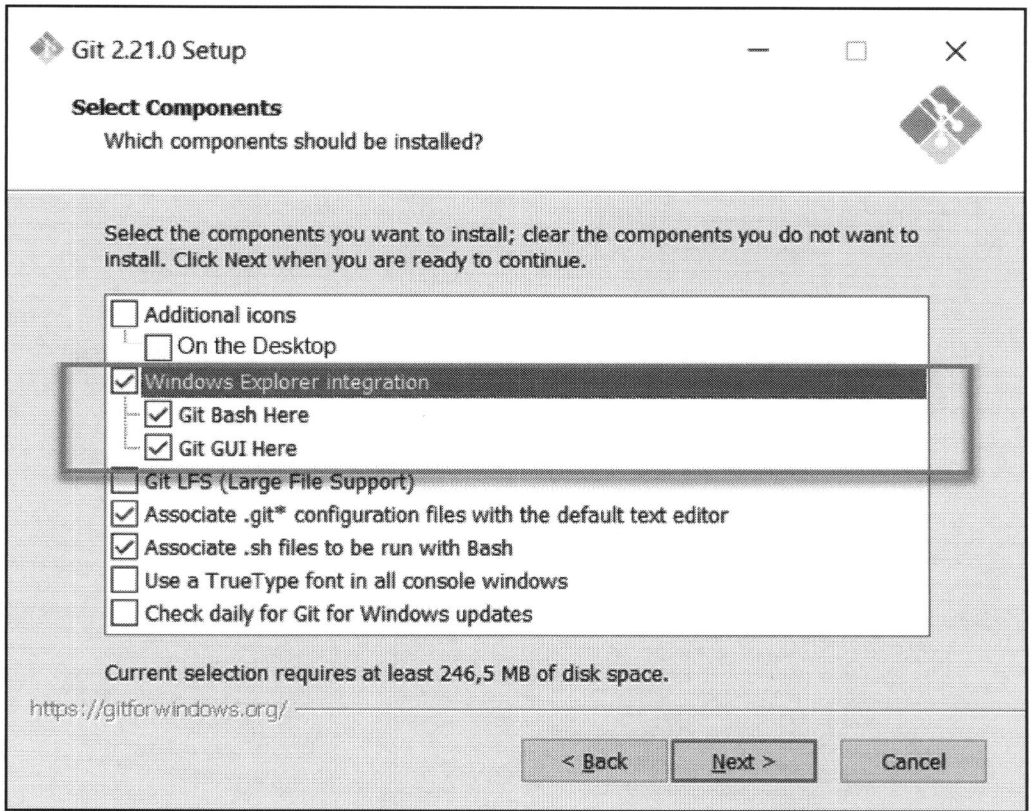

2. Choose your code editor IDE; in our case, we use **Visual Studio Code** by selecting the **Use Visual Studio Code as Git's default editor** option:

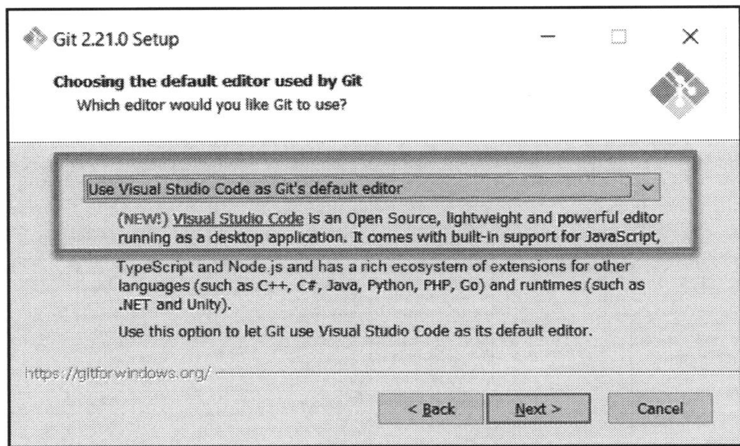

3. Choose the **PATH** environment variable option or we can leave the default choice proposed by the installer:

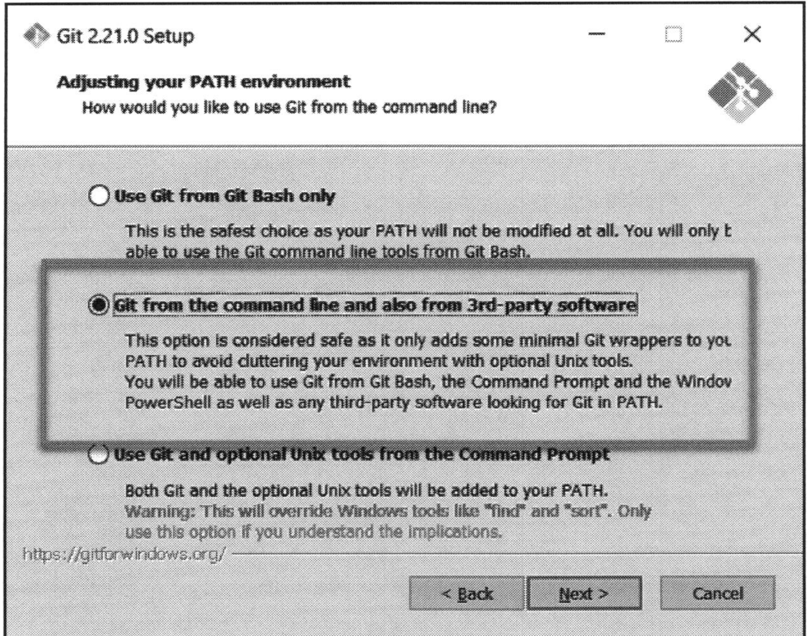

4. Choose the type of HTTPS transport that we will also leave by default with the **Use the OpenSSL library** option:

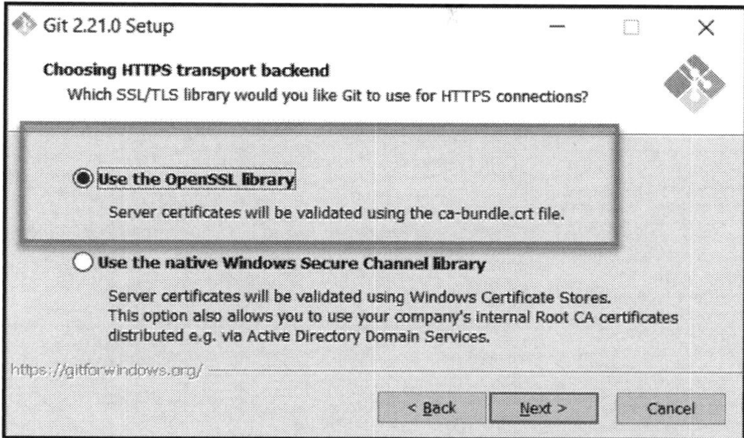

5. Choose the encoding of the end files; we will also select the default option, which archives in Unix format:

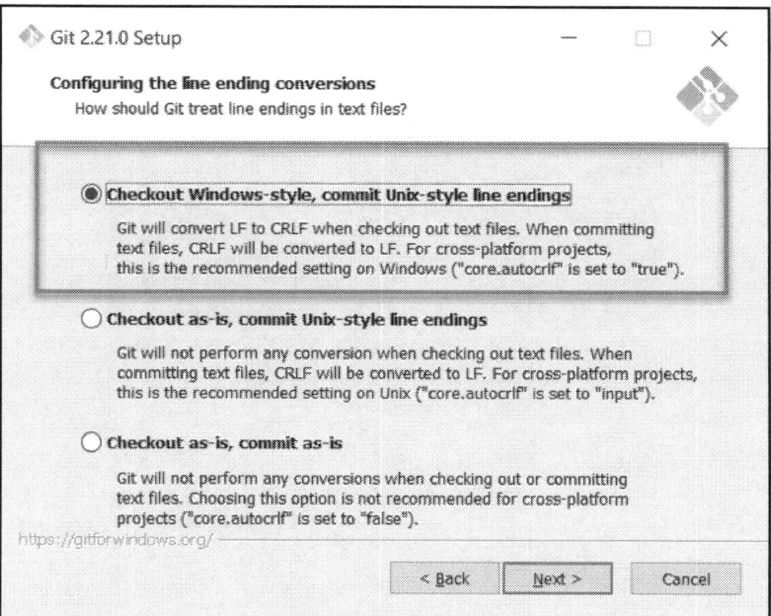

6. Then, finish the installation configuration by clicking on the **Install** button. At the end of the installation, the installation utility proposes to open Git Bash, which is a command-line Terminal Linux emulator dedicated to Git commands:

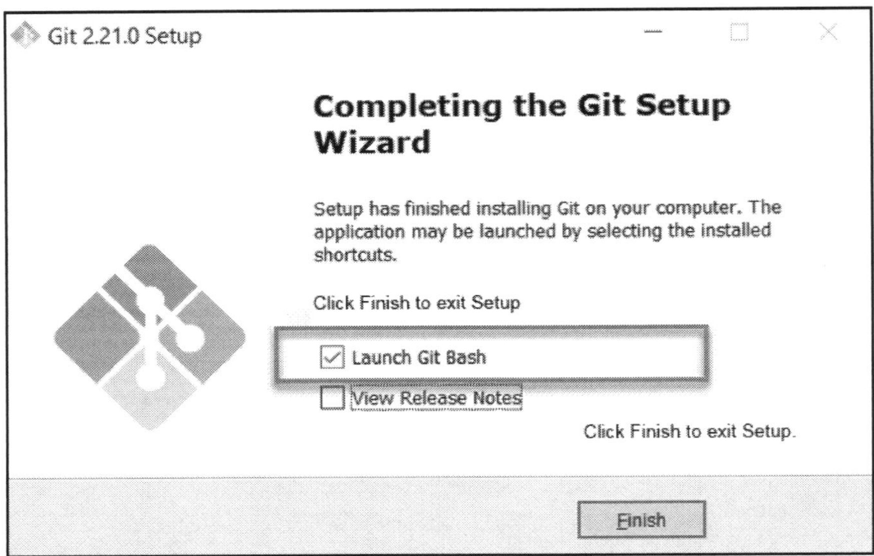

After the installation, we can immediately check the status of the Git installation directly in the Git Bash window by running the `git version` command, as follows:

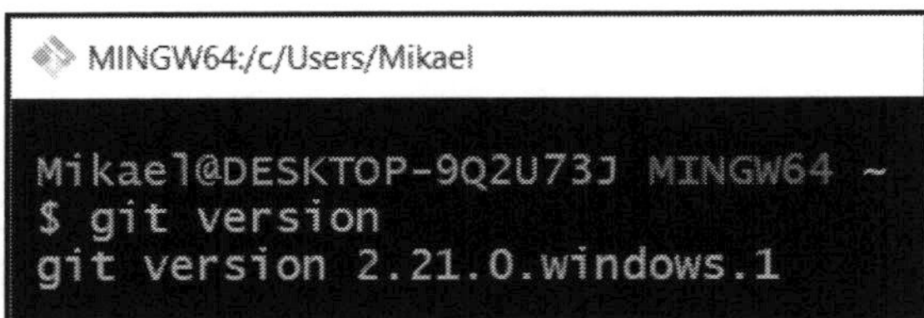

We can also install Git using an automatic script with **Chocolatey**, the Windows software package manager, which we already used in the previous chapters on Terraform and Packer. (As a reminder, the Chocolatey documentation is available here `https://chocolatey.org/`.)

To install Git for Windows with Chocolatey, we must execute the following command in an operating Terminal:

```
choco install -y git
```

And the result of this execution is displayed in the following screenshot:

```
PS C:\Windows\system32> choco install -y git
Chocolatey v0.10.11
Installing the following packages:
git
By installing you accept licenses for the packages.
Progress: Downloading git.install 2.21.0... 100%
Progress: Downloading git 2.21.0... 100%

git.install v2.21.0 [Approved]
git.install package files install completed. Performing other installation steps.
Using Git LFS
Installing 64-bit git.install...
git.install has been installed.
git.install installed to 'C:\Users\MikaelKRIEF\AppData\Local\Programs\Git'
  git.install can be automatically uninstalled.
Environment Vars (like PATH) have changed. Close/reopen your shell to
 see the changes (or in powershell/cmd.exe just type `refreshenv`).
 The install of git.install was successful.
  Software installed to 'C:\Users\MikaelKRIEF\AppData\Local\Programs\Git\'

git v2.21.0 [Approved]
git package files install completed. Performing other installation steps.
 The install of git was successful.
  Software install location not explicitly set, could be in package or
  default install location if installer.

Chocolatey installed 2/2 packages.
 See the log for details (C:\ProgramData\chocolatey\logs\chocolatey.log).
```

To install Git on a **Linux** machine, for Debian systems such as Ubuntu, we run the `apt-get` command with the following:

```
apt-get install git
```

Or for **CentOS** or Fedora, install Git with the `yum` command like this:

```
yum install git
```

For a **macOS** system, we can download and install Git using Homebrew (`https://brew.sh/`), which is the package manager dedicated to macOS, by executing this command in the Terminal:

```
brew install git
```

The installation of Git is finished, and we will now proceed to its configuration.

Configuration Git

Git configuration requires us to configure our username and email that will be used during code commit. To perform this configuration, we execute the following commands in a Terminal—either the one that is native to your OS or Git bash for Windows:

```
git config --global user.name "<your username>"
git config --global user.email "<your email>"
```

Then, we can check the configuration values by executing the following command:

```
git config --global --list
```

Refer to the following screenshot:

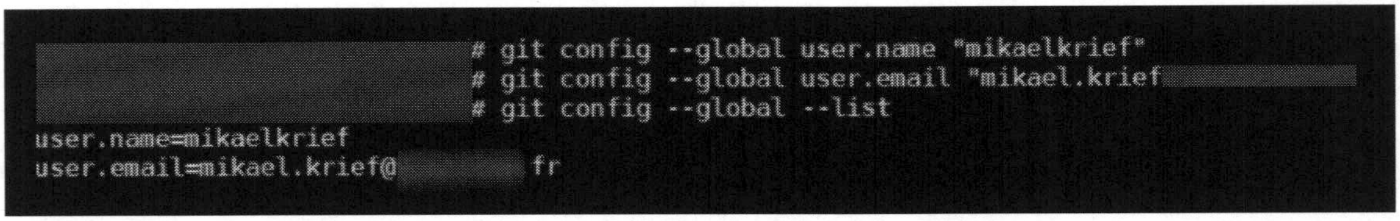

```
# git config --global user.name "mikaelkrief"
# git config --global user.email "mikael.krief
# git config --global --list
user.name=mikaelkrief
user.email=mikael.krief@        fr
```

Git is now configured and ready to use, but before using it, we will give an overview of its vocabulary.

Git vocabulary

Git is a tool that is very rich in objects and terminology and has its own concepts. Before using it, it is important to have some knowledge of its artifacts and the terms that compose it. Here is a brief list of this vocabulary:

- **Repository**: A repository is the basic element of Git; it is the storage space where the sources are tracked and versioned. There are **remote repositories** that centralize a team's code and allow team collaboration. There is also the **local repository**, which is a copy of the local repository on the local machine.
- **Clone**: Cloning is the act of making a local copy of a remote repository.
- **Commit**: A commit is a change made to one or more files, and the change is saved to the local repository. Each commit is unique and is identified by a unique number called `SHA-1`, by which code changes can be tracked.

- **Branch**: The code that is in the repository is stored by default in a `master` branch. A branch can create other branches that will be a replica of the master on which developers make changes, and that will allow us to work in isolation without affecting the master branch. At any time, we can merge one branch with another.
- **Merge**: This is the action that consists of merging the code of one branch with another.
- **Checkout**: This is the action that allows to switch from one branch to another.
- **Fetch**: This is the action of retrieving the code from the remote repository without merging it with the local repository.
- **Pull**: This is the action that consists of updating your local repository from the remote repository. A pull is equivalent to fetch and merge.
- **Push**: A push is the reverse action of a pull—it allows us to update the remote repository from the local repository.

These are the concepts to know when using Git and its workflow. Of course, this list is not exhaustive, and there are other important terms and notions; we can find information on several sites. Here is a small list:

- **GitHub vocabulary**: `https://help.github.com/en/articles/github-glossary`
- **Atlassian glossary**: `https://www.atlassian.com/git/glossary/terminology`
- **Linux academy**: `https://linuxacademy.com/blog/linux/git-terms-explained/`

With these concepts explained, we can now see how to use Git with its command lines.

Git command lines

With everything we've seen so far on Git, we can now start manipulating it. The best way to do so is to first learn to use Git on the command line; then, once the process is assimilated, we can use the graphical tools better.

Here is a presentation of the main Git command lines that are now part of (or should be part of) developers' daily lives. We will see their application in practice in the next section, *Understanding the Git process and GitFlow pattern*.

The first command is the one that allows us to retrieve the code from a remote repository.

Retrieving a remote repository

The first command line to know is the `clone` command that makes a copy of a remote repository to create a local repository. The command to execute is the following:

```
git clone <url of the remote repository>
```

The only mandatory parameter is the repository URL (any repository is identifiable by a unique URL). Once this command is executed, the content of the remote repository is downloaded to the local machine and the local repository is automatically created and configured.

Initializing a local repository

Note that `init` is the Git command that allows you to create a local repository. To do this in the directory that will contain your local repository, run the following simple command:

```
git init
```

This command creates a `.git` directory that contains all of the folders and configuration files of the local repository.

Configuring a local repository

After the `init` command, the new local repository must be configured by setting up the linked remote repository. To make this setting, we will add `remote` with the following command:

```
git remote add <name> <url of the remote>
```

The name passed as a parameter allows this remote repository to be identified locally; it is the equivalent of an alias. It is also possible to configure several remote devices on our local repository.

Adding a file for the next commit

Making a commit (which we will see next) is to archive our changes in our local repository. When we edit files, we can choose which ones will be included in the next commit; it's the staged concept. The other files not selected will be set aside for a later commit.

To add files to the next commit, we execute the `add` command, as follows:

```
git add <files path to add>
```

So, for example, if we want all of the files modified at the next commit, execute the `git add .` command.
We can also filter the files to be added with RegEx, such as `git add *.txt`.

Creating a commit

A commit is the Git entity that contains a list of changes made to files and that have been registered in the local repository. Making a commit, therefore, consists of archiving changes made to files that have been previously selected with the `add` command.

The command to create a commit is as follows:

```
git commit -m "<your commit message>"
```

The `-m` parameter corresponds to a message, or description, that we assign to this commit. The message is very important because we will be able to identify the reason for the changes in the files.

It is also possible to commit all files modified since the last commit, without having to execute the `add` command, by executing the `git commit -a  -m "<message>"` command.

Once the commit is executed, the changes are archived in the local repository.

Updating the remote repository

When we make commits, they are stored in the local repository and when we are ready to share them with the rest of the team for validation or deployment, we must publish them to the remote repository. To update a remote repository from commits made on a local repository, a `push` operation is performed with this command:

```
git push <alias> <branch>
```

The alias passed as a parameter corresponds to the alias of the remote repository configured in the Git configuration of the local repository (done by the `git remote add` command).

And the branch parameter is the branch to be updated—by default, it is the `master` branch.

Synchronizing the local repository from the remote

As we discussed previously, the Git command line is used to update the remote repository from the local repository. Now, to perform the reverse operation, that is, update the repository with all of the changes of the other members that have been pushed on the remote repository, we will perform a `pull` operation with this command:

```
git pull
```

The execution of this command leads to two operations:

1. Merging the local code with the remote code.
2. Committing to the local repository.

On the other hand, if we do not want to commit—to be able to make other changes, for example—instead of the `pull` command, we must execute `fetch` with this command:

```
git fetch
```

Its execution-only merges the local code from the remote code, and to archive it we will have to execute `commit` to update the local repository.

Managing branches

By default, when creating a repository, the code is placed in the main branch called **master**. In order to be able to isolate the developments of the master branch, for example, to develop a new feature, fix a bug, or even make technical experiments, we can create new branches from other branches and merge them together when we want to merge their code.

To create a branch from the current locally loaded branch, we execute the following command:

```
git branch <name of the desired branch>
```

To switch to another branch, we execute the following command:

```
git checkout <name of the branch>
```

This command changes the branch and loads the current working directory with the contents of that branch.

To merge a branch to the current branch, execute the `merge` command, as follows:

```
git merge <branch name>
```

With the name of the branch, we want to merge as a parameter.

Finally, to display the list of local branches, we execute the command branch:

```
git branch
```

Branch management is not easy to use from the command line. The graphical tools of Git, already mentioned, allow better visualization and management of the branches. We will see their uses in detail in the next section, which will deal with the Git process. That's all for the usual Git command lines, although there are many more to handle.

In the next section, we will put all of this into practice by applying the work and collaboration process with Git, and look at an overview of the GitFlow pattern.

Understanding the Git process and GitFlow pattern

So far, we have seen the fundamentals of a very powerful VCS, which is Git, with its installation, configuration, and some of its most common command lines. In this section, we will put all of this into practice with a case study that will show which Git process to apply throughout the life cycle of a project.

In this case study, for remote repositories, we will use **Azure Repos** (one of the services of the Azure DevOps services), which is a free Git cloud platform that can be used for personal or even business projects. To learn more about Azure DevOps, consult the documentation here: `https://azure.microsoft.com/en-us/services/devops/`. We will often talk about it in this book.

Let's first look at how collaboration with Git is constituted, then we will see how to isolate the code using branches.

Starting with the Git process

In this lab, we will explain the Git collaboration process with a team of two developers who start a new application development project.

Following are the steps of the Git process that we will discuss in detail in this section:

1. The first developer commits the code on the local repository and pushes it to the remote repository.
2. Then, the second developer gets the pushed code from the remote repository.
3. The second developer updates this code, creates a commit, and pushes the new version of the code to the remote repository.
4. Finally, the first developer retrieves the last version of the code in the local repository.

However, before we start working with Git commands, we will have a look at the steps to create and configure a Git repository in Azure Repos.

Creating and configuring a Git repository

To start with, we will create a remote Git repository in Azure DevOps that will be used to collaborate with other team members. Follow these steps:

1. In Azure DevOps, we will create a new project:

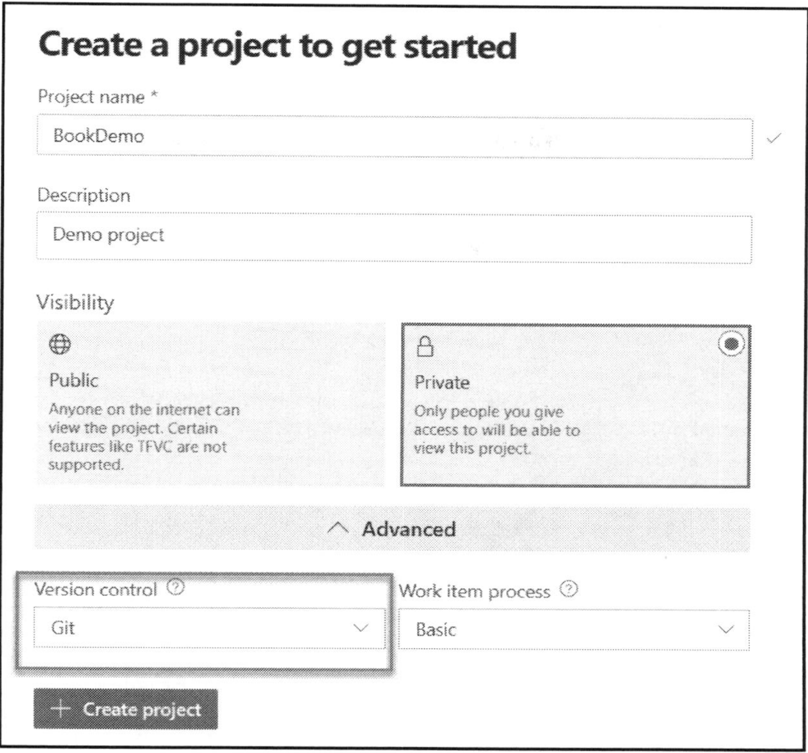

2. Enter a name and description for the project, and as we can see from the preceding screenshot, the **Version control** type is **Git**. Then, on the left-side menu, we click on the Azure **Repos** service as follows:

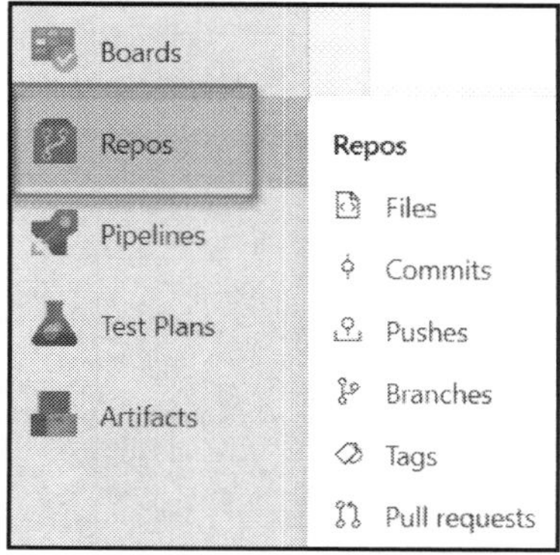

3. A default repository is already created, and it has the same name as the project:

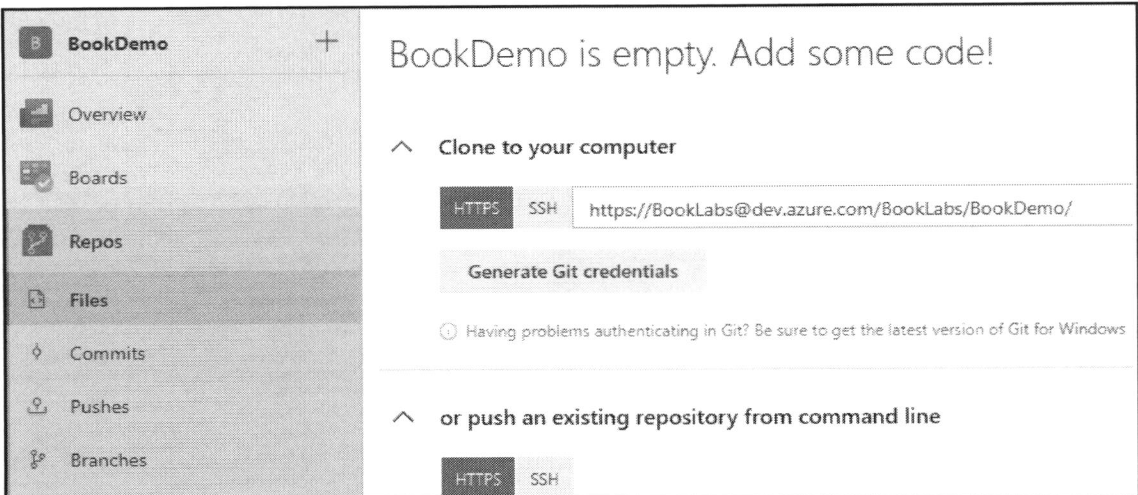

Now that the Git repository is created in Azure Repos, we will see how to initialize and configure our local working directory to work with this Git repository.

4. To initialize this repository, we will create a local directory and name it, for example, `bookProject`, which will contain the application code.
5. Once this directory is created, we will initialize a local Git repository by executing inside the `bookProject` folder the `git init` command, as follows:

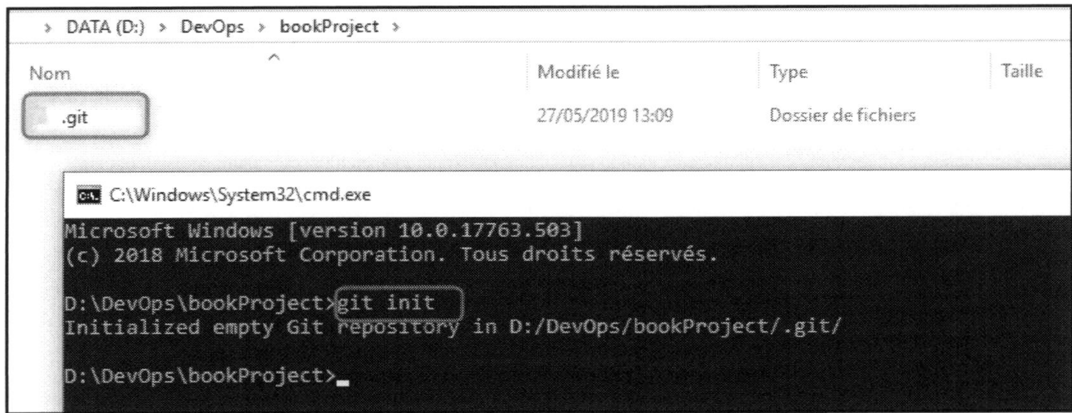

The result of the preceding command is a confirmation message and a new `.git` directory. This new directory will contain all of the information and configuration of the local Git repository.

6. Then, we will configure this local repository to be linked to the Azure DevOps remote repository. This link will allow the synchronization of local and remote repositories. For this, in Azure DevOps, we get the URL of the repository, which is found in the repository information, as shown in the following screenshot:

7. Run the `git remote add` command inside the `bookProject` folder that we created in *step 4*:

```
git remote add origin <url of the remote repository>
```

By executing the preceding command, we create an `origin` alias that will point to the remote repository URL.

This whole initialization and configuration procedure also applies to an empty directory, as in our case, or to a directory that already contains code that we want to archive.

So, we now have a local Git repository that we will work on. After creating and configuring our repository, we will start developing code for our application and collaborate using the Git process by starting with the commit of the code.

Committing the code

The first step in the process is the code commit, which allows you to store your code changes in your local Git repository, by following these steps:

1. In our directory, we will create a `Readme.md` file that contains the following example text:

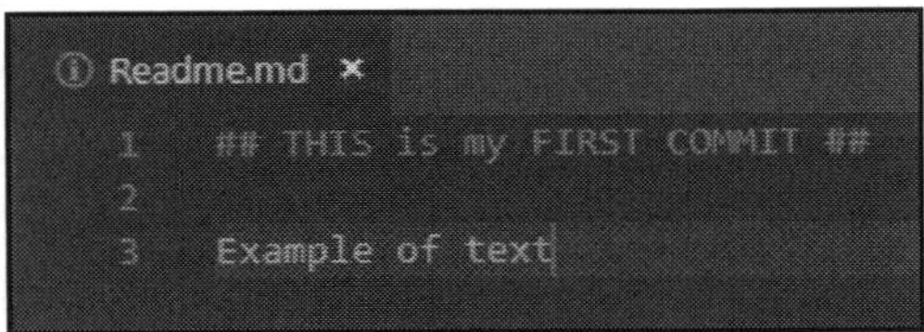

2. To make a commit of this file in our local repository, we must add it to the list of the next commit by executing the `git add .` command (the `.` character at the end indicates to include all of the files to modify, add, or delete).
3. We can also see the status of the changes that will be made to the local repository by running the `git status` command, as shown in the following screenshot:

```
\bookProject>git add .

\bookProject>git status
On branch master

No commits yet

Changes to be committed:
  (use "git rm --cached <file>..." to unstage)

        new file:   Readme.md
```

From the following execution, we can see that a `Readme.md` file is created and it will also integrate to commit.

The last operation to do is validate the change by archiving it in the local repository, and for this we execute the following command:

```
git commit -m "Add file readme.md"
```

We make a commit with a description, for keeping track of changes. Following is a screenshot of the preceding command:

```
D:\DevOps\bookProject>git commit -m "Add file readme.md"
[master (root-commit) 0ffca3e] Add file readme.md
 Committer: Mikael KRIEF <mikael.krief@younited-credit.fr>
your name and email address were configured automatically based
on your username and hostname. Please check that they are accurate.
You can suppress this message by setting them explicitly. Run the
following command and follow the instructions in your editor to edit
your configuration file:

    git config --global --edit

After doing this, you may fix the identity used for this commit with:

    git commit --amend --reset-author

 1 file changed, 3 insertions(+)
 create mode 100644 Readme.md
```

In this example, we also notice a Git message that gives information about the user's Git configuration.

Now that our local Git repository is up to date with our changes, we just have to archive them on the remote repository.

Archiving on the remote repository

To archive local changes and allow the team to work and collaborate on this code as well, we will push our commit in the remote repository by executing the command:

```
git push origin master
```

We indicate to the `push` command the alias and the branch of the remote repository. During the first execution of this command, we will be asked to authenticate to the Azure DevOps repository, as shown in the following screenshot:

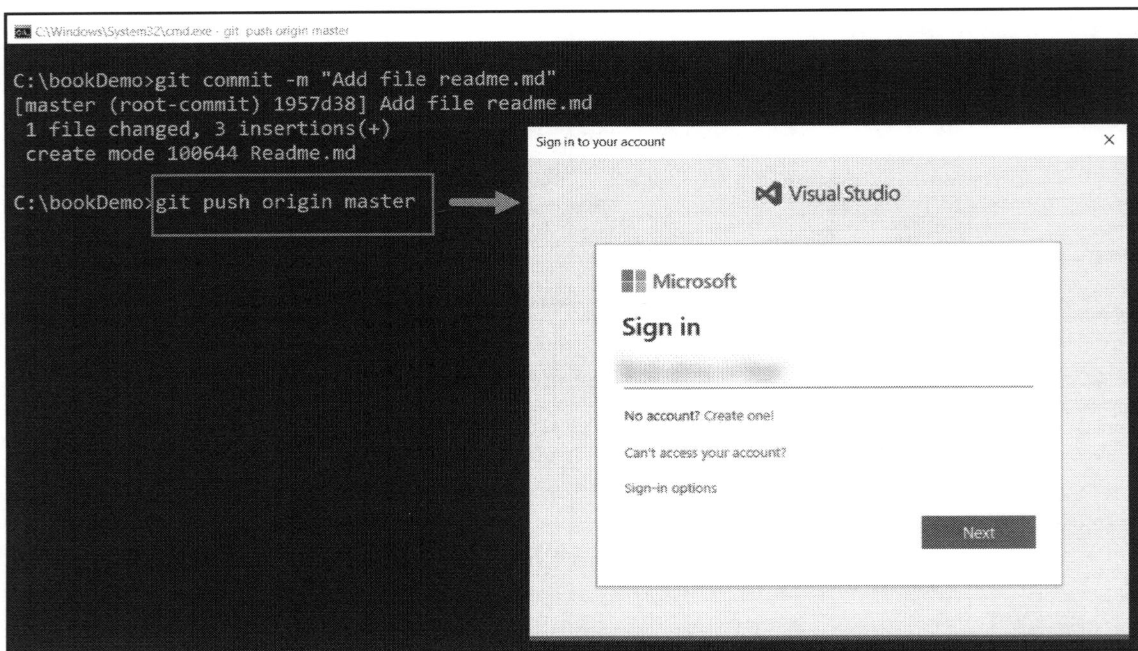

And after authentication, the push is executed as follows:

```
\bookProject>git push origin master
Enumerating objects: 3, done.
Counting objects: 100% (3/3), done.
Writing objects: 100% (3/3), 275 bytes | 275.00 KiB/s, done.
Total 3 (delta 0), reused 0 (delta 0)
remote: Analyzing objects... (3/3) (6 ms)
remote: Storing packfile... done (324 ms)
remote: Storing index... done (148 ms)
To https://dev.azure.com/BookLabs/BookDemo/_git/BookDemo
 * [new branch]      master -> master
```

The remote repository is up to date with our changes, and in the Azure Repos interface, we can see the code of the remote repository. The following screenshot shows the code in Azure Repos that contains the added file:

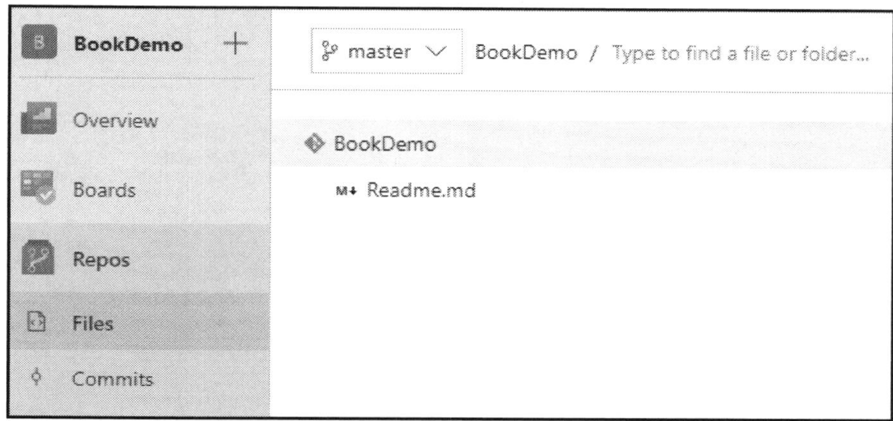

That's it: we have performed the first major step of the Git process, which consists of initializing a repository and pushing code into a remote repository. The next step is to make changes to the code by another team member.

Cloning the repository

When another member wants to retrieve the entire remote repository code for the first time, they perform a clone operation, and to do so they must execute this command:

```
git clone <repository url>
```

The execution of this command performs these actions:

1. Creates a new directory with the name of the repository.
2. Creates a local repository with its initialization and configuration.
3. Downloads the remote code.

The team member can therefore modify and make changes to the code.

The code update

When the code is modified and the developer updates their changes on the remote repository, they will perform exactly the same actions as when the remote repository was initialized, with the execution of the following commands:

```
git add .
git commit -m "update the code"
git push origin master
```

We added the files to the next commit, created the commit, and pushed the commit to the remote repository.

The last remaining step is the retrieval of updates by other members.

Retrieving updates

When one of the members updates the remote repository, it is possible to retrieve them and update our local repository with the changes by running the `git pull` command, as follows:

```
git pull origin master
```

The following screenshot shows the execution of this `git pull` command:

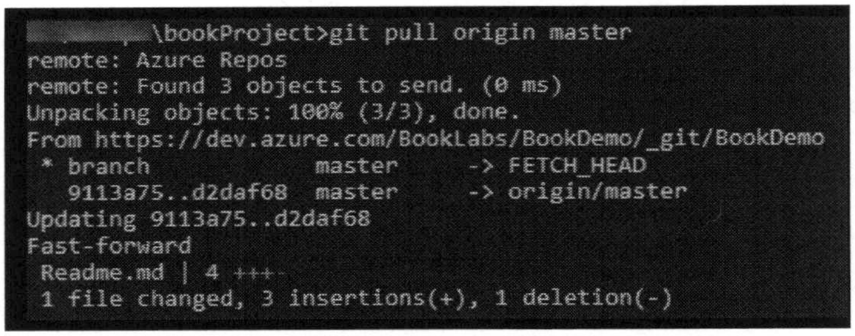

```
\bookProject>git pull origin master
remote: Azure Repos
remote: Found 3 objects to send. (0 ms)
Unpacking objects: 100% (3/3), done.
From https://dev.azure.com/BookLabs/BookDemo/_git/BookDemo
 * branch            master     -> FETCH_HEAD
   9113a75..d2daf68  master     -> origin/master
Updating 9113a75..d2daf68
Fast-forward
 Readme.md | 4 +++-
 1 file changed, 3 insertions(+), 1 deletion(-)
```

From the preceding execution, we indicate the `origin` alias and `master` branch to be updated in the local repository, and its execution displays the pushed commits. At the end of the execution of this command, our local repository is up to date.

For the rest of the process with the code update, it is the same as we saw earlier in *The code update* section.

We have just seen the simple process of using Git, but it is also possible to isolate its development with the use of branches.

Isolating your code with branches

When developing an application, we often need to modify part of the code without wanting to impact the existing stable code of the application. For this, we will use a feature that exists in all VCSes that allows us to create and manage branches.
The mechanics of using the branches are quite simple:

1. From a branch, we create another branch.
2. We develop this new branch.
3. To apply the changes made on this new branch to the original branch, a merge operation is performed.

Conceptually, a branch system is represented in this way:

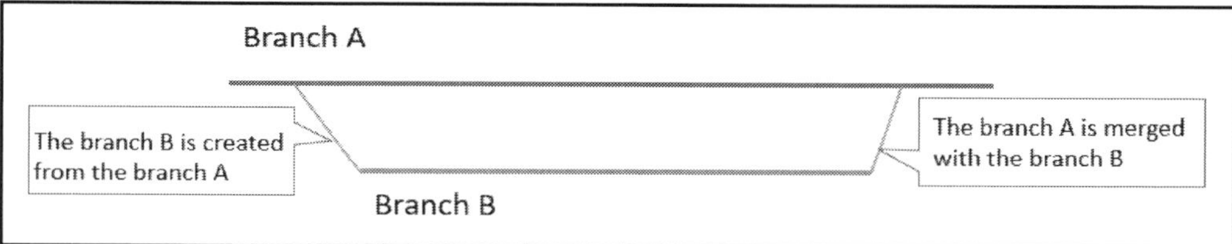

In this diagram, we see that branch B is created from branch A. These two branches will be modified by development teams for adding a new feature to our application. But in the end, branch A is merged with branch B, and hence the branch B code changes.

Now that we have seen the branch bases, let's look at how to use branches in Git by following these steps.

The purpose of this lab is to create a `Feature1` branch from the `master` branch. Then, after some changes on the `Feature1` branch, we will merge the code changes from the `Feature1` branch to the `master` branch:

1. First, create the `Feature1` branch on the local repository, from `master`, with the help of the following command:

```
git branch Feature1
```

   The execution of this command created a new branch, `Feature1`, which contains exactly the same code as the parent branch, `master`.

2. To load your working directory with the code of this branch, we execute the following command:

```
git checkout Feature1
```

From the following output, we can see the switch of the branch:

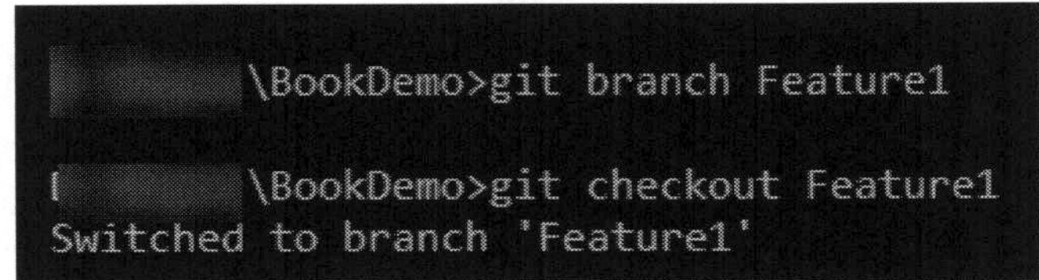

3. We can also see the list of branches of our local repository with the following command:

```
git branch
```

The output of this command is shown as follows:

```
\BookDemo>git branch
* Feature1
  master
```

During execution, we can see two branches and the active branch as well.

4. Now, we will make changes to our code on this branch. The update mechanism is identical to everything we have already seen previously, so we will execute the following commands:

```
git add .
git commit -m "Add feature 1 code"
```

5. For the `push` operation, we will specify the name of the `Feature1` branch in the parameter with this:

```
git push origin Feature1
```

The following screenshot is the execution that performs all of these steps:

```
\BookDemo>git add .

\BookDemo>git commit -m "Add feature 1 code"
[Feature1 0ed2a45] Add feature 1 code
 1 file changed, 3 insertions(+), 1 deletion(-)

\BookDemo>git push origin Feature1
Enumerating objects: 5, done.
Counting objects: 100% (5/5), done.
Delta compression using up to 4 threads
Compressing objects: 100% (2/2), done.
Writing objects: 100% (3/3), 306 bytes | 153.00 KiB/s, done.
Total 3 (delta 1), reused 0 (delta 0)
remote: Analyzing objects... (3/3) (97 ms)
remote: Storing packfile... done (188 ms)
remote: Storing index... done (92 ms)
To https://dev.azure.com/BookLabs/BookDemo/_git/BookDemo
 * [new branch]      Feature1 -> Feature1
```

With the preceding commands that we executed, we created a commit on the `Feature1` branch. We then executed the `push` command to publish the new branch as well as its commit in the remote repository.

6. On the other hand, in the Azure Repos interface, we can see our two branches in the **Branches** section, as shown in the following screenshot:

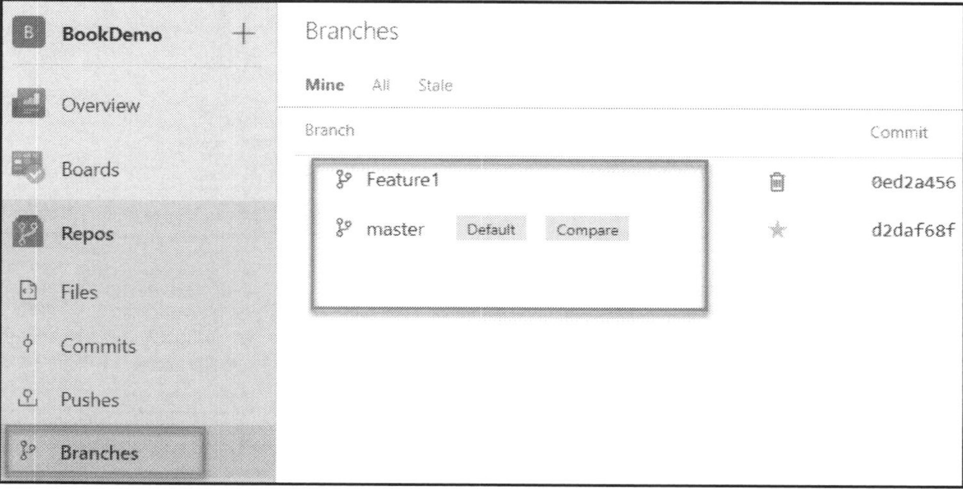

7. The last step in our process is to merge the code of the `Feature1` branch into that of the master branch, `master`. To perform this merge, we will first load our working directory with the `master` branch, then merge the `Feature1` branch to the current branch, which is `master`.

   To do this, we will execute the following commands:

   ```
   git checkout master
   git merge Feature1
   ```

   The execution of these commands is displayed as follows:

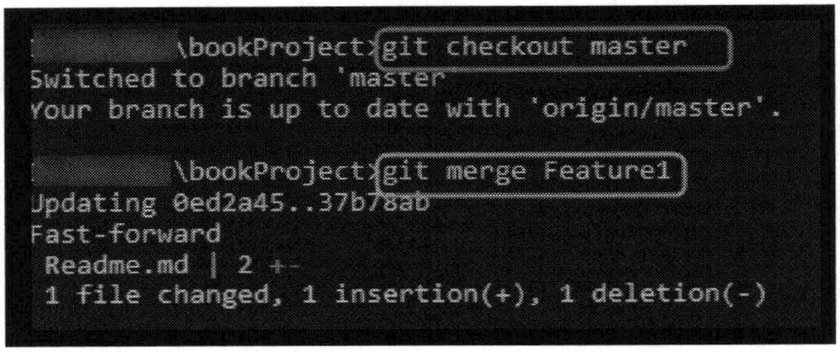

   ```
   \bookProject>git checkout master
   Switched to branch 'master'
   Your branch is up to date with 'origin/master'.

   \bookProject>git merge Feature1
   Updating 0ed2a45..37b78ab
   Fast-forward
    Readme.md | 2 +-
    1 file changed, 1 insertion(+), 1 deletion(-)
   ```

At the end of its execution, the code of the `master` branch, therefore, contains the changes made in the `Feature1` branch.

We have just seen the use of branches and their merges in Git; now, let's look at the GitFlow branch pattern and its utility.

Branching strategy with GitFlow

When we start using branches in Git, the question that often arises is: what is the right branch strategy to use? In other words, what is the key that we need to isolate our code—is it by environment, functionality, theme, or release?

This question has no universal answer, and the management of branches within a project depends on the context of the project. However, there are branch strategy patterns, which have been approved by several communities and a multitude of users that allow a better collaboration process in a project under Git.

I suggest we have a closer look at one of these branch patterns, which is GitFlow.

The GitFlow pattern

One of these patterns is **GitFlow**, developed by Nvie, which is very popular and has a very easy-to-learn Git workflow.

The branch diagram of GitFlow is as follows:

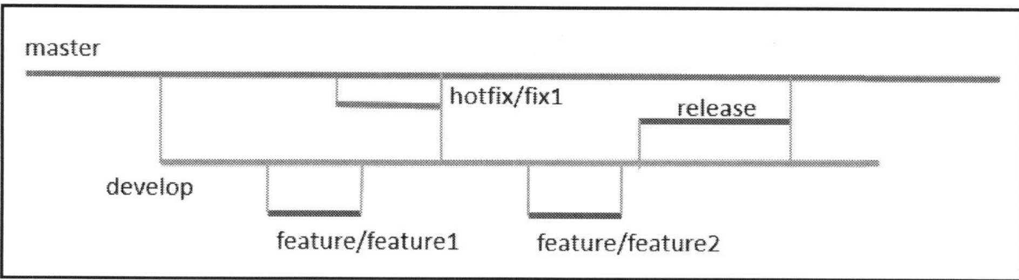

Let's look at the details of this workflow as well as the purpose of each of these branches in their order of creation:

1. First of all, we have the `master` branch, which contains the code that is currently in production. No developer is working directly on it.
2. From this `master` branch, we create a `develop` branch, which is the branch that will contain the changes to be deployed in the next delivery of the application.
3. For each of the application's functionalities, a `feature/<name of the feature>` branch is created (the / will hence create a feature directory) from the `develop` branch.
4. As soon as a feature is finished being coded, the branch of the feature is merged into the `develop` branch.
5. Then, as soon as we want to deploy all of the latest features developed, we create a `release` branch from `develop`.
6. The `release` branch content is deployed in all environments successively.
7. As soon as the deployment in production has taken place, the `release` branch is merged with the `master` branch, and, with this merge operation, the `master` branch contains the production code. Hence, the code that is on the `release` branch and features branch is no longer necessary, and these branches can be deleted.
8. If a bug is detected in production, we create a `hotfix/<bug name>` branch; then once the bug is fixed, we merge this branch into the `master` and `develop` branches to propagate the fix on the next branches and deployments.

After seeing the workflow of the GitFlow pattern, we'll discuss the tools that facilitate its implementation.

GitFlow tools

To help teams and developers to use GitFlow, Nvie created a Git override command-line tool that allows you to easily create branches according to the workflow step. This tool is available on the Nvie GitHub page here: `https://github.com/nvie/gitflow`.

On the other hand, there are also Git graphical tools that support the GitFlow model, such as GitKraken (`https://www.gitkraken.com/`) and SourceTree (`https://www.sourcetreeapp.com/`). These tools allow us to use the GitFlow process via a graphical interface, as shown in the following screenshot, with the use of SourceTree with a feature **Git-flow** that allows us to create a branch using the GitFlow pattern.

The following screenshot shows the configuration of SourceTree for a GitFlow pattern branch naming:

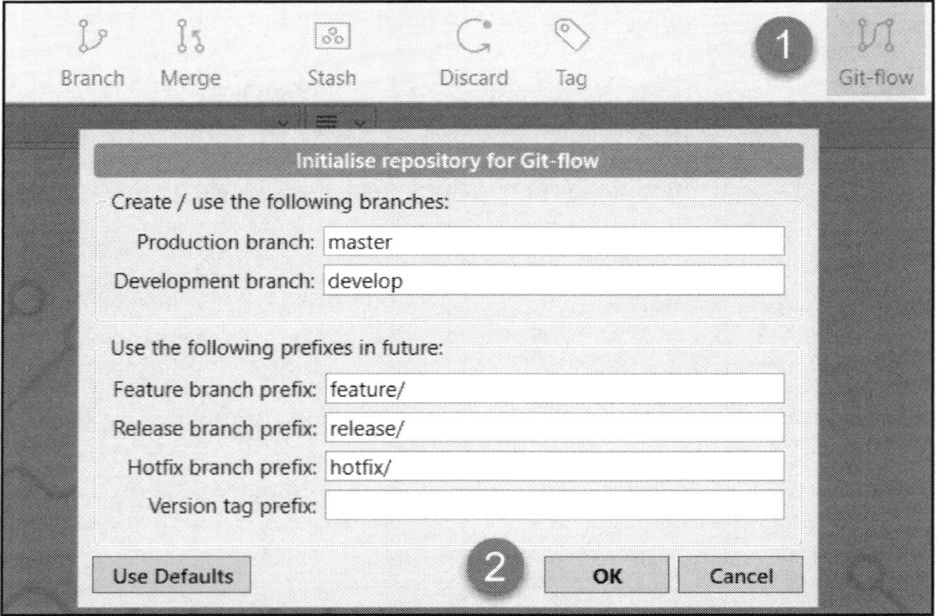

Then, after this configuration, we can create a branch easily with this name, as shown in this screenshot:

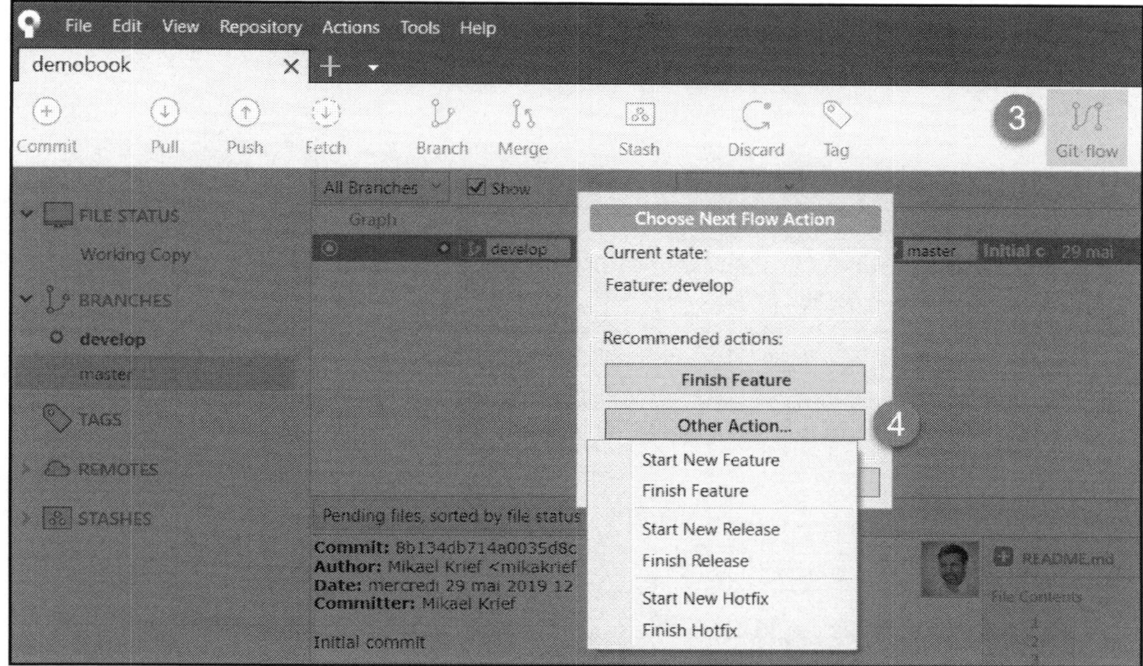

As shown in the preceding screenshots, we use SourceTree and the graphical interface that allows us to create branches intuitively.

For more documentation on GitFlow and its process, read this article: `https://jeffkreeftmeijer.com/git-flow/`.

To get started with GitFlow, I suggest you get used to small projects, then you will see that the mechanism is the same for larger projects. In this section, we saw the process of using Git's command lines with its workflow. Then, we talked about the management of development branches, and finally we went a little further into branches with GitFlow, which allows simple management of branches and a better development workflow with Git.

Summary

Git is today an essential tool for all developers; it allows us to use command lines or graphical tools and sharing and versioning code for better collaboration of team members.

In this chapter, we saw how to install Git on the different OSes and an overview of the main command lines. Then, through a small lab, we saw the Git workflow with the application of Git command lines. Finally, we presented the isolation of code with the implementation of branches and the use of the GitFlow pattern, which gives a simple model of branch strategy.

In the next chapter, we will talk about continuous integration and continuous deployment, which is one of the key practices of the DevOps culture.

Questions

1. What is Git?
2. Which command is used to initialize a repository?
3. What is the name of the Git artifact for saving part of the code?
4. Which Git command allows you to save your code in the local repository?
5. Which Git command allows you to send your code to the remote repository?
6. Which Git command allows you to update your local repository from the remote repository?
7. What Git mechanism is used for Git code isolation?
8. What is GitFlow?

Further reading

If you want to know more about Git, here are some resources:

- Git documentation: `https://www.git-scm.com/doc`
- Pro Git book: `https://git-scm.com/book/en/v2`
- Mastering Git book: `https://www.packtpub.com/application-development/mastering-git`
- Try Git: `https://try.github.io/`
- Learning Git branching: `https://learngitbranching.js.org/`
- Atlassian Git tutorials: `https://www.atlassian.com/git/tutorials`
- Git Flow cheat sheet: `https://danielkummer.github.io/git-flow-cheatsheet/`

6
Continuous Integration and Continuous Delivery

One of the main pillars of DevOps culture is the implementation of continuous integration and deployment processes, as we explained in `Chapter 1`, *DevOps Culture and Practices*.

In the previous chapter, we looked at the use of Git with its command lines and usage workflow, and in this chapter, we will look at the important role Git has in the CI/CD workflow.

Continuous integration (**CI**) is a process that provides rapid feedback on the consistency and quality of code to all members of a team. It occurs when each user's code commit retrieves and merges the code from a remote repository, compiles it, and tests it.

Continuous delivery (**CD**) is the automation of the process that deploys an application in different stages (or environments).

In this chapter, we will learn the principles of the CI/CD process as well as its practical use with different tools such as **Jenkins**, **Azure Pipelines**, and **GitLab CI**. For each of these tools, we will present the advantages, disadvantages, and the best practices, and look at a practical example of implementing a CI/CD pipeline.

You will learn about the concept of a CI/CD pipeline. After this, we will explore the package managers and their role in the pipeline.

Then, you will learn how to install Jenkins and finally build a CI/CD pipeline on Jenkins, Azure Pipelines, and GitLab CI.

This chapter covers the following:

- The CI/CD principles
- Using a package manager in the CI/CD process
- Jenkins CI/CD implementation

- Azure Pipelines for CI/CD
- Using GitLab CI

Technical requirements

The only requirement for this chapter is to have Git installed on your system, as detailed in the previous chapter.

The source code for this chapter is available at `https://github.com/PacktPublishing/Learning_DevOps/tree/master/CHAP06`.

Check out the following video to see the Code in Action:
`http://bit.ly/2pdXFrM`

The CI/CD principles

To implement a CI/CD pipeline, it is important to know the different elements that will be required to build an efficient and safe pipeline. In order to understand the principles of CI/CD, the following diagram shows the different steps of a CI/CD workflow, which we already saw in `Chapter 1`, *DevOps Culture and Practices*:

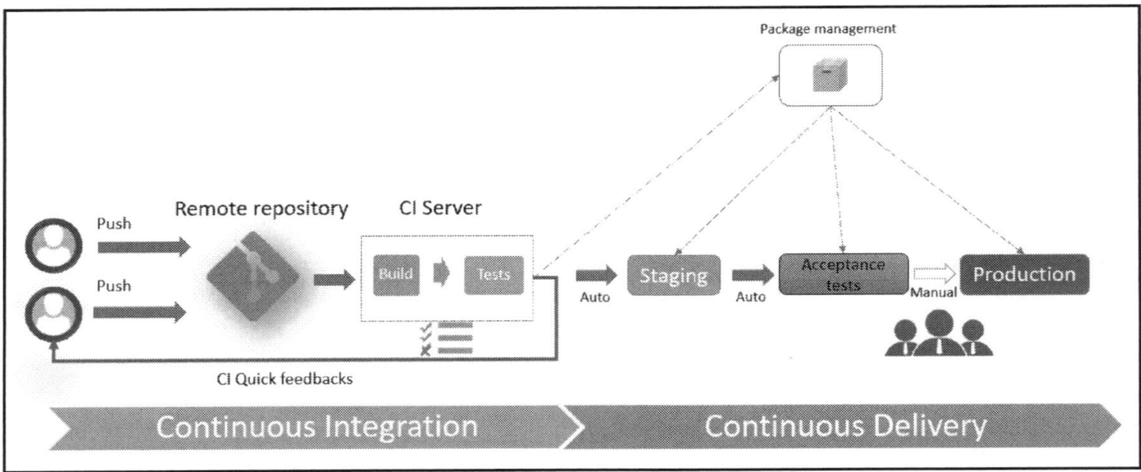

Let's look in detail at each of these steps in order to list the artifacts of the CI/CD process.

Continuous integration (CI)

The CI phase checks the code archived by the team members. It must be executed on each commit that has been pushed to the remote repository.

The setting up of a Git-type **SCV** is a necessary prerequisite that makes it possible to centralize the code of all the members of a team.

The team will have to decide on a code branch that will be used for continuous integration. For example, we can use the *master* branch, or the *develop* branch as part of GitFlow; it just needs to be an active branch that very regularly centralizes code changes.

In addition, CI is achieved by an automatic task suite that is executed on a server, following similar patterns executed on a developer's laptop that has the necessary tools for continuous integration; this server is called the **CI server**.

The CI server can be either of the on-premise type, installed in the company data center, such as Jenkins or TeamCity, or perhaps a cloud type that we don't have to worry about installing and maintaining, such as Azure Pipelines or GitLab CI.

The tasks performed during the CI phase must be automated and take into account all the elements that are necessary for the verification of the code.

These tasks are generally the compilation of code and the execution of unit tests with code coverage. We can also add static code analysis with SonarQube (or SonarCloud), which we will look at in `Chapter 10`, *Static Code Analysis with SonarQube*.

At the end of the verification tasks, in many cases, the CI generates an application package that will be deployed on the different environments (also called **stages**).

To be able to host this package, we need a **package manager**, also called a **repository manager**, which can be on-premise (installed locally) such as Nexus, Artifactory or ProGet, or a SaaS Solution such as Azure Pipelines, Azure Artifacts, or GitHub Package Registry. This package must also be neutral in terms of environment configuration and *must* be versioned in order to deploy the application in a previous version if necessary.

Continuous delivery (CD)

Once the application has been packaged and stored in a package manager during CI, the continuous delivery process is ready to retrieve the package and deploy it in different environments.

The deployment in each environment consists of a succession of automated tasks that are also executed on a remote server that has access to the different environments.

It is, therefore, necessary to involve Devs, Ops, and also the security team in the implementation of CI/CD tools and processes. It will, indeed, be this union of people with the tools and processes that will deploy applications on the different servers or cloud resources, respecting the network rules but also the company's security standards.

During the deployment phase, it is often necessary to modify the configuration of the application in the generated package in order for it to be adapted to the target environment. It is, therefore, necessary to integrate a **configuration manager** that is already present in common CI/CD tools such as Jenkins, Azure Pipelines, or Octopus Deploy. In addition, when there is a new configuration key, it is good practice for every environment, including production, to be entered with the involvement of the Ops team.

Finally, the triggering of a deployment can be done automatically, but for some environments that are more critical (for example, production environments), heavily regulated companies may have gateways that require a manual trigger with checks on the people authorized to trigger the deployment.

The different tools for setting up a CI/CD pipeline are as follows:

- A source control version
- A package manager
- A CI server
- A configuration manager

But let's not forget that all these tools will only be really effective in delivering added value to the product if development and operations team work together around them.

We have just looked at the principles of implementing a CI/CD pipeline. In the rest of the chapter, we will look at the practical implementation with different tools, starting with package managers.

Using a package manager

A package manager is a central repository to centralize and share packages, development libraries, tools, and software.

There are many public package managers, such as NuGet, npm, Maven, Bower, and Chocolatey, that provide frameworks or tools for developers in different languages and platforms.

The following screenshot is from the NuGet package manager, which publicly provides more than 150K .NET Frameworks:

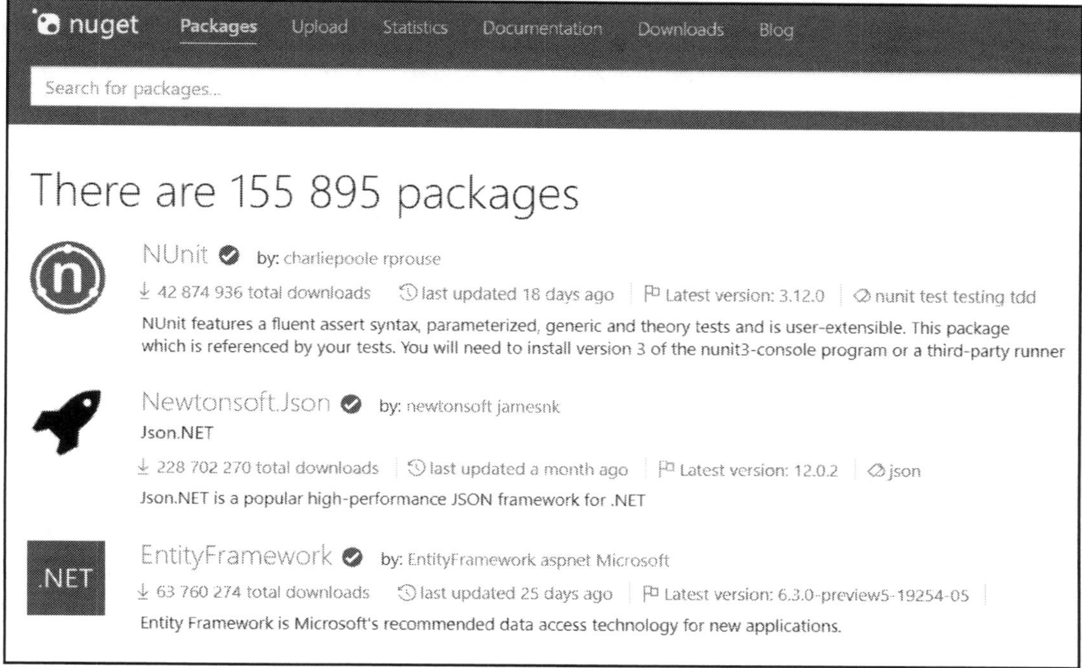

One of the advantages for the developer of using this type of package manager is that they don't have to store the packages with the application sources, but can make them a reference in a configuration file, so the packages will be automatically retrieved.

In an enterprise application, things are a little different because, although developers use packages from public managers, some elements that are generated in an enterprise must remain internal.

Indeed, it is often the case that frameworks (such as NuGet or npm libraries) are developed internally and cannot be exposed publicly. Moreover, as we have seen in the CI/CD pipeline, we need to make a package for our application and store it in a package manager that will be private to the company.

That's why looking at package managers such as NuGet and npm, which can be used within an enterprise or for personal needs, is suggested.

Private NuGet and npm repository

If you need to centralize your NuGet or npm packages, you can create your own local repository.

To create your **NuGet Server**, here is the Microsoft documentation: `https://docs.microsoft.com/en-us/nuget/hosting-packages/overview`.

For npm, we can also install it locally with the `npm local-npm` package, whose documentation is available here: `https://www.npmjs.com/package/local-npm`.

The problem with installing one repository per package type method is that we need to install and maintain a repository and its infrastructure for the different types of packages. This is why it is preferable to switch to universal repository solutions such as Nexus (Sonatype), ProGet, and Artifactory for on-premise solutions, and Azure Artifacts or MyGet for SaaS solutions.

To understand how a package manager works, we will look at Nexus Repository OSS and Azure Artifacts.

Nexus Repository OSS

Nexus Repository is a product of the Sonatype company (`https://www.sonatype.com/`), which specializes in DevSecOps tools that integrate security controls in the code of applications.

Nexus Repository exists in an open source/free version, and its documentation is available at `https://www.sonatype.com/nexus-repository-oss?smtNoRedir=1` and `https://help.sonatype.com/repomanager3`.

Before installing and using Nexus, please take into consideration the software and hardware requirements detailed in the requirements documentation here: `https://help.sonatype.com/repomanager3/system-requirements`.

For the installation and configuration steps, refer to the installation procedure, which is available here: `https://help.sonatype.com/repomanager3/installation`.

You can also use it via a Docker container (we will look at Docker in detail in the next chapter), and here is the documentation: `https://hub.docker.com/r/sonatype/nexus3/`.

Once Nexus Repository is installed, we must create a repository by following these steps:

1. In the **Repositories** section, click on the **Create repository** button.
2. Then choose the type of packages (for example, npm, NuGet, or Bower) that will be stored in the repository, as shown in the following screenshot:

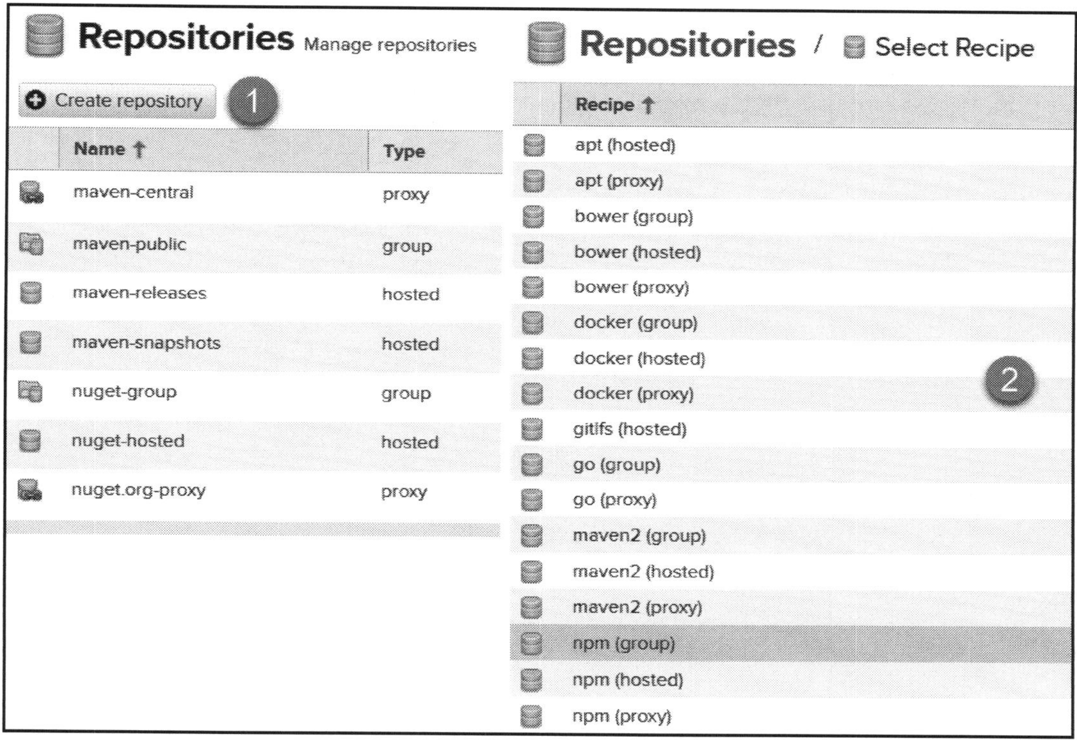

Nexus is a high-performance and widely used enterprise repository, but it requires effort to install and maintain it. This is not the case for SaaS package managers, such as Azure Artifacts, which we'll look at now.

Azure Artifacts

Azure Artifacts is one of the services provided by Azure DevOps. We already looked at it in the previous chapter, and we will also cover it again later, in the *Using Azure Pipelines* section of this chapter. It is hosted in the cloud, and therefore, allows managing private package feeds.

The packages supported today are NuGet, npm, Maven, Gradle, Python, and also universal packages. The main difference from Nexus is that in Azure Artifacts, the feed is not by package type, and one feed can contain different types of packages.

One of the advantages of Azure Artifacts is that it is fully integrated with other Azure DevOps services such as Azure Pipelines, which allows managing CI/CD pipelines, as we will see shortly.

In Azure Artifacts, there is also a type of package called **universal packages** that allows storing all types of files (called a package) in a feed that can be consumed by other services or users.

Here is an example of an Azure Artifacts feed containing several types of packages, in which we can see one NuGet package, one npm package, one universal package, one Maven package, and one Python package:

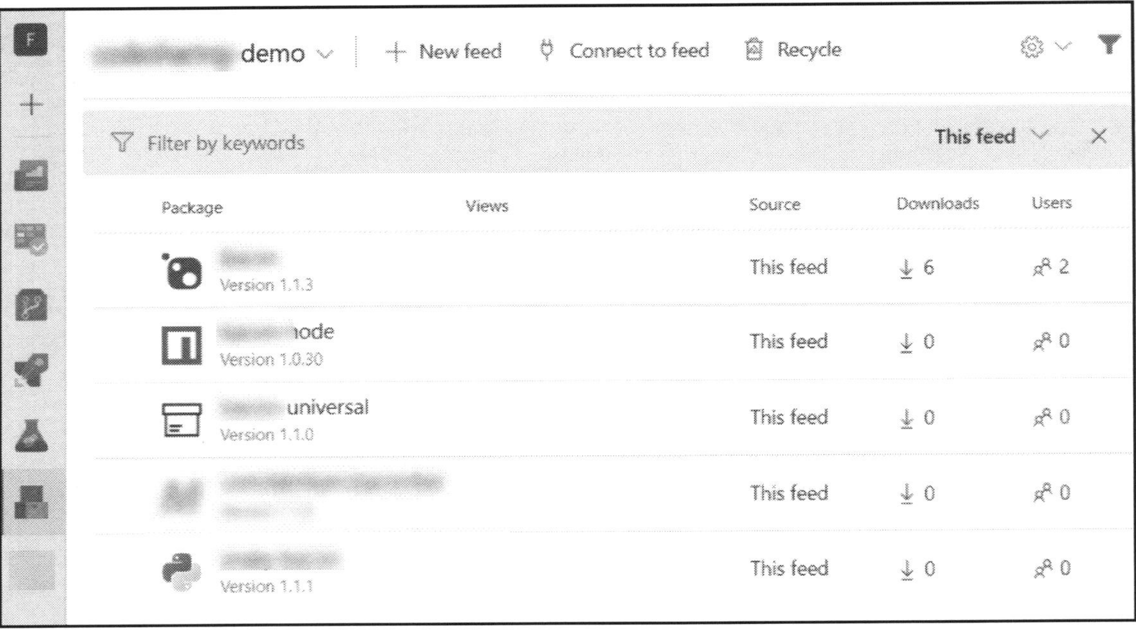

Azure Artifacts has the advantage of being in SaaS offering mode, so there is no installation or infrastructure to manage; for more information, the documentation is here: `https://azure.microsoft.com/en-us/services/devops/artifacts/`.

We finished this overview of package managers with the local NuGet Server, npm, Nexus, and Azure Artifacts. Of course, there are many other package manager tools that should be considered according to the company's needs.

After looking at package managers, we will now implement a CI/CD pipeline with a well-known tool called **Jenkins**.

Using Jenkins

Jenkins is one of the oldest continuous integration tools, initially released in 2011. It is open source and developed in Java.

Jenkins has become famous thanks to the large community working on it and its plugins. Indeed, there are more than 1,500 Jenkins plugins that allow you to perform all types of actions within your jobs. And if, despite everything, one of your tasks does not have a plugin, you can create it yourself.

In this section, we will looked at the installation and configuration of Jenkins, and will create a CI Jenkins job that will be executed during commit of a code that is in a Git repository.

The source code of the demo application is a Java project that is open source and available on the Microsoft GitHub here: `https://github.com/microsoft/MyShuttle2`. To be able to use it, you need to fork it into your GitHub account.

Before talking about the Jenkins job, we will see how to install and configure Jenkins.

Installing and configuring Jenkins

Jenkins is a cross-platform tool that can be installed on any type of support, such as VMs or even Docker containers. Its installation documentation is available here: `https://jenkins.io/doc/book/installing/`.

For our demo, and to quickly access the configuration of a CI/CD pipeline, we will use Jenkins on an Azure VM. In fact, Azure Marketplace already contains a VM with Jenkins and its prerequisites already installed.

These steps show how to create an Azure VM with Jenkins and its basic configuration:

1. To get all the steps to create an Azure VM with Jenkins already installed, read the documentation available here: `https://docs.microsoft.com/en-us/azure/jenkins/install-jenkins-solution-template`.

   The following screenshot shows Jenkins integration on Azure Marketplace:

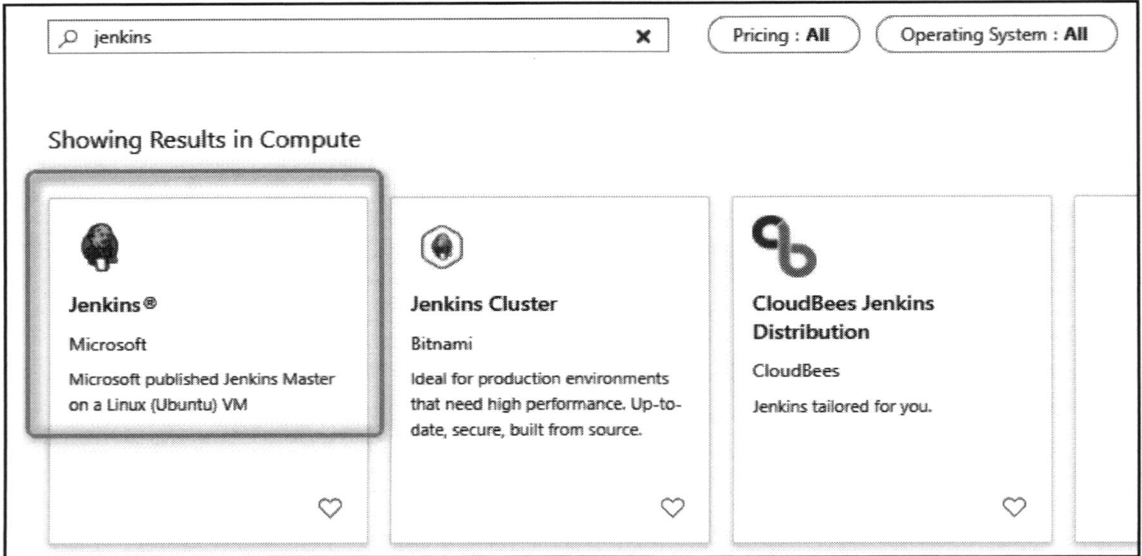

2. Once installed and created, we will access it in the browser by providing its URL in the Azure portal in the **DNS name** field, as shown in the following screenshot:

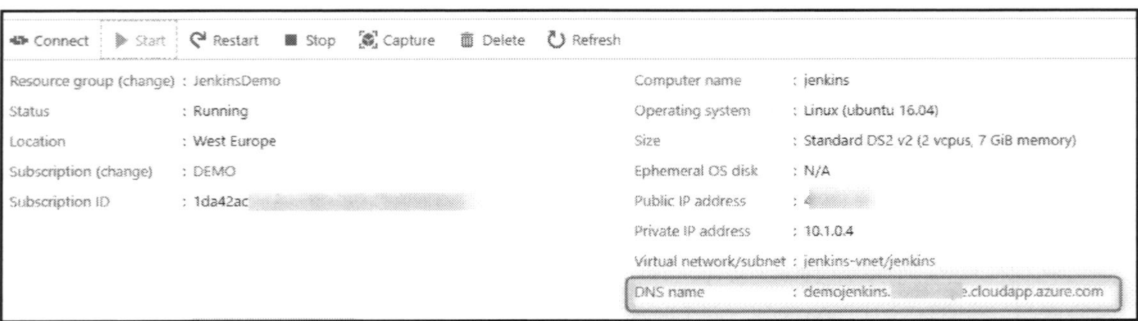

3. Follow the displayed instructions on the Jenkins home page to enable access to this Jenkins instance via secure SSL tunneling. For more details about this step, read the documentation at `https://docs.microsoft.com/en-us/azure/jenkins/install-jenkins-solution-template#connect-to-jenkins` and this article: `https://jenkins.io/blog/2017/04/20/secure-jenkins-on-azure/`.
4. Then, follow the configuration instructions on the **Unlock Jenkins** displayed on the Jenkins screen. Once the configuration is complete, we get Jenkins ready to create a CI job.

In order to use GitHub features in Jenkins, we also installed the **GitHub integration plugin** from the Jenkins plugin management by following the documentation: `https://jenkins.io/doc/book/managing/plugins/`.

The following screenshot shows the installation of the GitHub plugin:

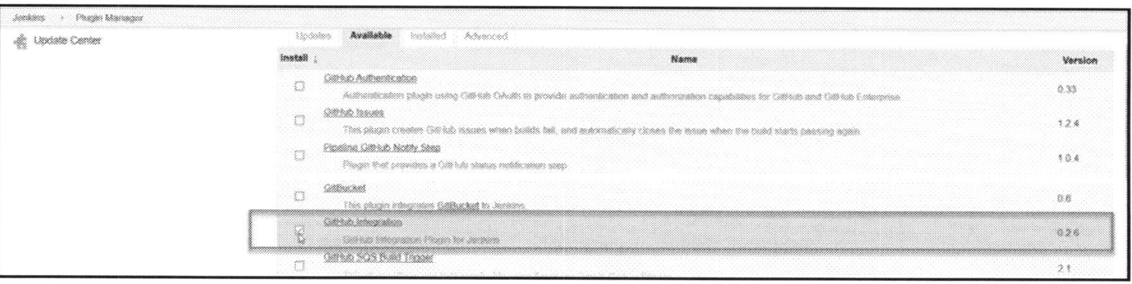

Now that we have installed the GitHub plugin in Jenkins, let's look at how to configure GitHub with a webhook for its integration with Jenkins.

Configuring a GitHub webhook

In order for Jenkins to run a new job, we must first create a webhook in the GitHub repository. This webhook will be used to notify Jenkins as soon as a new push occurs in the repository.

To do this, follow these steps:

1. In the GitHub repository, go to the **Settings** | **Webhooks** menu.
2. Click on the **Add Webhook** button.
3. In the **Payload URL** field, fill in the URL address of Jenkins followed by `/github-webhook/`, leave the secret input as it is, and choose the **Just the push event** option.

4. Validate the webhook.

   The following screenshot shows the configuration of a GitHub webhook for Jenkins:

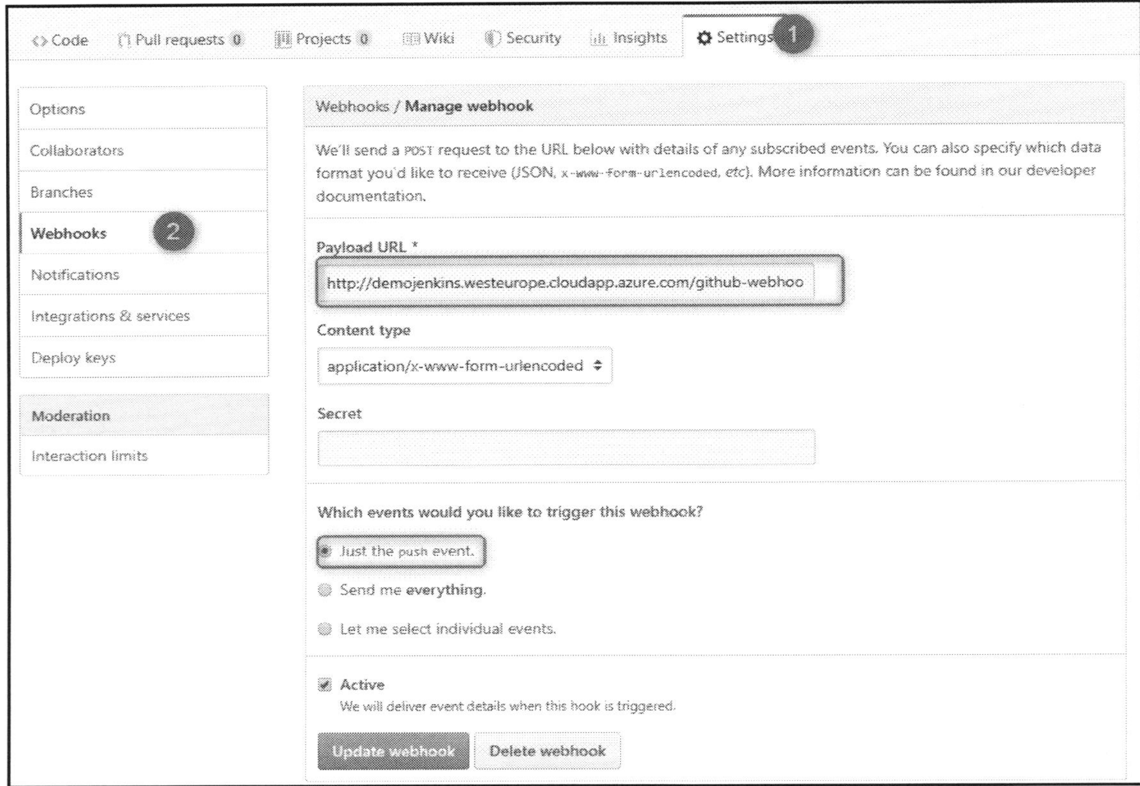

5. Finally, we can check on the GitHub interface, as shown in the following screenshot, that the webhook is well configured and that it communicates with Jenkins:

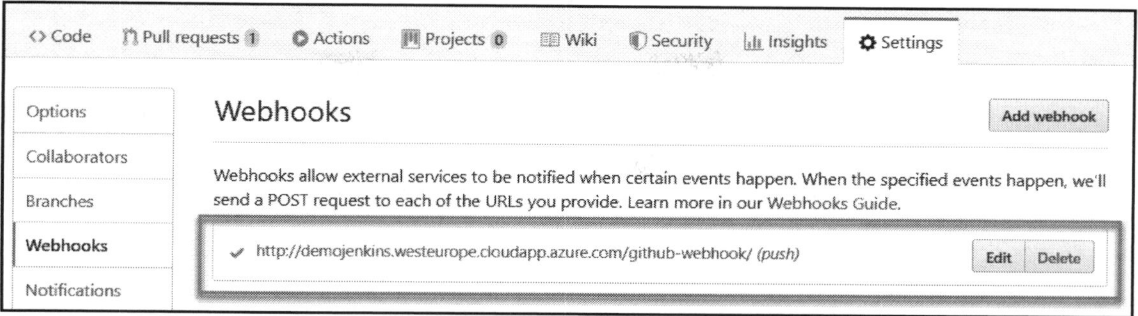

The configuration of GitHub is done. We will now proceed to create a new CI job in Jenkins.

Configuring a Jenkins CI job

To configure Jenkins, let's follow these steps:

1. First, we will create a new job, by clicking on **New Item** or on the **create new jobs** link, as shown in the following screenshot:

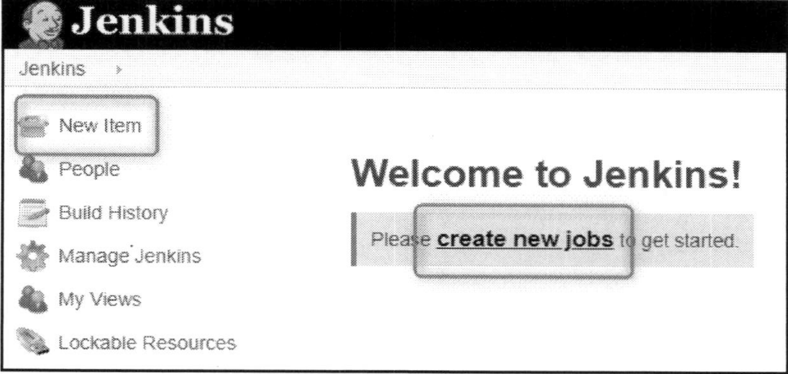

2. On the job configuration form, enter the name of the job, for example, `demoCI`, and choose the **Freestyle project** template, then validate that by clicking on **OK**, as shown in the following screenshot:

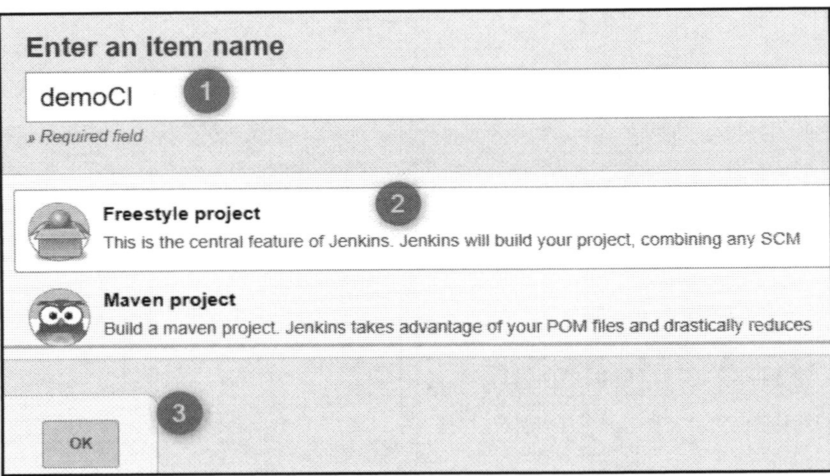

3. Then, we configure the job with the following parameters:
    - In the **GitHub project** input, we enter the URL of the GitHub repository as follows:

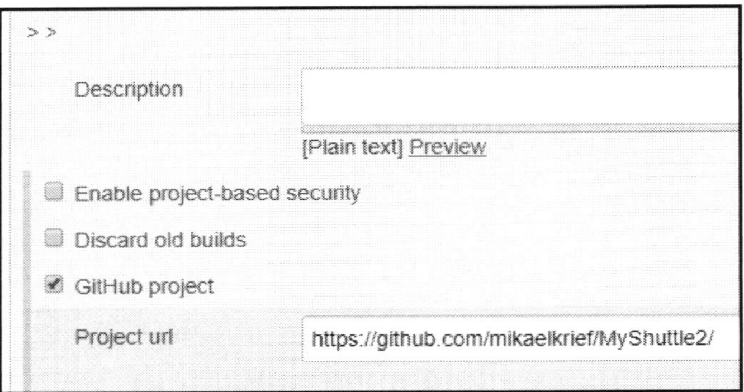

- In the **Source Code Management** section, enter the URL of the GitHub repository and the code branch, like this:

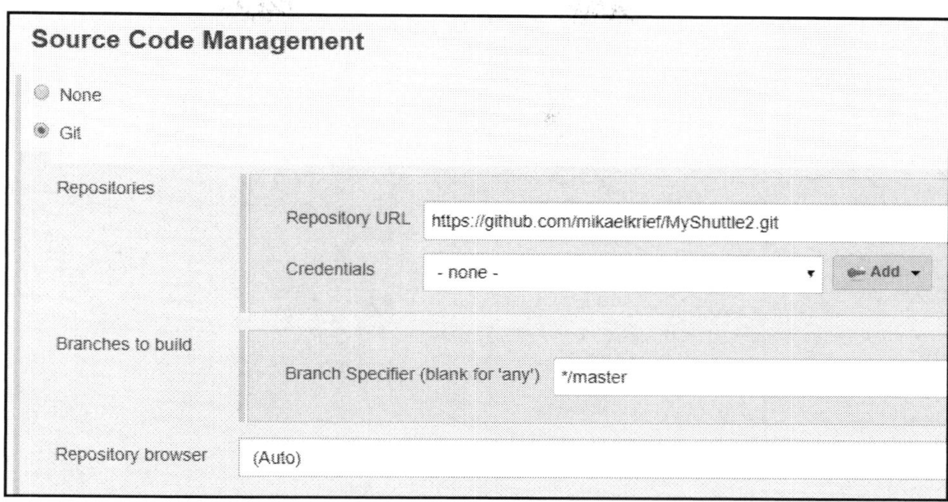

- In the **Build Triggers** section, check the **GitHub hook trigger for GITScm polling** box, like this:

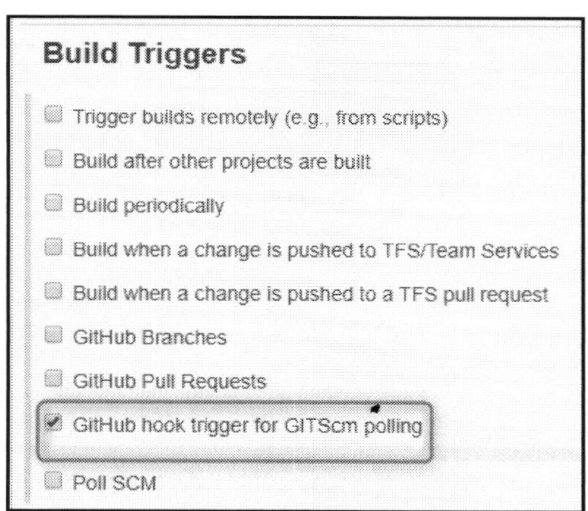

- In the **Build** section, in the actions drop-down list, for this lab we'll choose the **Execute shell** step. You can add as many actions as necessary to your CI (compilation, file copies, and tests).

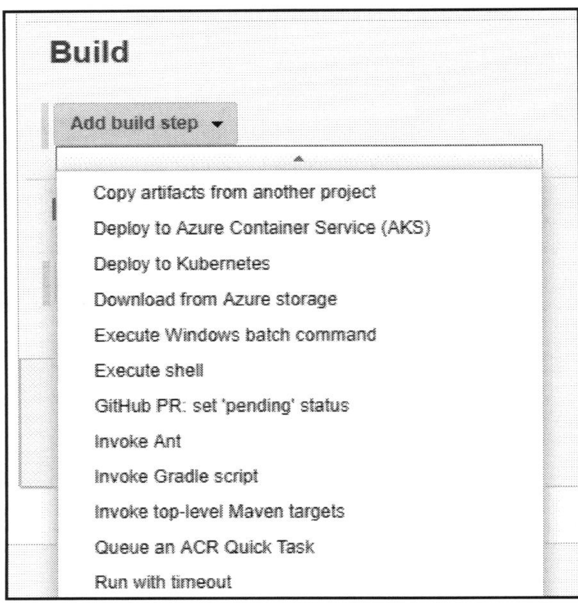

- Inside the textbox of the shell command, we enter the `printenv` command to be executed during the execution of the job pipeline:

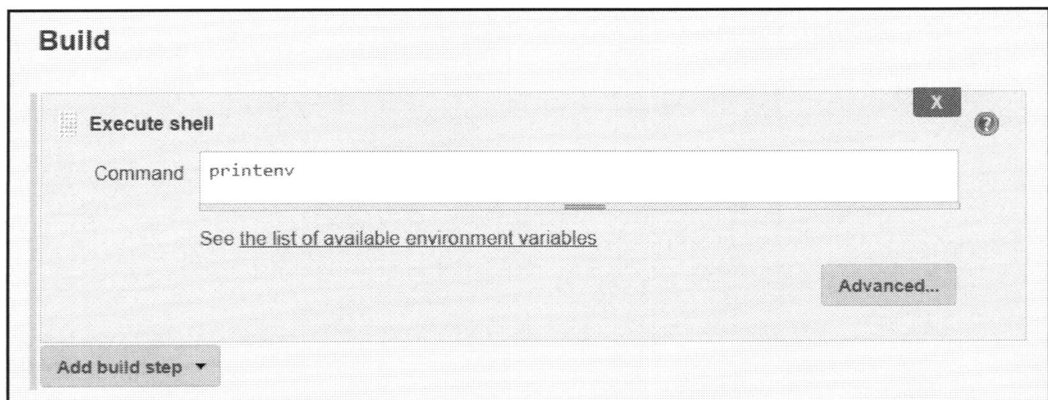

4. Then, we finish the configuration by clicking on **Apply** and then on the **Save** button.

Our Jenkins CI job has now been created and is configured to be triggered during a commit and to perform various actions.

We will now run it manually to test its proper functioning.

Executing the Jenkins job

To test job execution, we will perform these steps:

1. First, we will modify the code of our GitHub repository, for example, by modifying the `Readme.md` file.
2. Then, we, commit to the master branch directly from the GitHub web interface.
3. What we see in Jenkins, right after making this commit, is that the `DemoCI` job is queued up and running.

   The following screenshot shows the job execution queue:

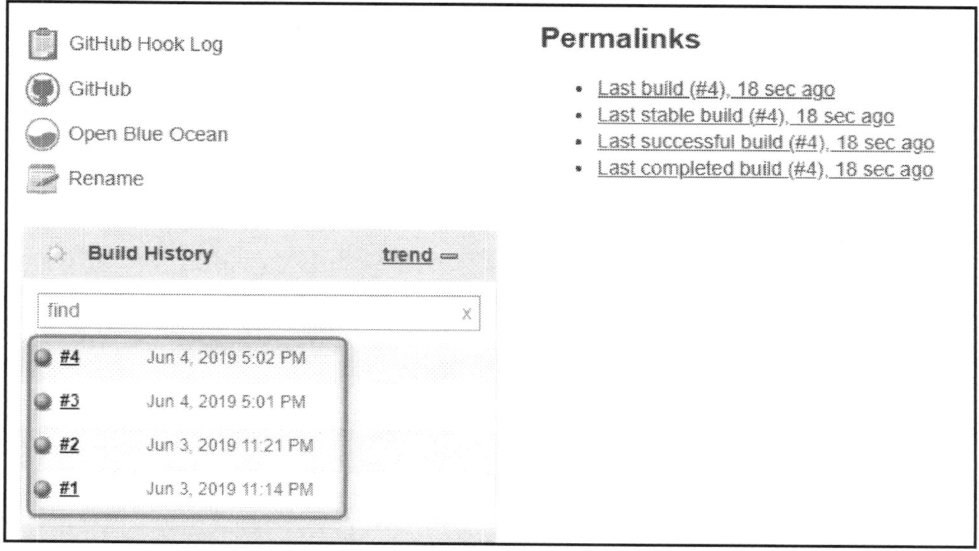

4. By clicking on the job, then on the link of the **Console Output** menu, we can see the job execution logs, as shown in the following screenshot:

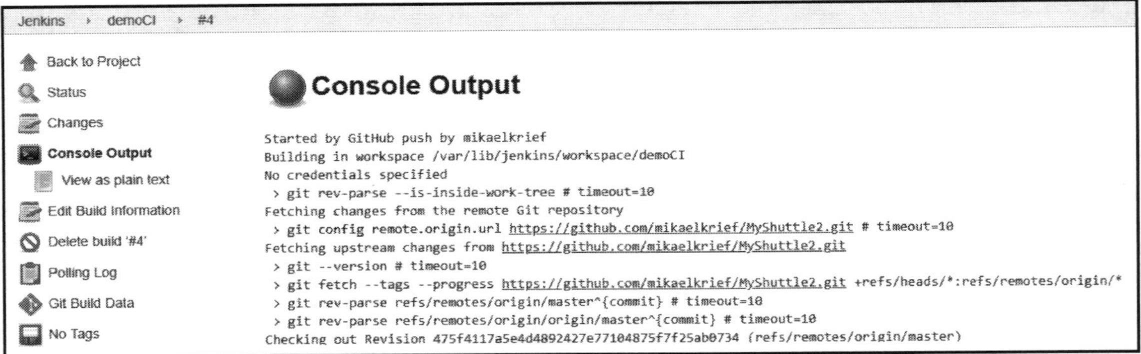

We have just created a CI job in Jenkins that runs during a commit of a Git repository (GitHub, in our example).

In this section, we have looked at the creation of a pipeline in Jenkins. Let's now look at how to create a CI/CD pipeline with another DevOps tool: Azure Pipelines.

Using Azure Pipelines

Azure Pipelines is one of the services offered by Azure DevOps. It was previously known as **Visual Studio Team Services** (**VSTS**).

Azure DevOps is a complete DevOps platform provided by Microsoft that is fully accessible via a web browser and requires no installation. It is very useful for following reasons:

- The DevOps tools manage its code in **Version Control System** (**VCS**)
- It manages a project in agile mode
- It deploys applications in a CI/CD pipeline, to centralize packages
- It performs manual test plans

Each of these features is combined into services that are summarized in this table:

Service name	**Description**	**Documentation link**
Azure Repos	It is a SCV, which we looked at in the previous chapter.	`https://azure.microsoft.com/en-us/services/devops/repos/`
Azure Boards	It is a service for project management in agile mode with sprints, backlogs, and boards.	`https://azure.microsoft.com/en-us/services/devops/boards/`
Azure Pipelines	It is a service that allows the management of CI/CD pipelines.	`https://azure.microsoft.com/en-us/services/devops/pipelines/`
Azure Artifacts	It is a private package manager.	`https://azure.microsoft.com/en-us/services/devops/artifacts/`
Azure Test Plans	It allows you to make and manage a manual test plan.	`https://azure.microsoft.com/en-us/services/devops/test-plans/`

Azure DevOps is free for up to five users. Beyond that, there is a license version with per user costs. For more information on licensing, refer to the product sheet at `https://azure.microsoft.com/en-us/pricing/details/devops/azure-devops-services/`, which also contains a calculator to get a cost estimate for your team.

There is also **Azure DevOps Server**, which is the same product as Azure DevOps, but it installs itself on premises. To find out the differences between these two products, read the documentation here: `https://docs.microsoft.com/en-us/azure/devops/user-guide/about-azure-devops-services-tfs?view=azure-devops`.

To register with Azure DevOps and create an account, called an organization, we need either a Microsoft live account or a GitHub account, and to follow these steps:

1. In your browser, go to this URL: `https://azure.microsoft.com/en-us/services/devops/`.
2. Click on the **Signup** button.
3. On the next page, choose the account to use (either **Live** or **GitHub**).
4. As soon as we register, the first step suggested is to create an organization with a unique name of your choice and the Azure location, for example, `BookLabs` for the name of the organization, and `West Europe` for the location.
5. In this organization, we will now be able to create projects with our CI/CD pipeline, as we learned in `Chapter 5`, *Managing Your Source Code with Git.*

In this lab, we will show how to set up an end-to-end CI and CD pipeline, starting with the use of Azure Repos to version our code. Then, in Azure Pipeline, we will look at the CI process and end with the automatic deployment of the application in the release.

Versioning of the code with Git in Azure Repos

As we have mentioned, the first prerequisite for setting up a continuous integration process is to have the application code versioned in an SVC, and we will do this in Azure Repos by following these steps:

1. To start our lab, we will create a new project; this operation has already been covered in `Chapter 5`, *Managing Your Source Code with Git*, in the *Starting with the Git Process* section.
2. Then, in Azure Repos, we will import code from another Git repository by using the **Import repository** option of the repository menu, as shown in the following screenshot:

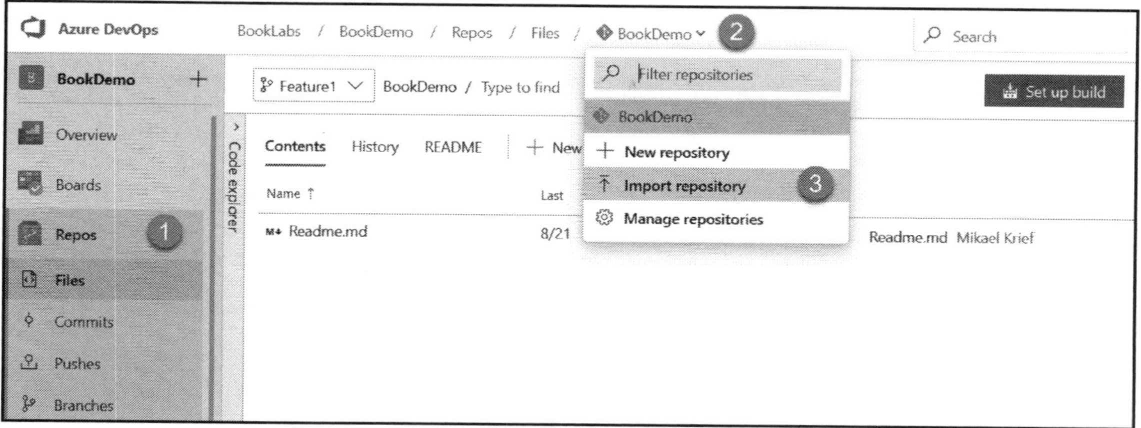

3. Once the **Import a Git repository** window opens, enter the URL of the Git repository whose sources we want to import. In our lab, we'll import the sources found in the `https://github.com/mikaelkrief/DemoAspNetApp.git` repository, as can be seen in the following screenshot:

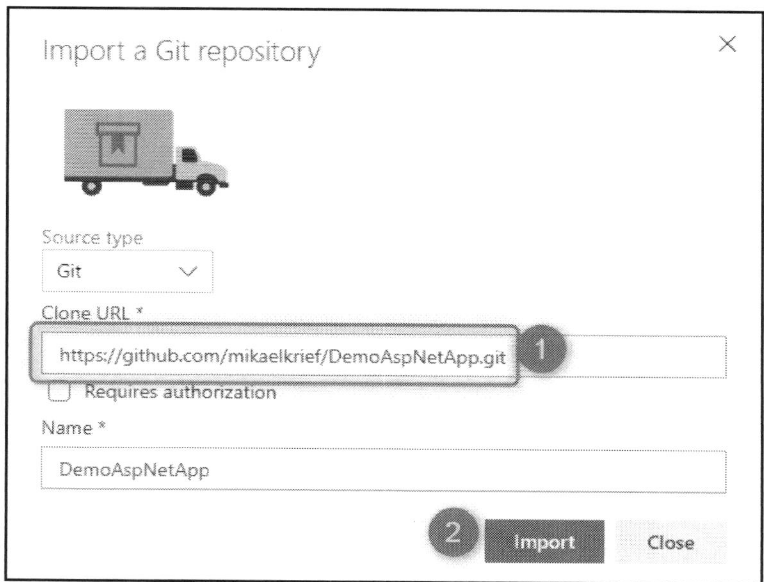

4. We'll click on the **Import** button, then we'll see that the code is imported into our repository. The following screenshot shows the code imported into our Azure Repos repository, which is an ASP.NET code application, and its unit tests:

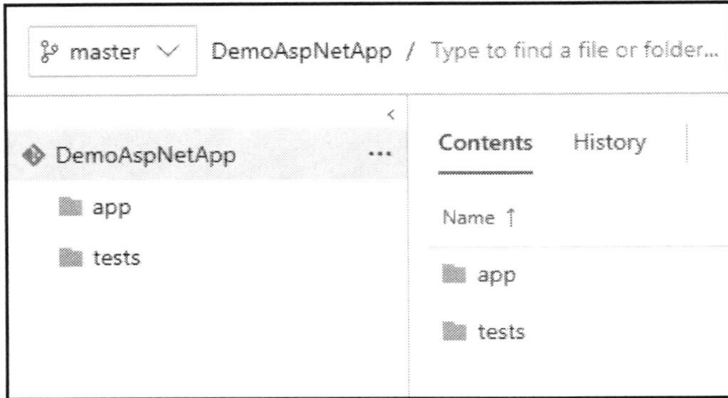

Now that we have the code in Azure Repos, we'll set up a CI pipeline that will check and test the code at each user commit.

Creating the CI pipeline

We will create a CI pipeline in Azure Pipelines by following these steps:

1. To create this pipeline, open the **Pipelines** | **Builds** menu.
2. Then, click on the **New pipeline** button, as shown in the following screenshot:

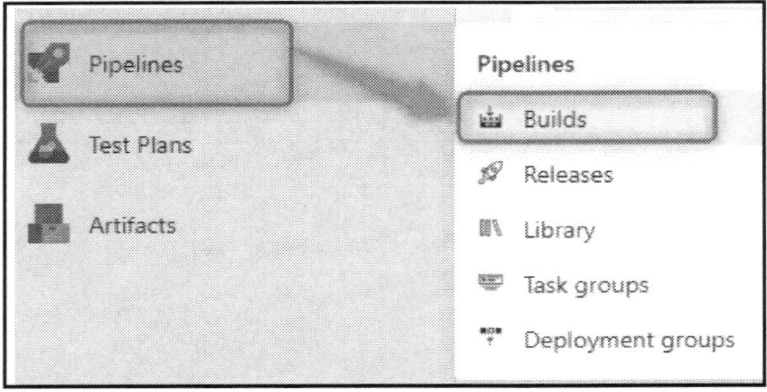

3. For the configuration mode, we choose the **Use the classic editor** option:

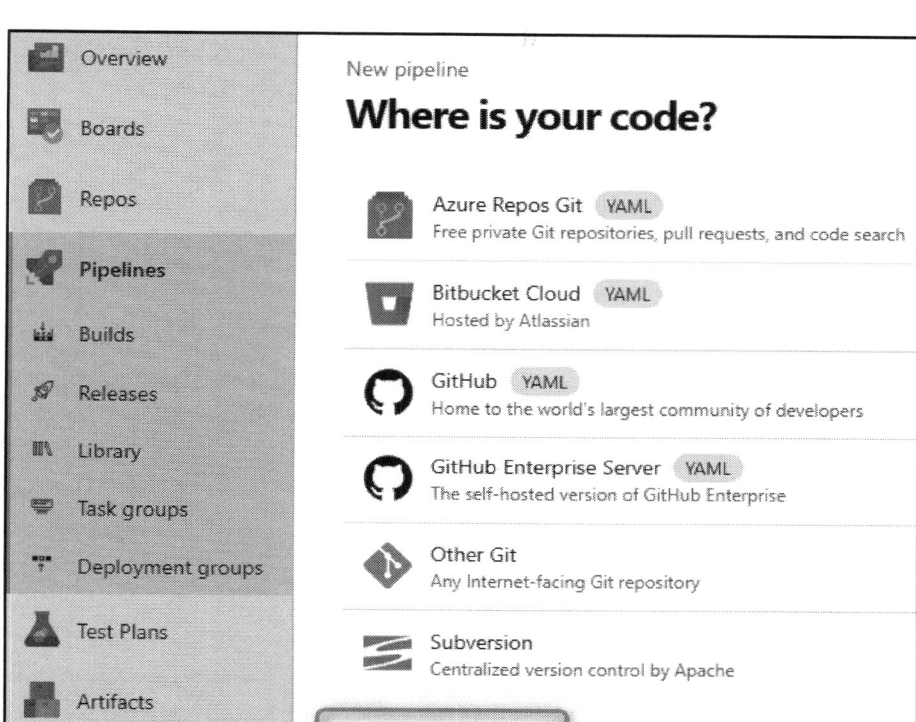

In Azure Pipelines, there is the choice of either the *classic editor* mode, which allows us to configure the pipeline via a graphical interface, or the *YAML pipeline* mode, which involves using a YAML file that describes the configuration of the pipeline.

In this lab, we will use the *classic editor* mode, so that we can visualize the different options and configuration steps.

4. The first configuration step of the pipeline consists of selecting the repository that contains the application's sources.

Today, Azure Pipelines supports several types of Git, such as Azure Repos, GitHub, Bitbucket, and Subversion, and so on. We will therefore select **Azure Repos Git** in the repository that contains the imported sources, as shown in the following screenshot:

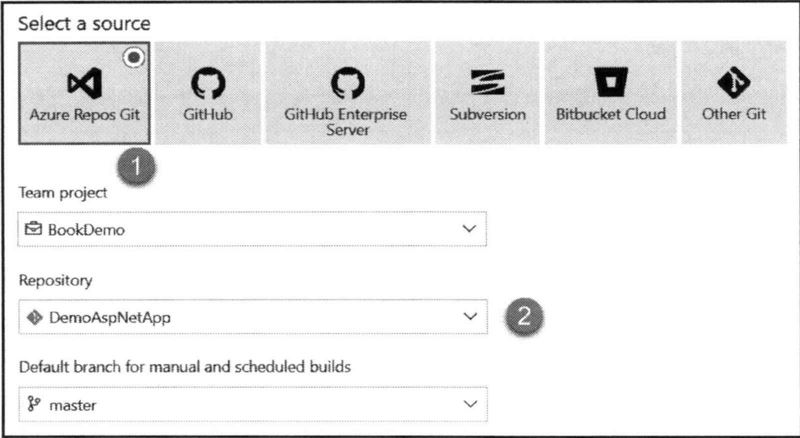

5. Azure Pipelines proposes to select a build template that will contain all the preconfigured build steps; there is also the possibility to start from an empty template.

Since our project is an ASP.NET core application, we will choose the **ASP.NET Core** template, as shown in the following screenshot:

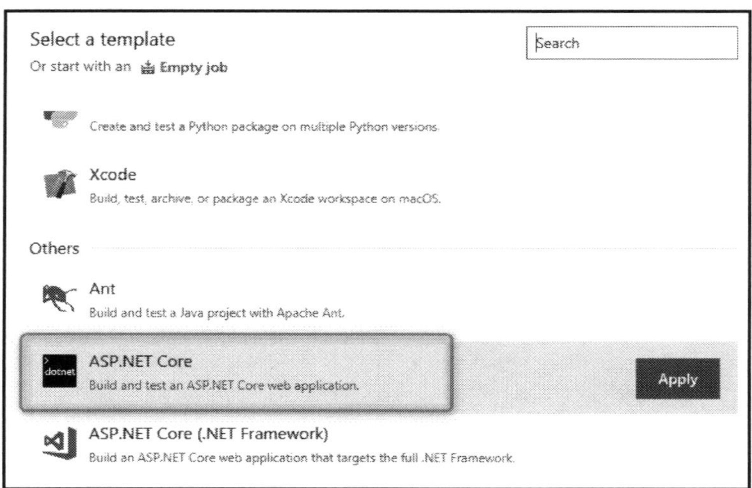

Once the template has been chosen, we reach the configuration page of the build definition.

The configuration of the build definition consists of four main sections:

- The variables
- The steps
- The triggers
- The options

6. We configure the **Variables** section, which allows us to fill in a list of variables in a key form, creating a value that can be used in the steps.

   The following screen shows the **Variables** tab of our build definition:

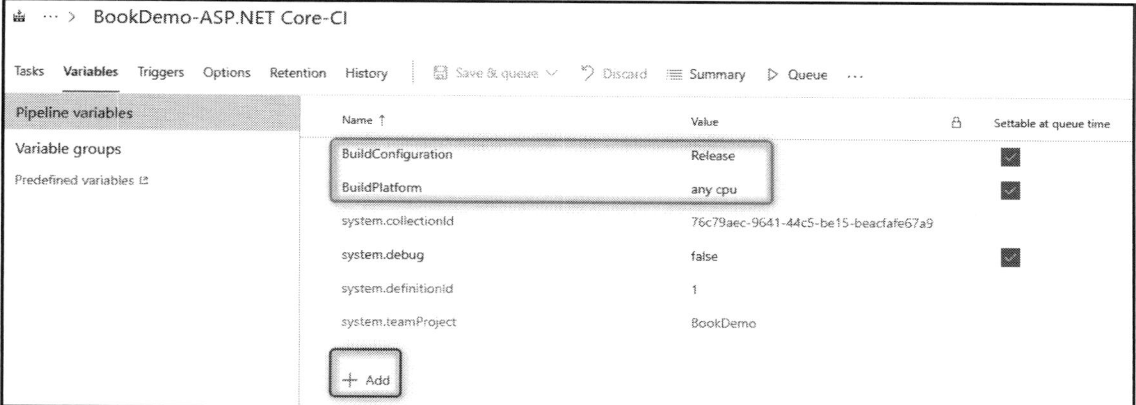

   We notice, when we navigate to the **Variables** tab, we see the **BuildConfiguration** and **BuildPlatform** variables, which are already prefilled by the template, and the **+ Add** button, which allows us to add other variables if we wish.

The documentation on the variables is here: `https://docs.microsoft.com/en-us/azure/devops/pipelines/process/variables?view=azure-devopstabs=classic%2Cbatch.`

7. We configure the **Tasks** tab, which contains the configuration of all the steps to be performed in the build.

 The following is a screenshot of this tab:

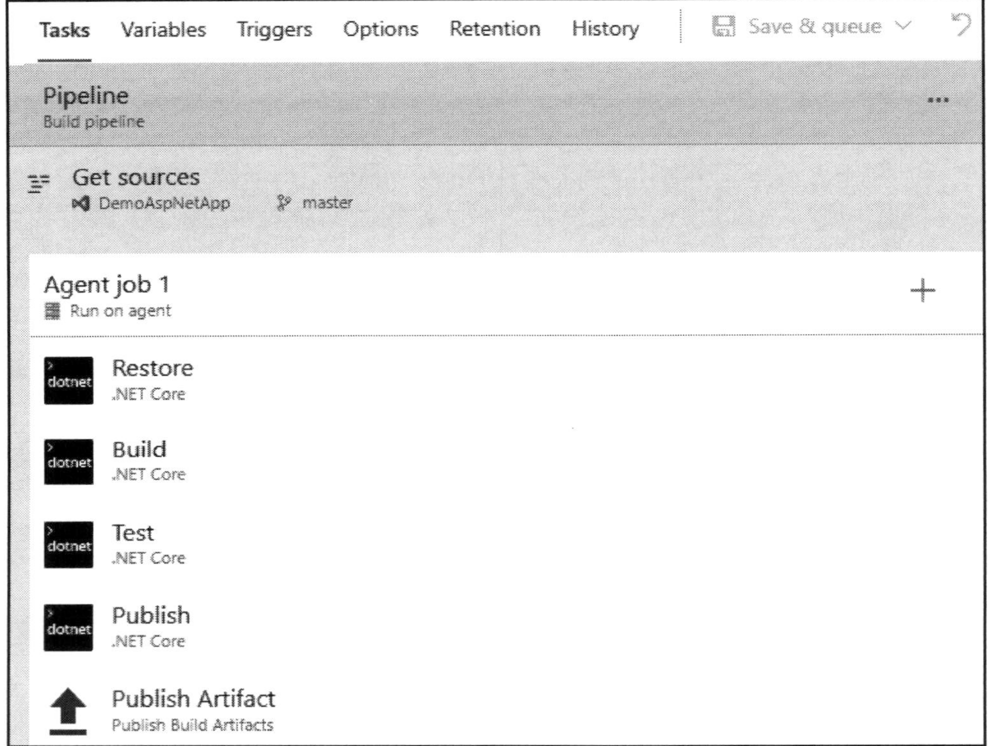

 In the preceding screenshot, the first part is **Pipeline**, which allows us to configure the name of the build definition as well as the agent to use.

 Indeed, in Azure DevOps, pipelines are executed on agents that are installed on VMs or containers.

 Azure DevOps offers free agents from multiple OSes, called a hosted agent, but it is also possible to install your own agents, and they are called **self-hosted**.

To learn more about hosted and self-hosted agents, refer to the documentation: `https://docs.microsoft.com/en-us/azure/devops/pipelines/agents/agents?view=azure-devops`.

The following screenshot shows the configuration of the **Pipeline** section:

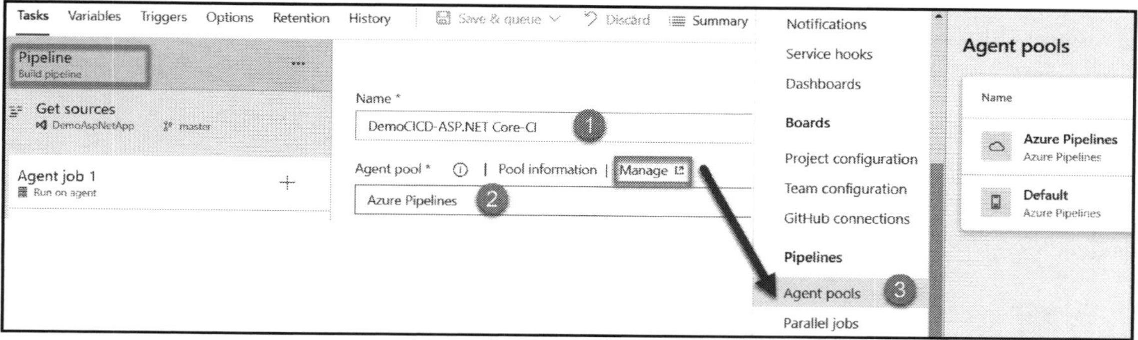

8. Then, we have the **Get sources** phase, which contains the configuration of the sources that we did at the beginning; it is, however, possible to modify it here:

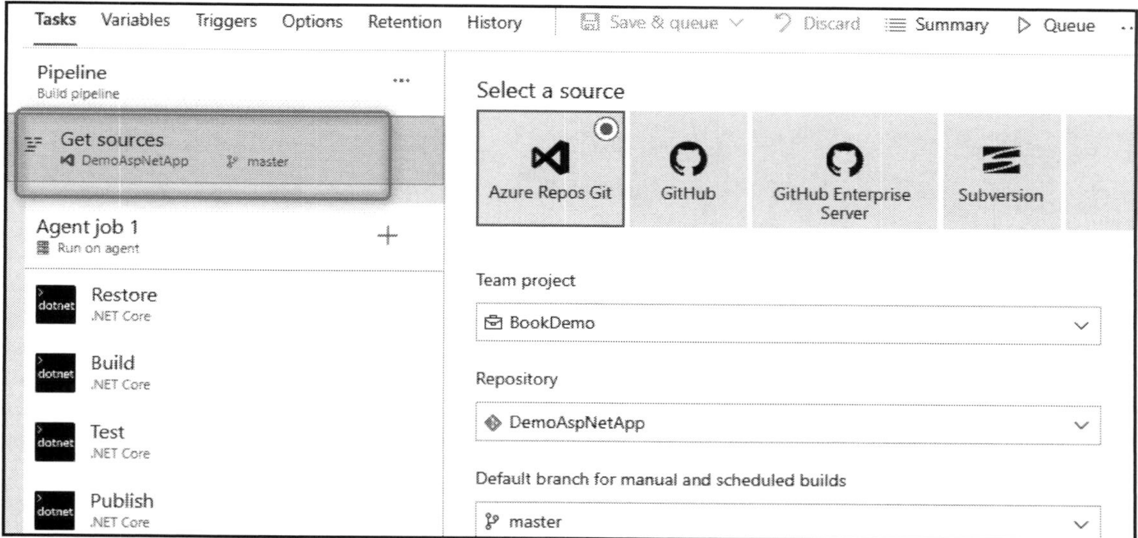

9. We have the **Agent job** part, which contains the ordered list of tasks to be performed in the pipeline. Each of these tasks is configured in panel on the right.

We can add tasks by clicking on the **+** button, and select them from the Azure Pipelines catalog, as follows:

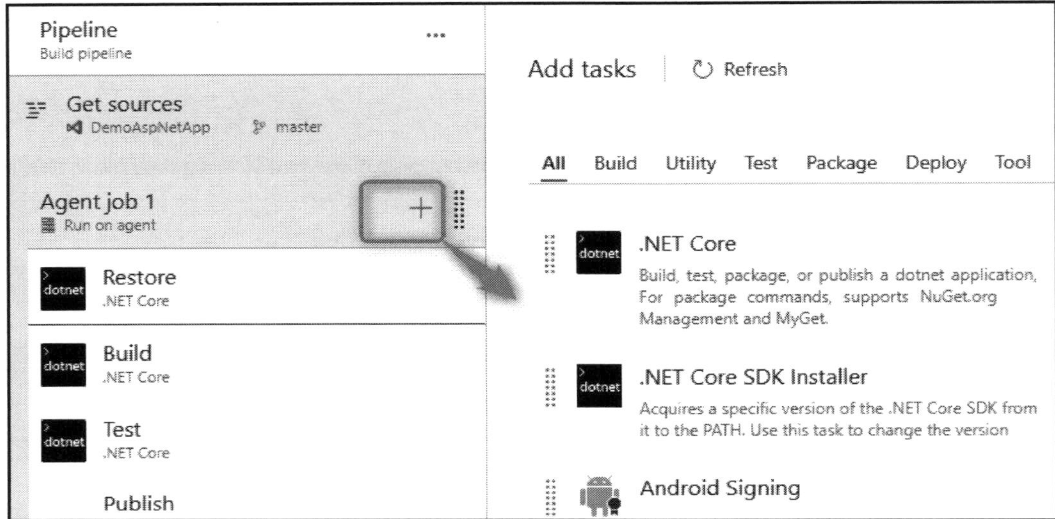

By default, Azure Pipelines contains a very rich catalog of tasks; the list is available here: `https://docs.microsoft.com/en-us/azure/devops/pipelines/tasks/index?view=azure-devops`.
We can also install tasks that can be found in the Azure Marketplace: `https://marketplace.visualstudio.com/search?target=AzureDevOpscategory=Azure%20PipelinessortBy=Downloads`. If necessary, you can also create your tasks for your needs by following the documentation here: `https://docs.microsoft.com/en-us/azure/devops/extend/get-started/node?view=azure-devops`.

Let's define the five tasks in the CI pipeline:

Step/task	**Description**
Restore	Restores the packages referenced in the project.
Build	Builds the project and generates binaries.
Test	Runs unit tests.
Publish	Creates a ZIP package that contains the binary files of the project.
Publish Build Artifacts	Defines an artifact that is our ZIP of the application, which we will publish in Azure DevOps, and which will be used in the deployment release, as seen in the previous, *Use package manager*, section.

10. The last important configuration of our CI pipeline is the configuration of the build trigger in the **Triggers** tab to enable continuous integration, as shown in the following screenshot:

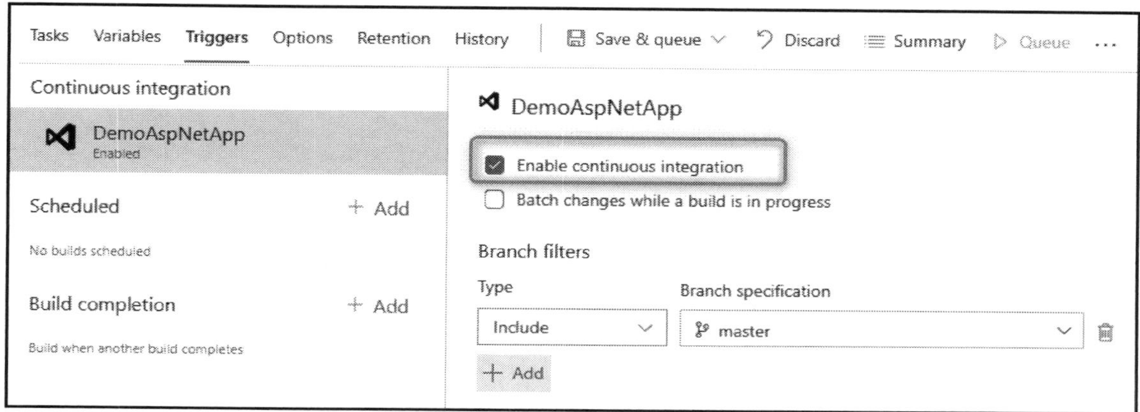

11. The configuration of our CI or Build pipeline is complete; we validate and test its execution for the first time by clicking on the **Save & queue** button:

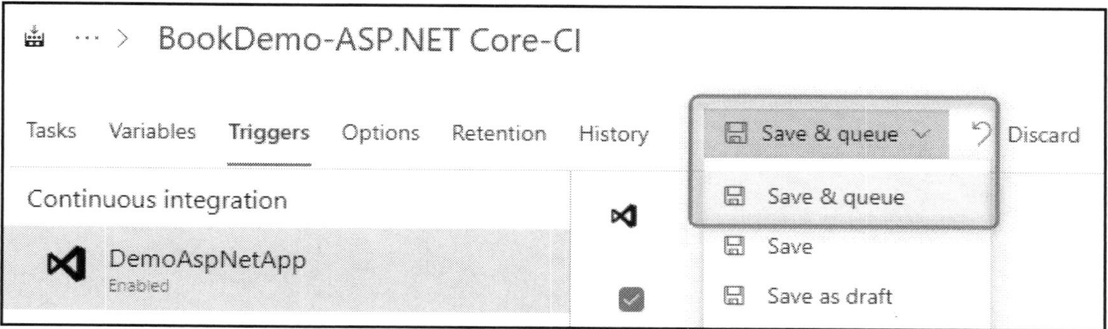

12. At the end of the execution of the build, we have some information, which helps us to analyze the status of the pipeline:

    - The following screenshot shows the execution logs:

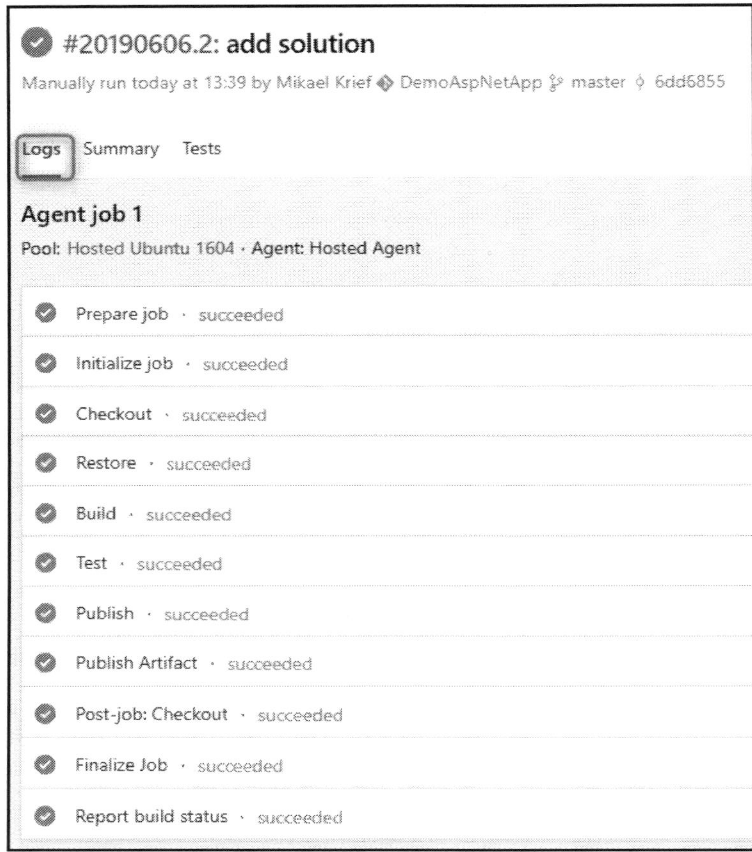

This displays the details of the execution of each task defined in the pipeline.

- The results of the execution of the unit tests are as follows:

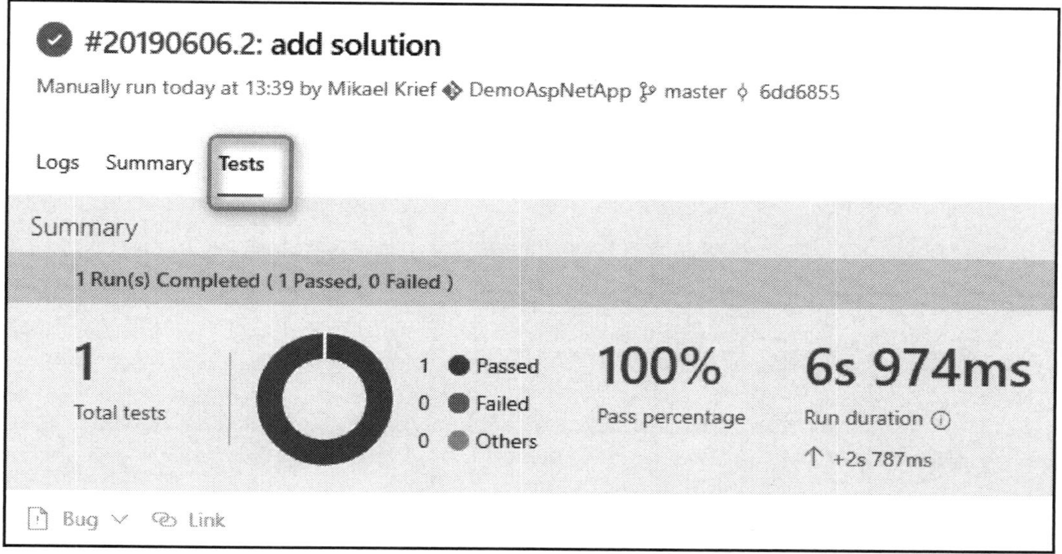

The preceding screenshot is a report of the executions of unit tests with some important metrics such as the number of passed/failed tests and the test execution time.

- The following screenshot shows the published artifacts:

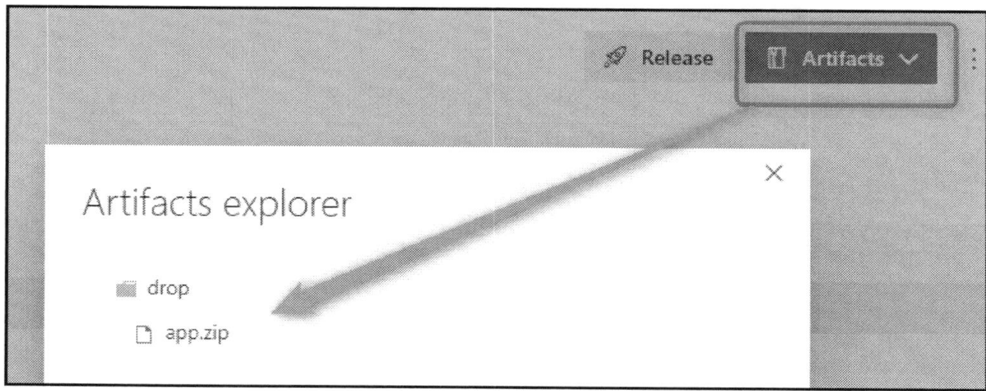

This provides the possibility to explore or download the published artifacts defined in the **Publish Build Artifact** task of the pipeline.

Now we know our CI build is configured and operational, we will create and configure a deployment release for the CD pipeline.

Creating the CD pipeline: the release

In Azure Pipelines, the element that allows deployment in the different stages or environments is called the release. We will now create a release definition that will deploy our build-generated artifacts to an Azure web app by following these steps:

1. To create the release definition, we go to the **Releases** menu and click on **New pipeline**, as follows:

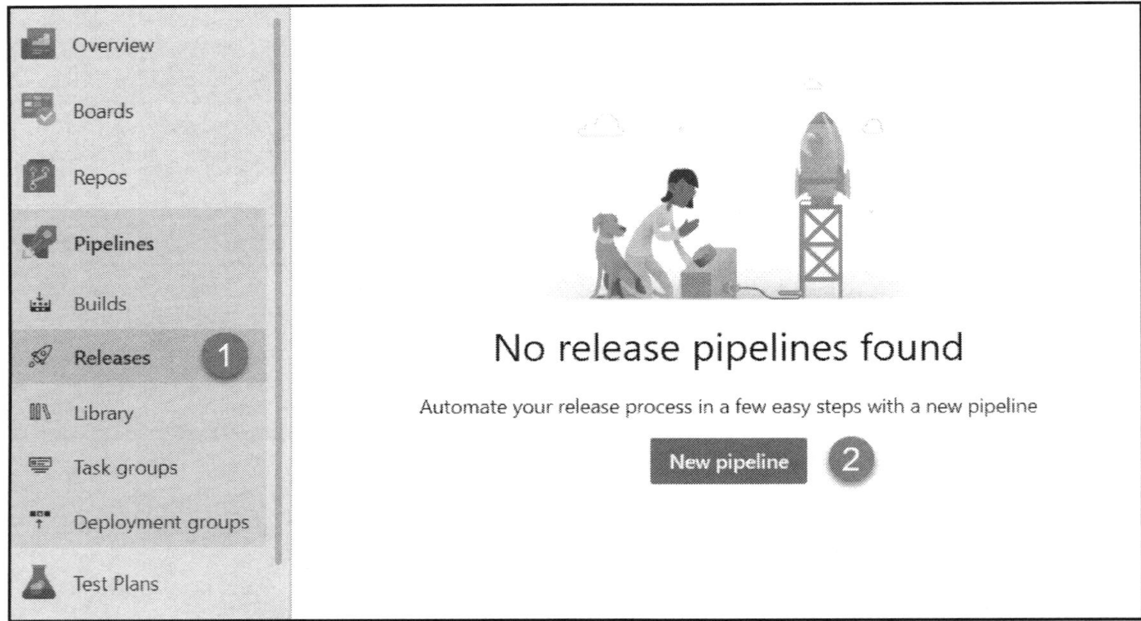

2. As for the build, the first step of the configuration is to choose a template already configured. For this lab, we will choose the **Azure App Service deployment** template:

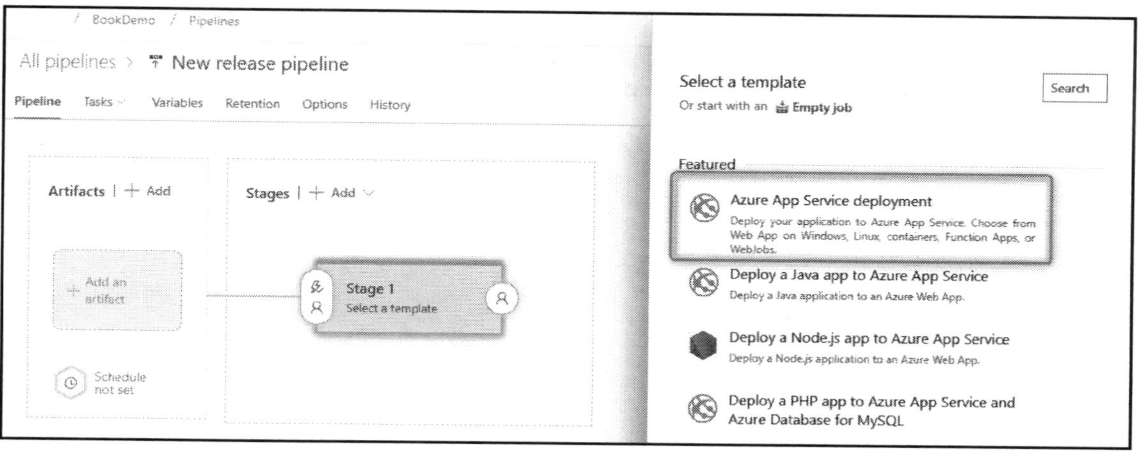

3. Then, in the next window, the first stage is named, for example, `CI` as the continuous integration environment:

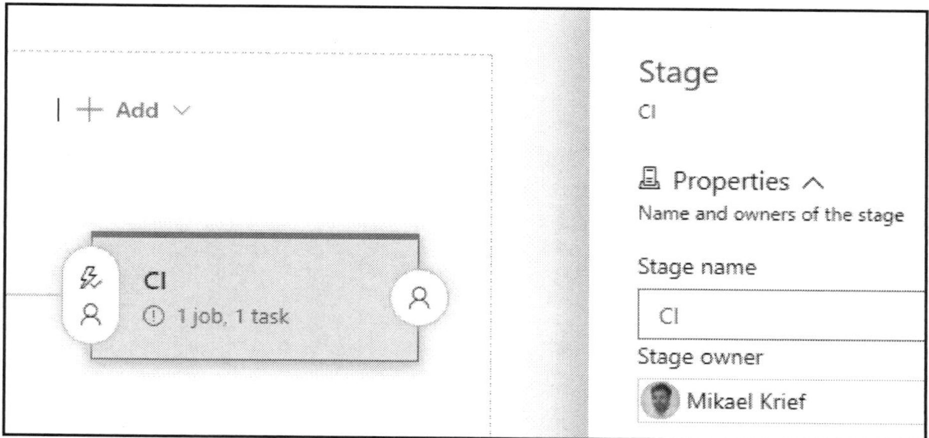

4. We configure the entry point of the release in the artifacts part by adding an artifact that is the build definition previously created in the *Creating the CI pipeline* section, as follows:

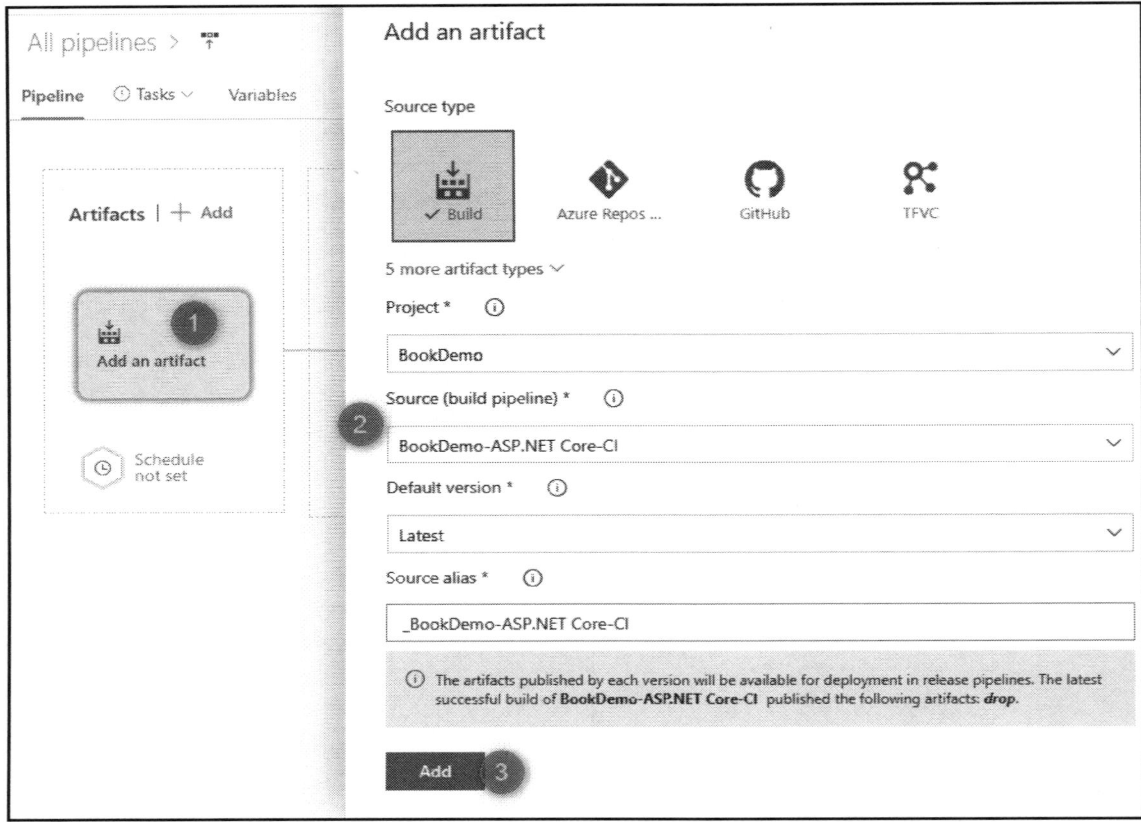

5. We configure the automatic release trigger for each successful build execution:

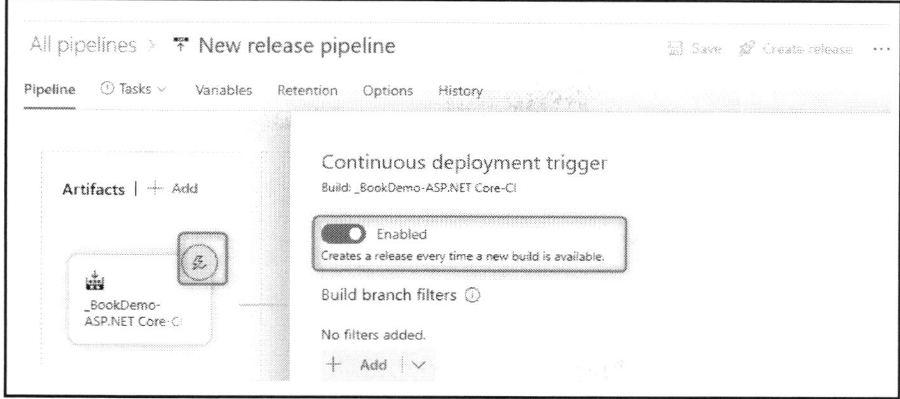

6. Now, we'll configure the steps that will be executed in the CI stage; by clicking on the stage, we get exactly the same configuration window as the build:

    - The agent's choice over the **Run on agent** section
    - The configuration of the steps with their parameters

In our case, we see the deployment task in an Azure web app that was already present in the template. We first fill in the parameters that are located in the CI header because they are shared for the CI course, as shown in the following screenshot:

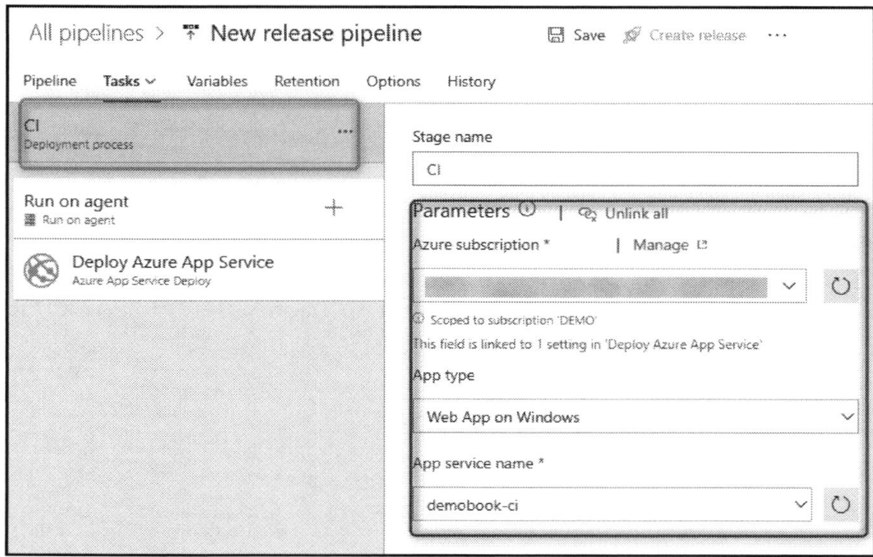

We will fill the following parameters:

- The connection to your Azure subscription
- The name of the web app in which we want to deploy our ZIP

The web app must already be created before deploying the application in it. If it is not created, you can use the Azure CLI `az webapp create` command, documented at `https://docs.microsoft.com/en-us/cli/azure/webapp?view=azure-cli-latest#az-webapp-create`, or the PowerShell `New-AzureRmWebApp` command, documented at `https://docs.microsoft.com/en-us/powershell/module/azurerm.websites/new-azurermwebapp?view=azurermps-6.13.0`.

Then, in the parameters of the Azure deployment task, we have nothing to modify.

7. We just have to rename the release with a name that simply describes what it does, and then we save it, as follows:

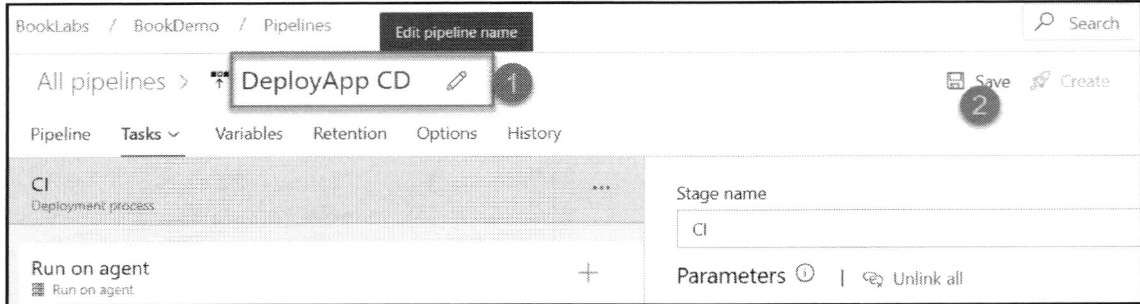

8. Now, we complete our definition of the release with the deployment of the other environments (or stages), which are, for our example, `QA` and `PROD`. To simplify the manipulation, we will clone the CI environment settings in our release, and change the name of the **app service name** settings to the name of the web app like this. The following screenshot shows the clone environment action:

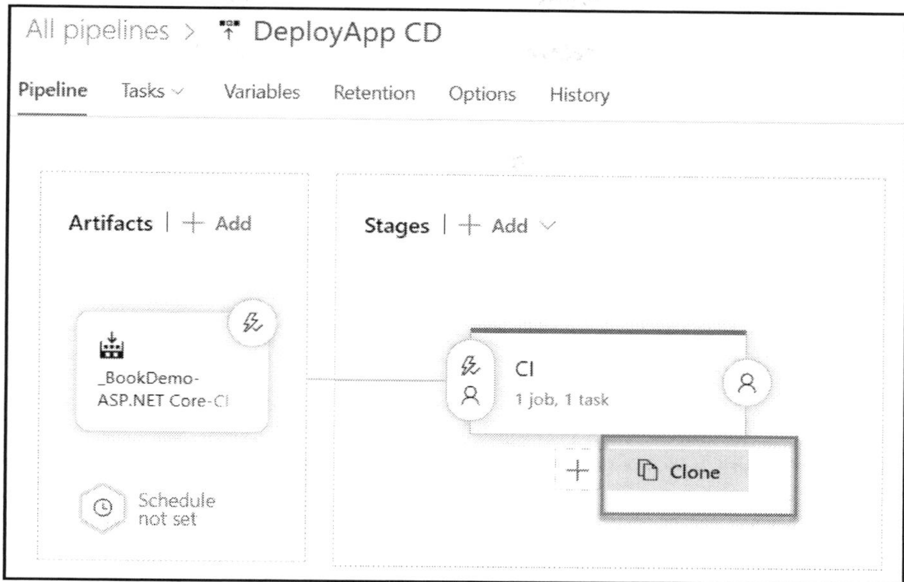

And the following screenshot shows the app service name settings with the web app name of the new environment:

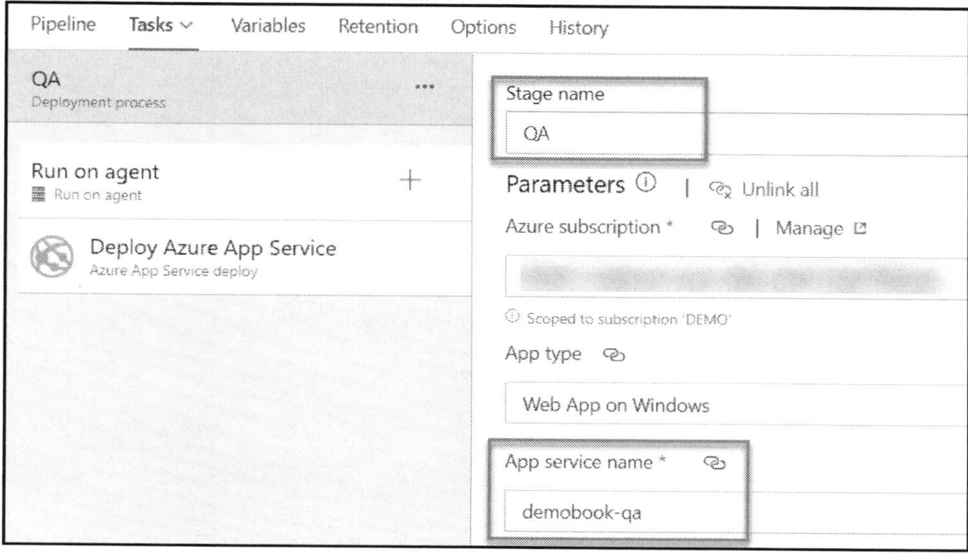

9. We finally get the release definition as follows:

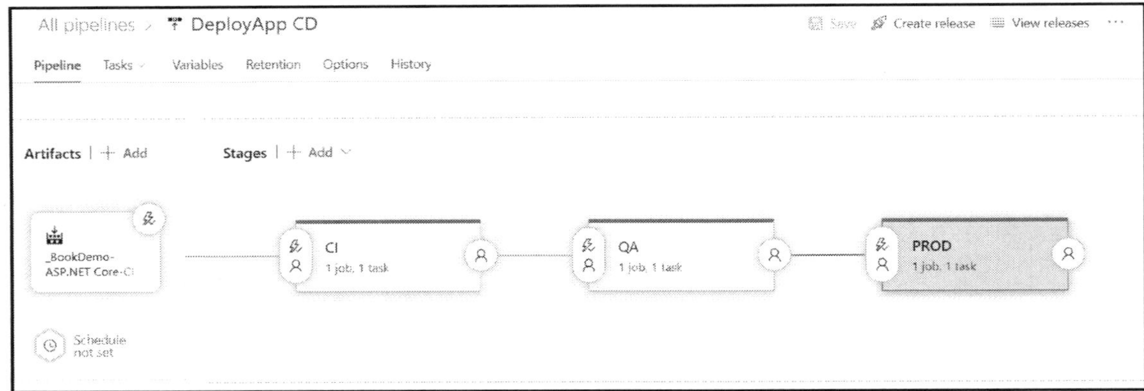

10. To trigger the deployment of our application, we will create a new release by clicking on the **Create release** button:

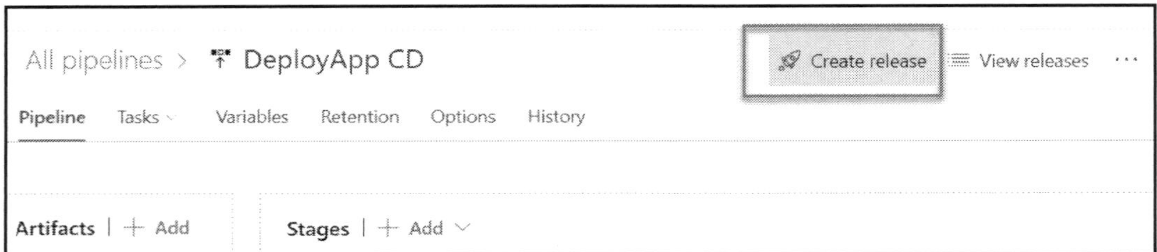

11. At the end of its execution, we can see its deployment status, as shown in the following screenshot:

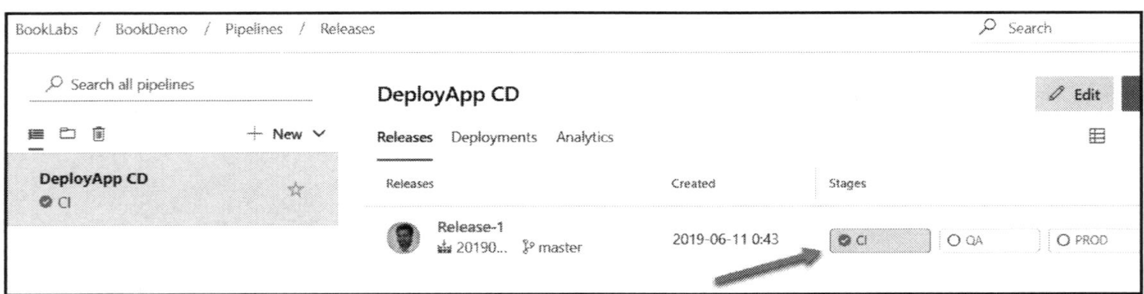

In this screenshot, we see that the deployment in the integration environment was successfully completed.

By following the steps in this lab, we have created a CI and CD pipeline with Azure DevOps, which has many other very interesting features. To learn more, consult the documentation here: `https://docs.microsoft.com/en-us/azure/devops/pipelines/?view=azure-devops`.

Let's look at creating a CI pipeline using GitLab CI.

Using GitLab CI

In the previous sections of this chapter, we learned how to create CI/CD pipelines with Jenkins and Azure Pipelines.

Now let's look at a lab using another DevOps tool that is gaining popularity: **GitLab CI**.

GitLab CI is one of the services offered by GitLab (`https://about.gitlab.com/`), which, like Azure DevOps, is a cloud platform with the following:

- A source code manager
- A CI/CD pipeline manager
- A board for project management

The other services it offers are listed here: `https://about.gitlab.com/features/`.

GitLab has a free price model with additional services that are subject to a charge; the price grid is available at `https://about.gitlab.com/pricing/` and is shown as follows:

This table highlights the differences between Azure DevOps and GitLab: `https://about.gitlab.com/devops-tools/azure-devops-vs-gitlab.html`.

In this lab, we'll see the following:

1. Authentication at GitLab
2. Creating a new project and versioning its code in GitLab
3. The creation and execution of a CI pipeline in GitLab CI

Authentication at GitLab

Creating a GitLab account is free and can be done either by creating a GitLab account or using external accounts, such as Google, GitHub, Twitter, or Bitbucket.

To create a GitLab account, we need to go to `https://gitlab.com/users/sign_in#register-pane` and choose the type of authentication.

The following screenshot shows the GitLab authentication form:

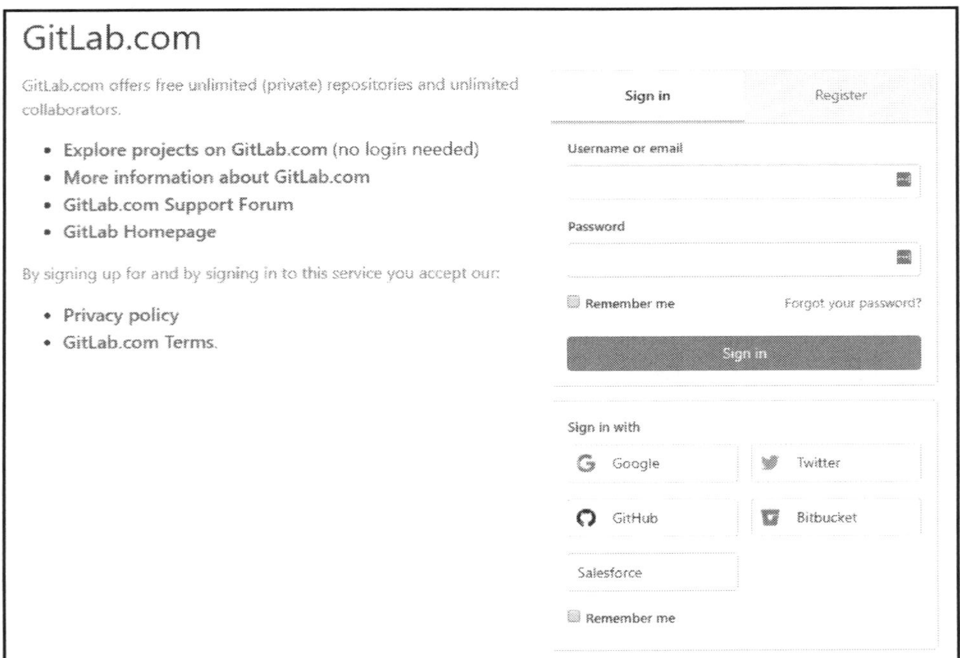

Once your account has been created and authenticated, you will be taken to the home page of your account, which offers all the functionalities shown, as in the following screenshot:

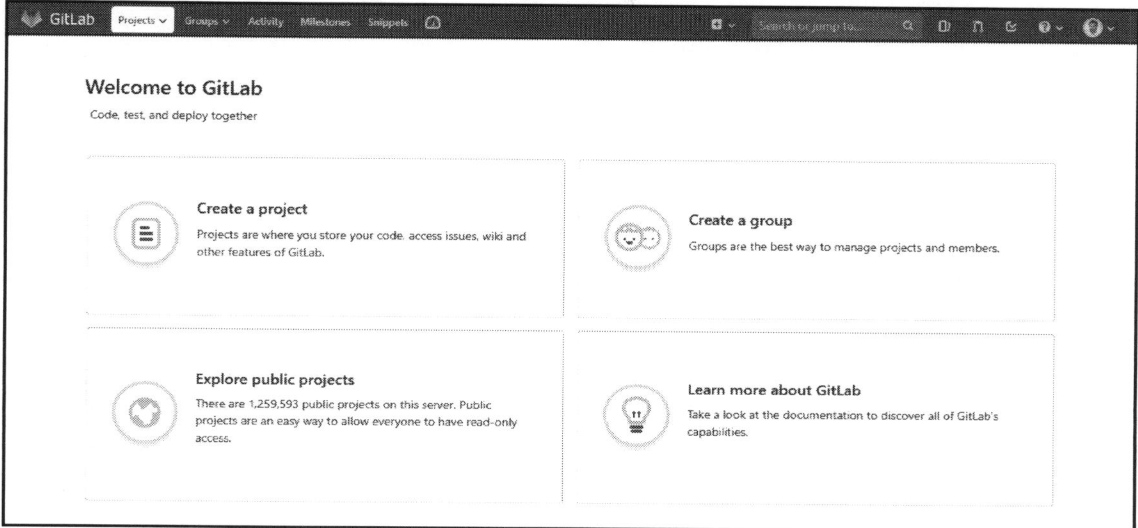

Now that we are done with the authentication, let's go ahead and create a new project.

Creating a new project and managing your code source

To create a new project in GitLab, follow these steps:

1. Click on **Create a project** on the home page:

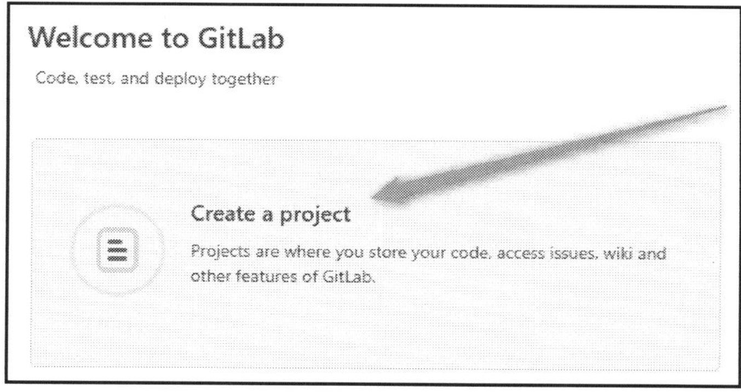

2. Then, we can choose a few options:
    - To create an empty project (without code)—the form asks you to enter its name:

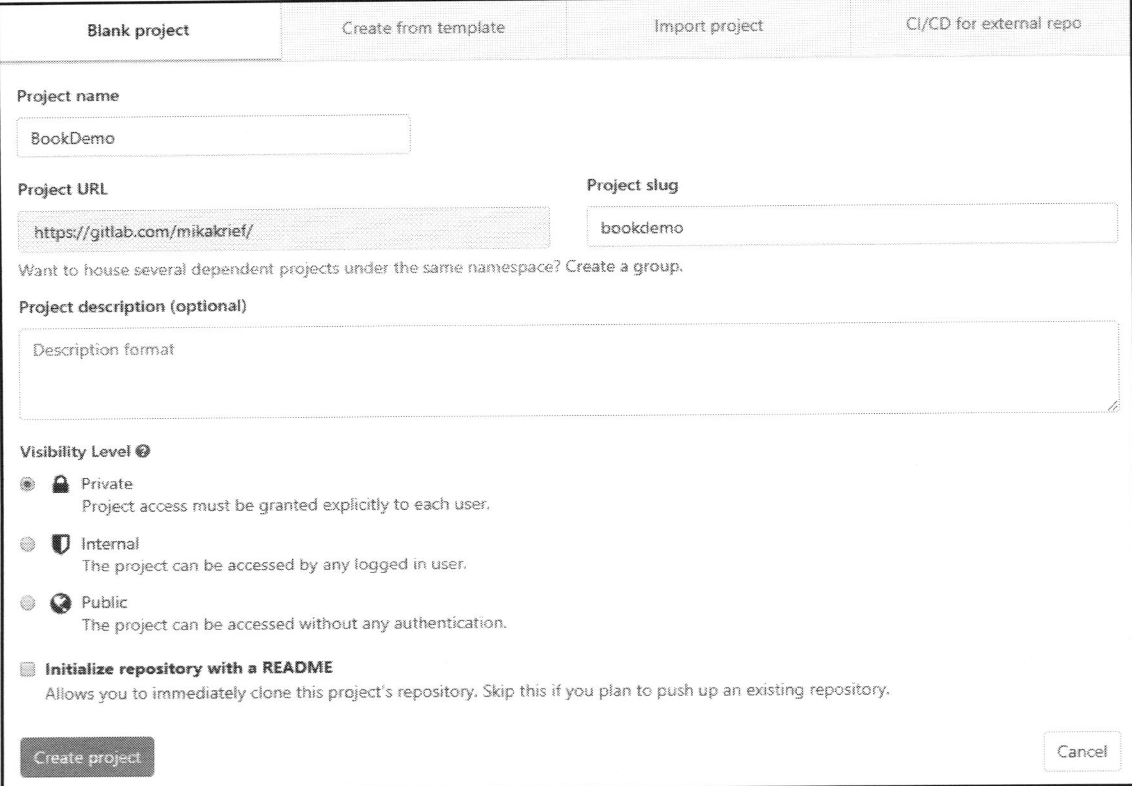

- To create a new project from a built-in template project, as follows:

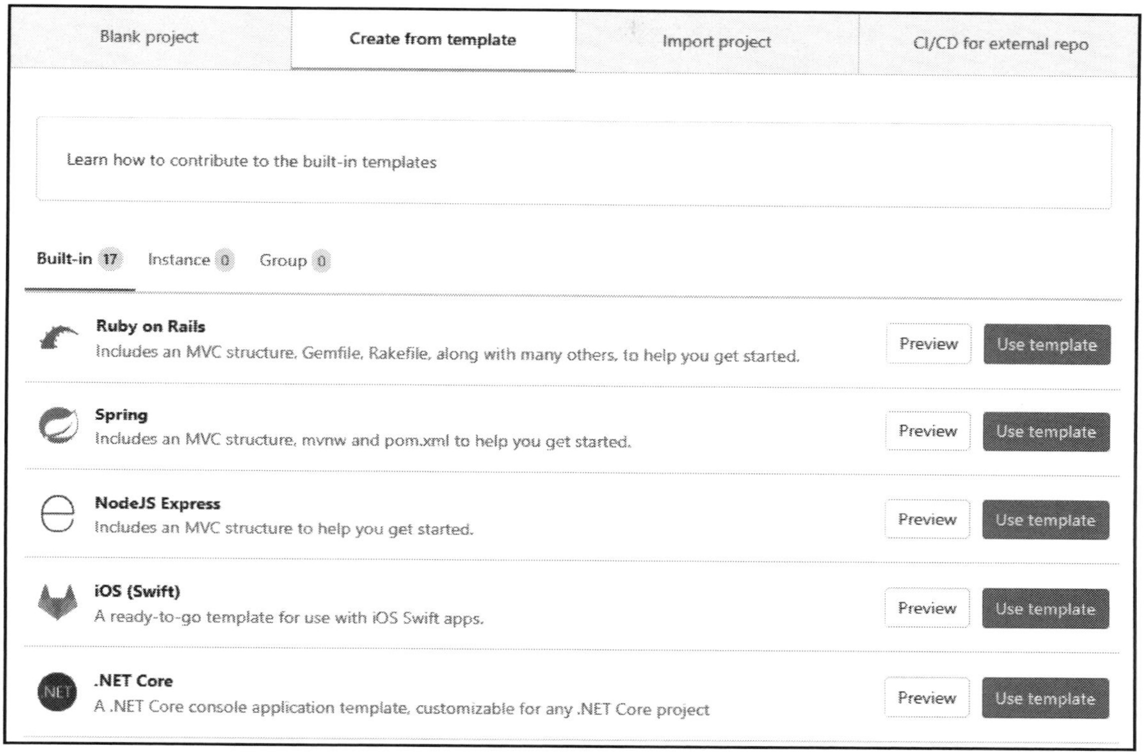

Blank project | **Create from template** | Import project | CI/CD for external repo

Learn how to contribute to the built-in templates

Built-in 17 Instance 0 Group 0

Ruby on Rails
Includes an MVC structure, Gemfile, Rakefile, along with many others, to help you get started. Preview Use template

Spring
Includes an MVC structure, mvnw and pom.xml to help you get started. Preview Use template

NodeJS Express
Includes an MVC structure to help you get started. Preview Use template

iOS (Swift)
A ready-to-go template for use with iOS Swift apps. Preview Use template

.NET Core
A .NET Core console application template, customizable for any .NET Core project Preview Use template

- To import code from an internal or external repository of another SVC platform, as shown in the following screenshot:

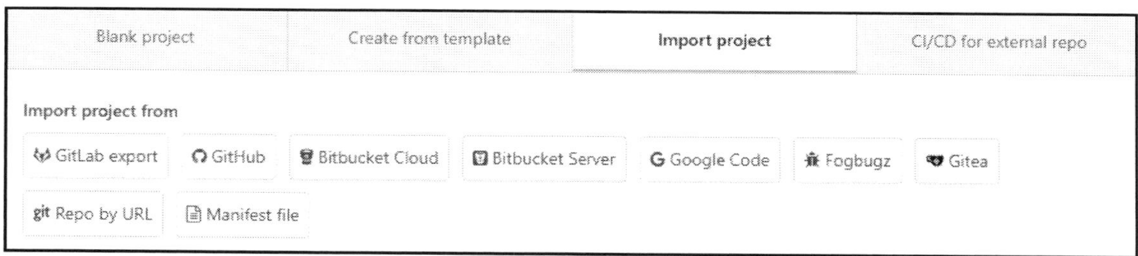

- The code to import is located in an external SVC repository, as shown in the following screenshot:

In our case, for this lab, we will start with the first option, which is an empty project, and as we saw in the form, we choose a project name such as `BookDemo`, then we validate it by clicking on the **Create a project** button.

3. Once the project is created, we'll have a page that indicates the different Git commands to execute to push its code.
4. To do this, on our local disk, we will create a new `gitlabdemo` directory and then clone the content of our example, which can be found at `https://github.com/PacktPublishing/Learning_DevOps/tree/master/CHAP06`, in the `gitlabdemo` directory.
5. Then, we will execute the following commands in a terminal to push the code into the repository, as seen in detail in `Chapter 5`, *Managing Your Source Code with Git*:

```
git init
git remote add origin <git repo Url>
git add .
git commit -m "Initial commit"
git push -u origin master
```

During execution, an identification window will ask us for your GitLab account ID because it is a private project. Once logged into your account on the GitLab web portal, your username will be available on your account page, `https://gitlab.com/profile/account`.

Once these commands have been executed, we'll obtain a remote GitLab repository with our lab code.

The following screenshot shows the remote GitLab repository:

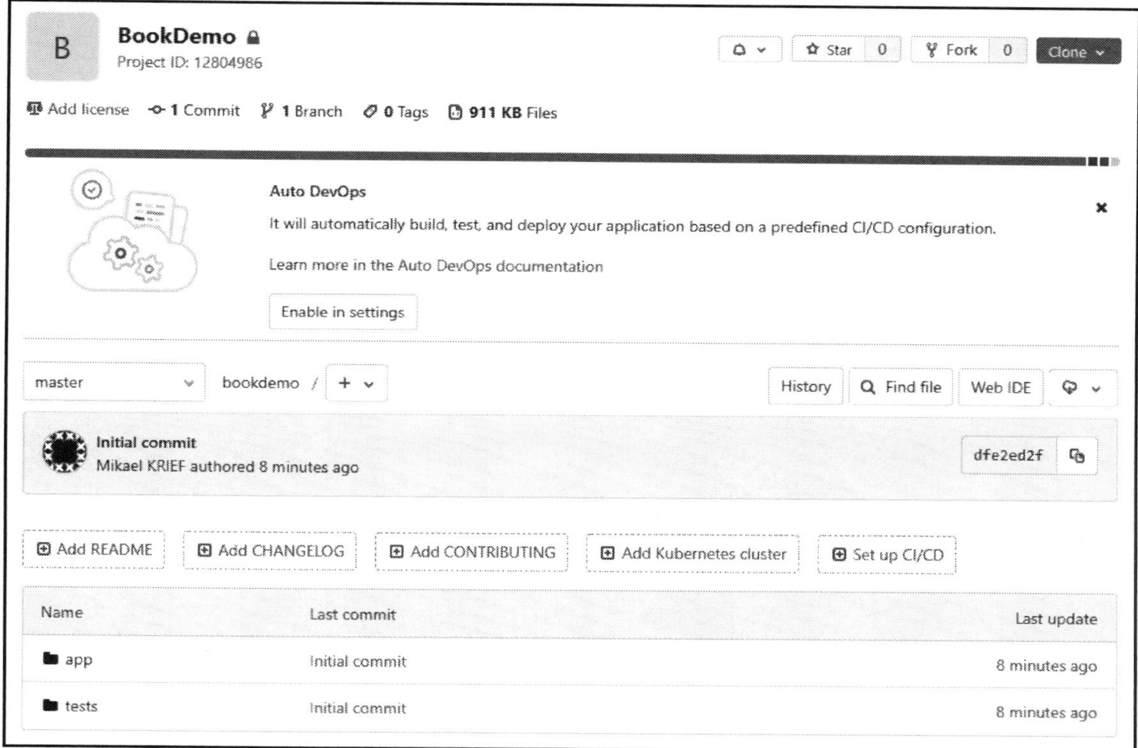

The code of our application has been deposited in GitLab, and we can now create our CI process with GitLab.

Creating the CI pipeline

In GitLab CI, the creation of a CI pipeline (and CD) is not done via a graphical interface, but with a YAML file at the root of the project.

This method, which consists of describing the process of a pipeline in a file that is located with the code, can be called **Pipeline as Code** in the same way as the IaC:

1. To create this pipeline, we will create, at the root of the application code, a `.gitlab-ci.yml` file with the following content code:

```
image: microsoft/dotnet:latest
stages:
    - build
    - test

variables:
    BuildConfiguration: "Release"

build:
    stage: build
    script:
        - "cd app"
        - "dotnet restore"
        - "dotnet build --configuration $BuildConfiguration"
test:
    stage: test
    script:
        - "cd tests"
        - "dotnet test --configuration $BuildConfiguration"
```

The code of this file is also available here: `https://github.com/PacktPublishing/Learning_DevOps/blob/master/CHAP06/gitlab-ci-demo.yml`.

We can see at the beginning of this code that we use a `microsoft/dotnet:latest` Docker image that will be mounted in a container and in which the actions of the pipeline will be executed.

Then, we define two stages: one for the build and one for the test execution, as well as a `BuildConfiguration` variable that will be used in the scripts.

Finally, we describe each of the stages of the scripts to be executed in their respective directories. These .NET core scripts are identical to the ones we saw in the *Using Azure Pipelines* section.

The full documentation on the format and syntax of this `.gitlab-ci.yml` file is available here: `https://docs.gitlab.com/ee/ci/yaml/`.

2. Then, we will commit and push this file into the remote repository.
3. Just after pushing the code, we can see that the CI process has been triggered.

Our CI pipeline was, therefore, triggered when the code was pushed into the repository, so let's now look at how to see the details of its execution.

Accessing the CI pipeline execution details

To access the execution details of the executed CI pipeline, follow these steps:

1. In the GitLab CI menu, we go to **CI / CD** | **Pipelines**, and we will see the list of pipeline executions, as shown in the following screenshot:

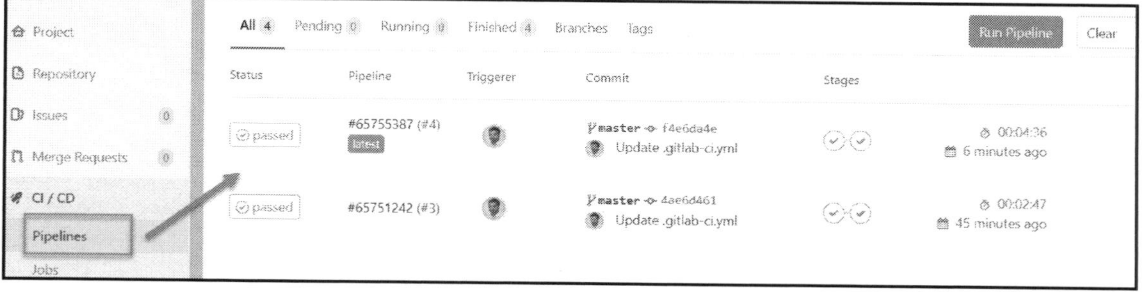

2. To display the details of the pipeline, we click on the desired pipeline execution:

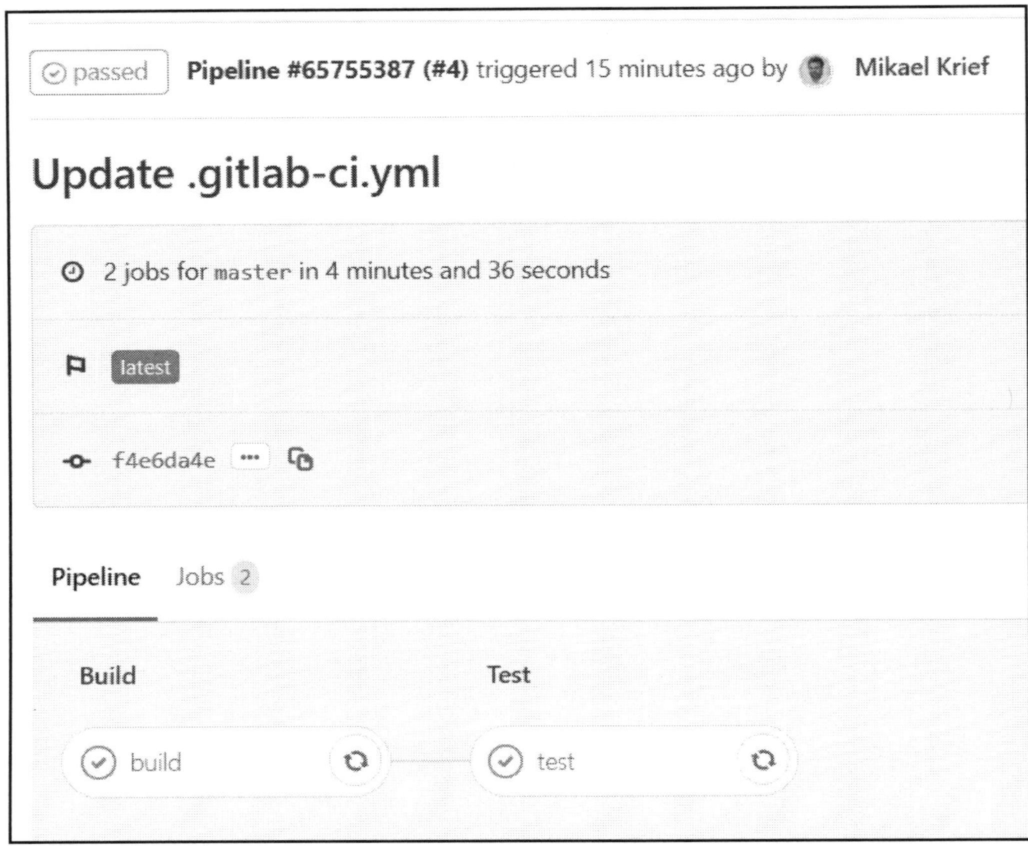

3. We can see the execution status as well as the two stages that you defined in the pipeline YAML file. To view the details of the execution logs for a stage, we click on the stage, as shown in the following screenshot:

Mikael Krief > BookDemo > Jobs > #229390643

passed **Job #229390643** triggered 19 minutes ago by Mikael Krief

```
Running with gitlab-runner 11.11.2 (ac2a293c)
  on docker-auto-scale ed2dce3a
Using Docker executor with image microsoft/dotnet:latest ...
Pulling docker image microsoft/dotnet:latest ...
Using docker image sha256:08663b8eaa01a928bf4b22c6d7892a5306dc76a40d34e9449465d7f8d0c5ec38 for microsoft/dotnet:latest ...
Running on runner-ed2dce3a-project-12804986-concurrent-0 via runner-ed2dce3a-srm-1560285231-e19116a0...
Initialized empty Git repository in /builds/mikakrief/bookdemo/.git/
Fetching changes...
Created fresh repository.
From https://gitlab.com/mikakrief/bookdemo
 * [new branch]      master     -> origin/master
Checking out f4e6da4e as master...

Skipping Git submodules setup
$ cd app
$ dotnet restore
  Restore completed in 13.32 sec for /builds/mikakrief/bookdemo/app/app.csproj.
```

We can see the execution of the scripts written in the pipeline YAML file.

In this section, we have seen the implementation of a CI pipeline in GitLab CI with the initialization of a remote repository and the creation of a YAML file for configuring the pipeline as well as the execution of the pipeline.

Summary

In this chapter, we looked at one of the most important topics in DevOps: the CI/CD process. We started with a presentation of the principles of continuous integration and continuous delivery. Then, we focused on package managers, looking at NuGet, npm, Nexus, and Azure Artifacts.

Finally, we saw how to implement and execute an end-to-end CI/CD pipeline using three different tools: Jenkins, Azure Pipelines, and GitLab CI. For each of them, we looked at the archiving of the application source code and the creation of the pipeline and its execution.

After reading this chapter, we should be able to create a pipeline for continuous integration and delivery with source code management as the source. In addition, we will be able to choose and use a package manager to centralize and distribute our packages.

In the next chapter, we will talk about Docker and the containerization of applications, which enables not only better isolation of applications on the host system but also an improvement in the CI/CD pipelines.

Questions

1. What are the prerequisites for implementing the CI pipeline?
2. When will the CI be triggered?
3. What is the purpose of a package manager?
4. Which type of packages are stored in a NuGet package manager?
5. Which platform is Azure Artifacts integrated into?
6. Is Jenkins a cloud service?
7. In Azure DevOps, what is the name of the service that allows the management of CI/CD pipelines?
8. What are the three services offered by GitLab?
9. In GitLab CI, which element allows you to build a CI pipeline?

Further reading

If you would like to find out more about CI/CD pipelines, here are some books:

- Hands-On Continuous Integration and Delivery: `https://www.packtpub.com/virtualization-and-cloud/hands-continuous-integration-and-delivery`
- Continuous Integration, Delivery, and Deployment: `https://www.packtpub.com/application-development/continuous-integration-delivery-and-deployment`
- Mastering Jenkins: `https://www.packtpub.com/application-development/mastering-jenkins`
- Azure DevOps Server 2019 Cookbook: `https://www.packtpub.com/networking-and-servers/azure-devops-server-2019-cookbook-second-edition`
- Mastering GitLab 12: `https://www.packtpub.com/cloud-networking/mastering-gitlab-12`

3
Section 3: Containerized Applications with Docker and Kubernetes

In this section, we present the basic usage of Docker including how to create and run a container from a Docker file. Then, we expose the role of Kubernetes and how to deploy a more complex application on Kubernetes.

We will have the following chapters in this section:

- `Chapter 7`, *Containerizing Your Application with Docker*
- `Chapter 8`, *Managing Containers Effectively with Kubernetes*

7
Containerizing Your Application with Docker

In the last few years, one technology in particular has been making headlines on the net, on social networks, and at events—Docker.

Docker is a containerization tool, which became open source in 2013. It allows you to isolate an application from its host system so that the application becomes portable and code tested on a developer's workstation can be deployed to production with fewer concerns about execution runtime dependencies. We'll talk a little about application containerization.

A container is a system that embeds an application and its dependencies. Unlike a VM, a container contains only a light operating system with only the elements required for the OS, such as system libraries, binaries, and code dependencies.

To learn more about the differences between VMs and containers, and why containers will replace VMs in the future, I suggest you read this blog article: `https://blog.docker.com/2018/08/containers-replacing-virtual-machines/`.

The principal difference between VMs and containers is that each VM that is hosted on a hypervisor contains a complete OS and is therefore completely independent of the guest OS that is on the hypervisor.

Containers don't contain a complete OS – only a few binaries—but they are dependent on the guest OS, using its resources (CPU, RAM, and network).

In this chapter, we will see how to install Docker on different platforms, how to create a Docker image, and how to register it in Docker Hub. Finally, we'll discuss an example of a CI/CD pipeline that deploys a Docker image in **Azure Container Instances** (**ACI**).

This chapter covers the following topics:

- Installing Docker
- Creating a Docker file
- Building and running a container on a local machine
- Pushing an image to Docker Hub
- Deploying a container in ACI with a CI/CD pipeline

Technical requirements

For this chapter, there are no technical prerequisites; however, in the last part of this chapter, the *CI/CD pipeline for the container* section, we will discuss Terraform and the CI/CD pipeline, which were explained in `Chapter 2`, *Provisioning Cloud Infrastructure with Terraform*, and `Chapter 6`, *Continuous Integration and Continuous Delivery*.

All of the source code for the scripts included in this chapter is available here: `https://github.com/PacktPublishing/Learning_DevOps/tree/master/CHAP07`.

Check out the following video to see the Code in Action:
`http://bit.ly/2WbYlKD`

Installing Docker

Docker's **community edition** (**CE**) is free and is very well suited to developers and small teams – it's what we'll use in this book.

If Docker is to be used throughout a company, it is better to use Docker Enterprise, which is not free. The documentation is at: `https://docs.docker.com/ee/supported-platforms/`.

Docker is a cross-platform tool that can be installed on Windows, Linux, or macOS and is also natively present on some cloud providers, such as AWS and Azure.

To operate, Docker needs the following elements:

- **The Docker client**: This allows you to perform various operations on the command line.
- **The Docker daemon**: This is Docker's engine.
- **Docker Hub**: This is a public (with a free option available) registry of Docker images.

Before installing Docker, we will first create an account on Docker Hub.

Registering on Docker Hub

Docker Hub is a public space called a **registry**, containing more than 2 million public Docker images that have been deposited by companies, communities, and even individual users.

To register on Docker Hub and list Docker images, perform the following steps:

1. Go to `https://hub.docker.com/` and click on the **Sign up for Docker Hub** button:

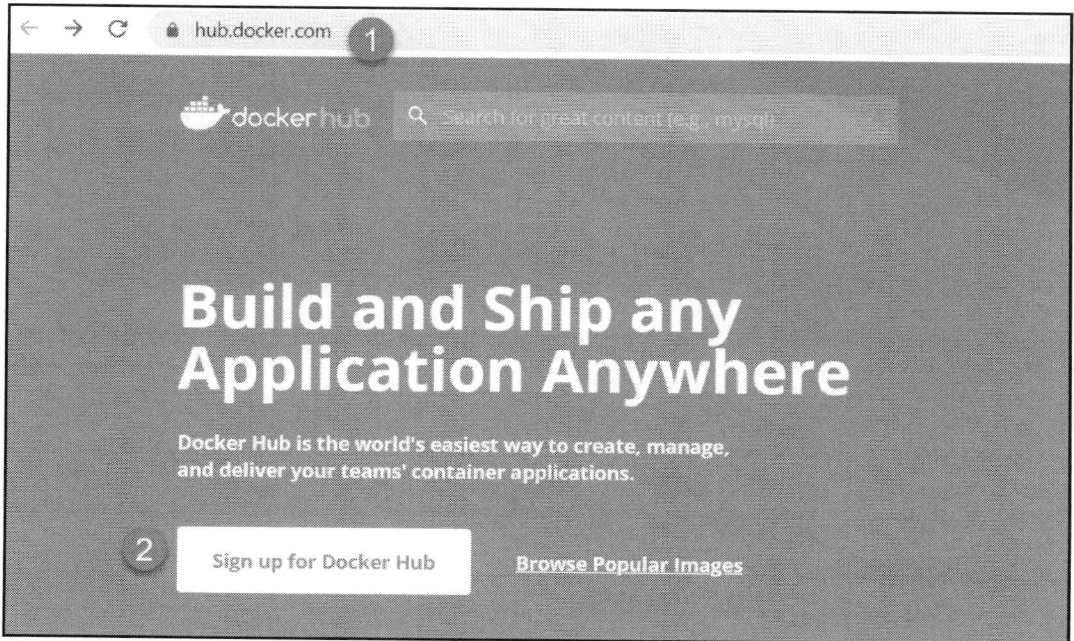

2. Fill in the form with a unique ID, an email, and a password.
3. Once your account is created, you can then log in to the site, and this account will allow you to upload custom images and download Docker Desktop.
4. To view and explore the images available from Docker Hub, go to the **Explore** section:

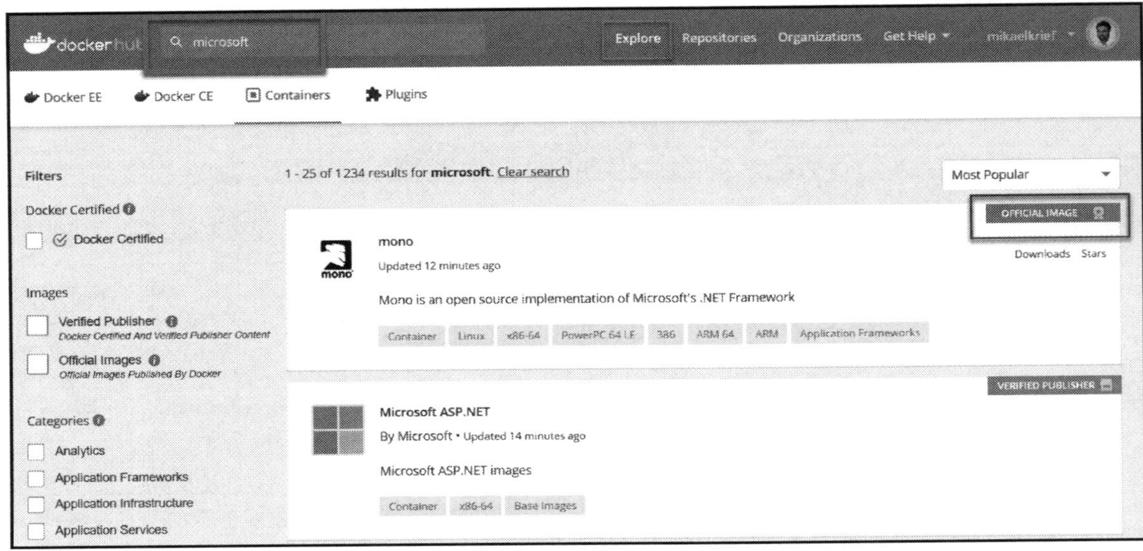

A list of Docker images is displayed with a search filter to search for official images or images from verified publishers, as well as images certified by Docker.

Having created an account on Docker Hub, we will now look at installing Docker.

Docker installation

We'll now discuss the installation of Docker on Windows in detail. To install Docker on a Windows machine, it is necessary to first check the hardware requirements, which are as follows:

- Windows 10 64 bit with at least 4 GB of RAM
- A virtualization system (such as Hyper-V) enabled. You can refer to this documentation in the event of any problems: `https://docs.docker.com/docker-for-windows/troubleshoot/#virtualization-must-be-enabled`.

To install Docker Desktop, which is the name of the Docker installer for Windows and macOS, follow these steps:

1. First, download Docker Desktop by clicking on the **Get Docker** button from Docker Hub at `https://hub.docker.com/editions/community/docker-ce-desktop-windows` and log in if you are not already connected to Docker Hub. The following screen shows the **Get Docker** button on the download page:

2. Once that's downloaded, click on the downloaded EXE file.
3. Then, take the single configuration step, which is a choice between using Windows or Linux containers:

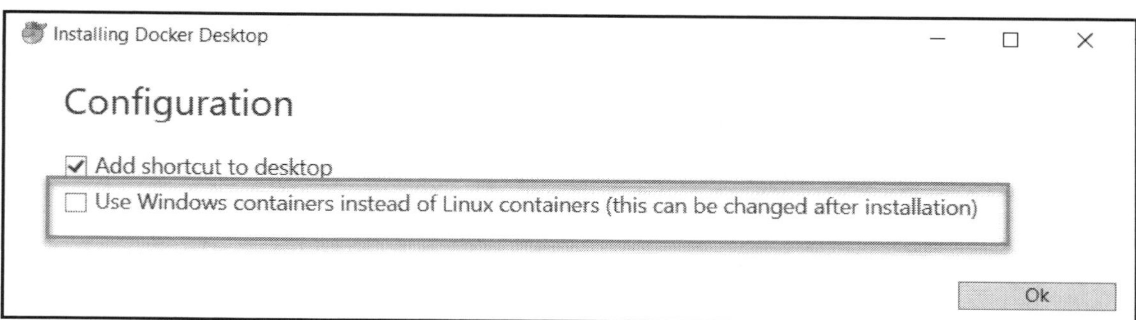

In our case, we will leave this as the default option to use Linux containers.

4. Once the installation is complete, we'll get a confirmation message and a button to close the installation:

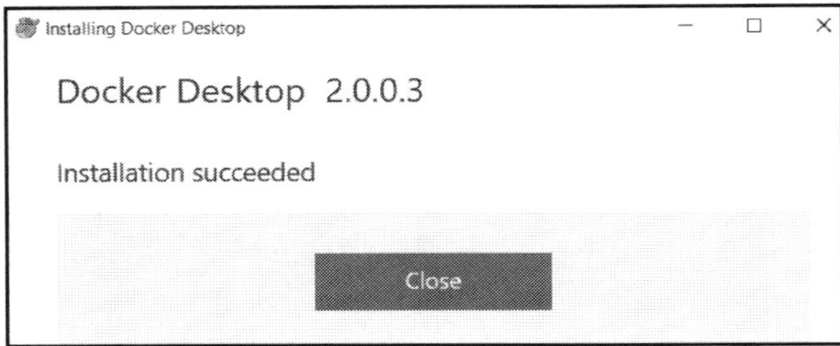

5. Finally, to start Docker, launch the Docker Desktop program. An icon will appear in the notification bar indicating that Docker is starting. It will then ask you to log in to Docker Hub via a small window. The startup steps of Docker Desktop are shown in the following screenshot:

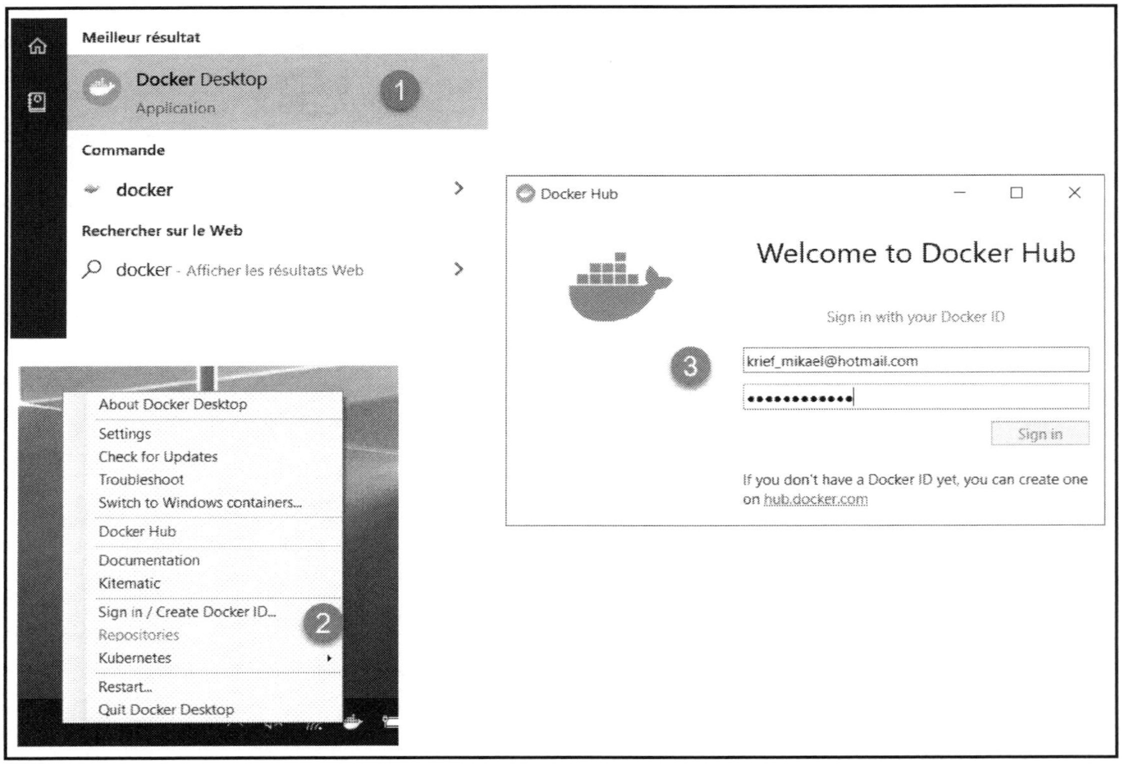

That's it – we've installed and started Docker on Windows.

To install Docker on another OS, you can read the documentation for each of them at `https://docs.docker.com/install/`. Then, from the menu options, we choose the desired target OS, as shown in the following screenshot:

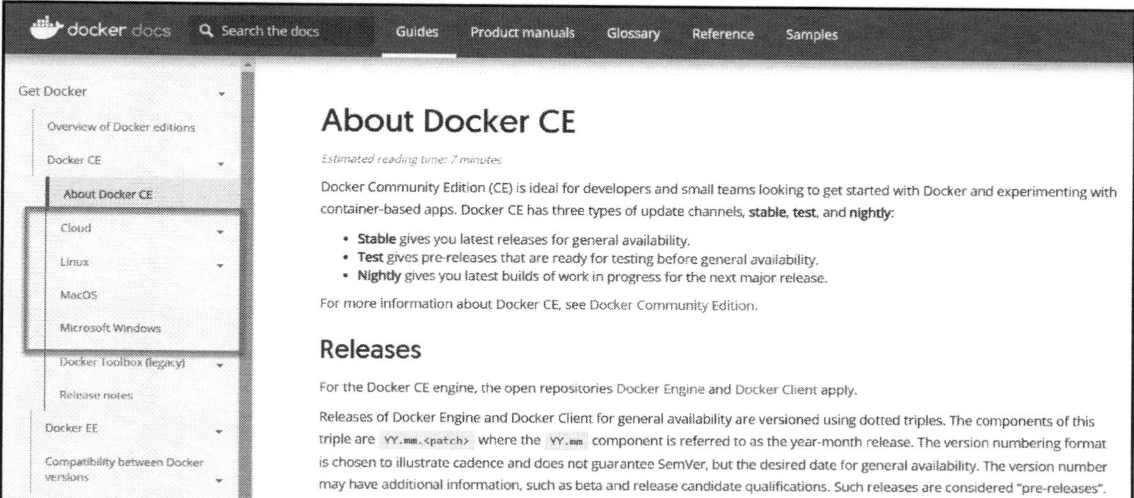

To check your Docker installation, open the Terminal window (it will also work on a Windows PowerShell Terminal), then execute the following command:

```
docker --help
```

You should see something like this:

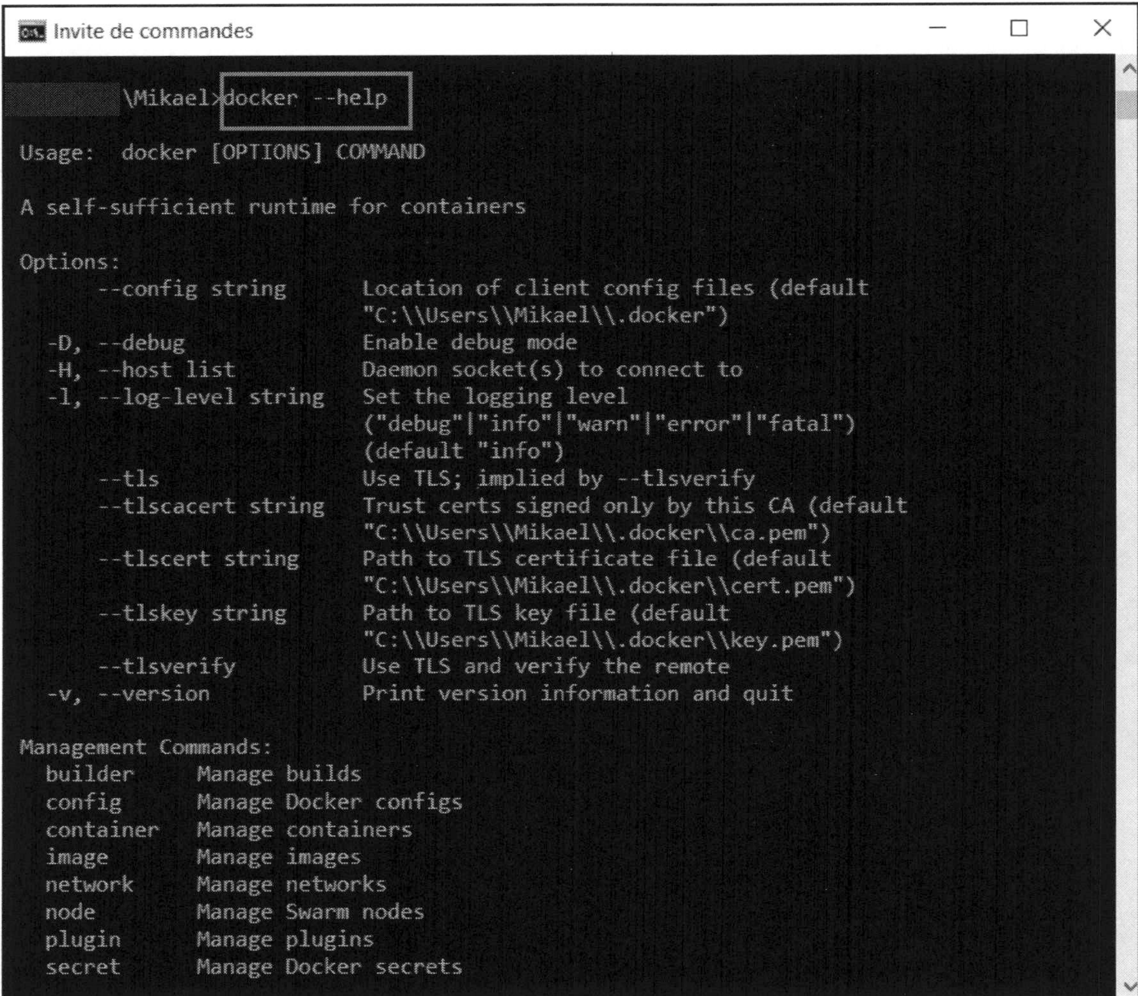

```
\Mikael>docker --help

Usage:  docker [OPTIONS] COMMAND

A self-sufficient runtime for containers

Options:
      --config string      Location of client config files (default
                           "C:\\Users\\Mikael\\.docker")
  -D, --debug              Enable debug mode
  -H, --host list          Daemon socket(s) to connect to
  -l, --log-level string   Set the logging level
                           ("debug"|"info"|"warn"|"error"|"fatal")
                           (default "info")
      --tls                Use TLS; implied by --tlsverify
      --tlscacert string   Trust certs signed only by this CA (default
                           "C:\\Users\\Mikael\\.docker\\ca.pem")
      --tlscert string     Path to TLS certificate file (default
                           "C:\\Users\\Mikael\\.docker\\cert.pem")
      --tlskey string      Path to TLS key file (default
                           "C:\\Users\\Mikael\\.docker\\key.pem")
      --tlsverify          Use TLS and verify the remote
  -v, --version            Print version information and quit

Management Commands:
  builder     Manage builds
  config      Manage Docker configs
  container   Manage containers
  image       Manage images
  network     Manage networks
  node        Manage Swarm nodes
  plugin      Manage plugins
  secret      Manage Docker secrets
```

As you can see in the preceding screenshot, the command displays the different operations available in the Docker client tool.

Before looking in detail at the execution of Docker commands, it is important to have an overview of Docker's concepts.

An overview of Docker's elements

Before executing Docker commands, we will discuss some of Docker's fundamental elements, which are Dockerfiles, containers, and volumes.

First of all, it is important to know that a **Docker image** is a basic element of Docker and consists of a text document called a **Dockerfile**, containing the binaries and application files we want to containerize.

A **container** is an instance that is executed from a Docker image. It is possible to have several instances of the same image within a container that the application will run. Finally, a **volume** is storage space that is physically located on the host OS (that is, outside the container) and it can be shared across multiple containers if required. This space will allow the storage of persistent elements (files or databases).

To manipulate these elements, we will use command lines, which will be discussed as we progress through this chapter.

In this section, we discussed Docker Hub, including the different steps for creating an account. Then, we looked at the steps for installing Docker Desktop locally, and finally, we finished with an overview of Docker elements.

We will now start working with Docker, and the first operation we will look at is the creation of a Docker image from a Dockerfile.

Creating a Dockerfile

A basic Docker element is a file called a **Dockerfile**, which contains step-by-step instructions for building a Docker image.

To understand how to create a Dockerfile, we'll look at an example that allows us to build a Docker image that contains an Apache web server and a web application.

Let's start by writing a Dockerfile.

Writing a Dockerfile

To do this, we will first create an HTML page that will be our web application. So, we create a new `appdocker` directory and an `index.html` page in it, which includes the example code that displays welcome text on a web page:

```
<html>
  <body>
    <h1>Welcome to my new app</h1>
    This page is test for my demo Dockerfile.<br />
    Enjoy ...
  </body>
</html>
```

Then, in the same directory, we create a **Dockerfile** (without an extension) with the following content, which we will detail right after:

```
FROM httpd:latest
COPY index.html /usr/local/apache2/htdocs/
```

To create a Dockerfile, start with the `FROM` statement. The required `FROM` statement defines the base image, which we will use for our Docker image – any Docker image is built from another Docker image. This base image can be saved either in Docker Hub or in another registry (for example, Artifactory, Nexus Repository, or Azure Container Registry).

In our code example, we use the Apache `httpd` image tagged the latest version, `https://hub.docker.com/_/httpd/`, and so we use the `FROM httpd:latest` Dockerfile instruction.

Then, we use the `COPY` instruction to execute the image construction process. Docker copies the local `index.html` file (which we just created) into the `/usr/local/apache2/htdocs/` directory of the image.

The source code for this Dockerfile and the HTML page can be found here: `https://github.com/PacktPublishing/Learning_DevOps/tree/master/CHAP07/appdocker`.

We have just looked at the `FROM` and `COPY` instructions of the Dockerfile, but there are other instructions, which there is a short overview of in the following section.

Dockerfile instructions overview

We previously mentioned that a Dockerfile file is comprised of instructions, and we also looked at a concrete example with the `FROM` and `COPY` instructions. There are other instructions that will allow you to build a Docker image – here is an overview of the principal instructions that can be used:

- `FROM`: This instruction is used to define the base image for our image, as shown in the example detailed in the *Writing a Dockerfile* section.
- `COPY` and `ADD`: These are used to copy one or more local files into an image. The `Add` instruction supports an extra two functionalities, to refer to a URL and to extract compressed files.

For more details about the differences between them, you can read this article: `https://nickjanetakis.com/blog/docker-tip-2-the-difference-between-copy-and-add-in-a-dockerile`.

- `RUN` and `CMD`: This instruction takes a command as a parameter that will be executed during the construction of the image. The `Run` instruction creates a layer so that it can be cached and versioned. The `CMD` instruction defines a default command to be executed during the call to run the image. The `CMD` instruction can be overwritten at runtime with an extra parameter provided.

  You can write the following example of the `RUN` instruction in a Dockerfile to execute the `apt-get` command:

  ```
  RUN apt-get update
  ```

  With this instruction, we update the `apt` packages that are already present in the image and create a layer. We can also use the `CMD` instruction in the following example, which, during execution, will display a `docker` message:

  ```
  CMD "echo docker"
  ```

- `ENV`: Allows you to instantiate environment variables that can be used to build an image. These environment variables will persist throughout the life of the container, as follows:

  ```
  ENV myvar=mykey
  ```

- `WORKDIR`: This instruction gives the execution directory of the container, as follows:

```
WORKDIR usr/local/apache2
```

That was an overview of Dockerfile instructions. There are other instructions that are commonly used, such as `EXPOSE`, `ENTRYPOINT`, and `VOLUME`, which you can find on the official documentation: `https://docs.docker.com/engine/reference/builder/`.

We have just seen that the writing of a Dockerfile is performed with different instructions, such as `FROM`, `COPY`, and `RUN`, which are used to create a Docker image. Let's now look at how to run Docker in order to build a Docker image from a Dockerfile, and run that image locally to test it.

Building and running a container on a local machine

So far, we have discussed Docker elements, along with an example of a Dockerfile that is used to containerize a web application, so we now have all the elements to run Docker.

The execution of Docker is performed by these different operations:

- Building a Docker image from a Dockerfile
- Instantiating a new container locally from this image
- Testing our locally containerized application

Let's take a deep dive into each operation.

Building a Docker image

To build a Docker image from our previously created Dockerfile that contains the following instructions:

```
FROM httpd:latest
COPY index.html /usr/local/apache2/htdocs/
```

We'll go to a Terminal to head into the directory that contains the Dockerfile, and then execute the `docker build` command with the following syntax:

```
docker build -t demobook:v1 .
```

The `-t argument` indicates the name of the image and its tag. Here, in our example, we call our image `demobook` and add the `v1` tag.

The `.` (dot) at the end of the command specifies that we will use the files in the current directory. The following screenshot shows the execution of this command:

```
PS                        \CHAP07\appdocker> docker build -t demobook:v1 .
Sending build context to Docker daemon  3.072kB
Step 1/2 : FROM httpd:latest
latest: Pulling from library/httpd
8d691f585fa8: Pull complete
8eb779d8bd44: Pull complete
574add29ec5c: Pull complete
9ccffbf4a714: Pull complete
166e14b82905: Pull complete
Digest: sha256:649bd29cc9284f06cf1a99726c4e747a83679e04eea3311b55022dd247026138
Status: Downloaded newer image for httpd:latest
 ---> 66a97eeec7b8
Step 2/2 : COPY index.html /usr/local/apache2/htdocs/
 ---> 808234df59cf
Successfully built 808234df59cf
Successfully tagged demobook:v1
SECURITY WARNING: You are building a Docker image from Windows against a non-Windows Docker host. All files and directories added to build context will have '-rwxr-xr-x' permissions. It is recommended to double check and reset permissions for sensitive files and directories.
```

Executing the `docker build` command downloads the base image indicated in the Dockerfile from Docker Hub, and then Docker executes the various instructions that are mentioned in the Dockerfile.

Note that if during the first execution of the `docker build` command, you get the `Get https://registry-1.docker.io/v2/library/httpd/manifests/latest: unauthorized: incorrect username or password` error, then execute the `docker logout` command, then restart the `docker build` command, as indicated in this article: `https://medium.com/@blacksourcez/fix-docker-error-unauthorized-incorrect-username-or-password-in-docker-f80c45951b6b`.

At the end of the execution, we obtain a locally stored Docker `demobook` image.

The Docker image is stored in a local folder system depending on your OS. For more information about the location of Docker images, read this article: `http://www.scmgalaxy.com/tutorials/location-of-dockers-images-in-all-operating-systems/`.

We can also check that the image is successfully created by executing the following Docker command:

```
docker images
```

Here's its output:

```
PS \Learning_DevOps\CHAP07\appdocker> docker images
REPOSITORY          TAG          IMAGE ID          CREATED            SIZE
demobook            v1           a121d88f6e18      23 minutes ago     140MB
httpd               latest       e77c77f17b46      6 days ago         140MB
```

This command displays the list of Docker images on my local machine, and we see the `demobook` image we just created as well as the basic `httpd` image that was downloaded from DockerHub. So, the next time the image is built, the `httpd` image will not need to be downloaded again.

Now that we have created the Docker image of our application, we will instantiate a new container of this image.

Instantiating a new container of an image

To instantiate a container of our Docker image, we will execute the `docker run` command in our Terminal with the following syntax:

```
docker run -d --name demoapp -p 8080:80 demobook:v1
```

The `-d` parameter indicates that the container will run in the background. In the `-name` parameter, we indicate the name of the container we want. In the `-p` parameter, we indicate the desired port translation; that is, in our example, port `80` of the container will be translated to port `8080` on our local machine. And finally, the last parameter of the command is the name of the image and its tag.

The execution of this command is shown in the following screenshot:

```
PS \Learning_DevOps\CHAP07\appdocker> docker run -d --name demoapp -p 8080:80 demobook:v1
381b476d62e568f382f251e0834fd8c69f713eb14ea41c95e5cd7004afdbb879
```

At the end of its execution, this command displays the ID of the container, and the container runs in the background. It is also possible to display the list of containers running on the local machine, by executing the following command:

```
docker ps
```

The following screenshot shows the execution with our container:

```
PS \Learning_DevOps\CHAP07\appdocker> docker ps
CONTAINER ID    IMAGE         COMMAND              CREATED          STATUS          PORTS                   NAMES
381b476d62e5    demobook:v1   "httpd-foreground"   3 minutes ago    Up 3 minutes    0.0.0.0:8080->80/tcp    demoapp
```

After the execution of each container, we have its shortcut ID, its associated image, its name, its execution command, as well as its translation port information displayed.

So, we have built a Docker image and instantiated a new container of that image locally. We will now see how to run a web application that is in the local container.

Testing a container locally

Everything that runs in a container remains inside it – this is the principle of container isolation. However, with the port translation that we did previously, with the `run` command, you can test your container on your local machine.

To do this, open a web browser and enter `http://localhost:8080` with `8080`, which represents the translation port indicated in the command, and here is the result:

We can see the content of our `index.html` page displayed.

In this section, we looked at the different Docker commands that can be used to build a Docker image, then we instantiated a new container from that image, and finally, we tested it locally.

In the next section, we will see how to publish a Docker image in Docker Hub.

Pushing an image to Docker Hub

The goal of creating a Docker image that contains an application is to be able to use it on servers that contain Docker and host the company's applications, just like a VM.

In order for an image to be downloaded to another computer, it must be saved in a Docker image registry. As already mentioned in this chapter, there are several Docker registries that can be installed on-premise, as in the case for Artifactory and Nexus Repository.

If you want to create a public image, you can push it (or upload it) to Docker Hub, which is Docker's public (and free) registry. We will now see how to upload the image we created in the previous section to Docker Hub. To do this, it is a requirement to have an account on Docker Hub, which we created just before installing Docker Desktop.

To push a Docker image to Docker Hub, perform the following steps:

1. **Sign in to Docker Hub**: Log in to Docker Hub using the following command:

   ```
   docker login -u <your dockerhub login>
   ```

   When executing the command, it will ask you to enter your Docker Hub password and indicate that you are connected, as shown in the following screenshot:

   ```
   PS \Learning_DevOps\CHAP07\appdocker> docker login -u mikaelkr
   Password:
   Login Succeeded
   ```

2. **Retrieving the image ID**: The next step consists of retrieving the ID of the image that has been created, and to do so we will execute the `docker images` command to display the list of images with their ID.

   The following screenshot shows the retrieval of the ID of the image:

   ```
   PS \Learning_DevOps\CHAP07\appdocker> docker images
   REPOSITORY     TAG        IMAGE ID        CREATED         SIZE
   demobook       v1         a121d88f6e18    6 hours ago     140MB
   httpd          latest     e77c77f17b46    6 days ago      140MB
   ```

3. **Tag the image for Docker Hub**: With the ID of the image we retrieved, we will now tag the image for Docker Hub. To do so, the following command is executed:

   ```
   docker tag <image ID> <dockerhub login>/demobook:v1
   ```

   The following screenshot shows the execution of this command on the created image:

```
PS \Learning_DevOps\CHAP07\appdocker> docker tag a121d88f6e18 mikaelkrief/demobook:v1
```

4. **Push the image Docker in the Docker Hub**: After tagging the image, the last step is to push the tagged image to Docker Hub.

   For this purpose, we will execute the following command:

   ```
   docker push docker.io/<dockerhub login>/demobook:v1
   ```

   The following screenshot shows its execution:

```
PS \Learning_DevOps\CHAP07\appdocker> docker push docker.io/mikaelkrief/demobook:v1
The push refers to repository [docker.io/mikaelkrief/demobook]
e5df7a05d9b7: Pushed
6c4a74a82dc9: Mounted from library/httpd
9cba8b480e83: Mounted from library/httpd
25797e1a8e3f: Mounted from library/httpd
d2583584487e: Mounted from library/httpd
cf5b3c6798f7: Mounted from library/httpd
v1: digest: sha256:ffe4e6e67b8bf200a1c86d42a00730491ded2e63279ddbaeb7e7ffdf3b56cd89 size: 1574
```

We can see from this execution that the image is uploaded to Docker Hub.

To view the pushed image in Docker Hub, we connect to the Docker Hub web portal at `https://hub.docker.com/` and see that the image is present, as shown in the following screenshot:

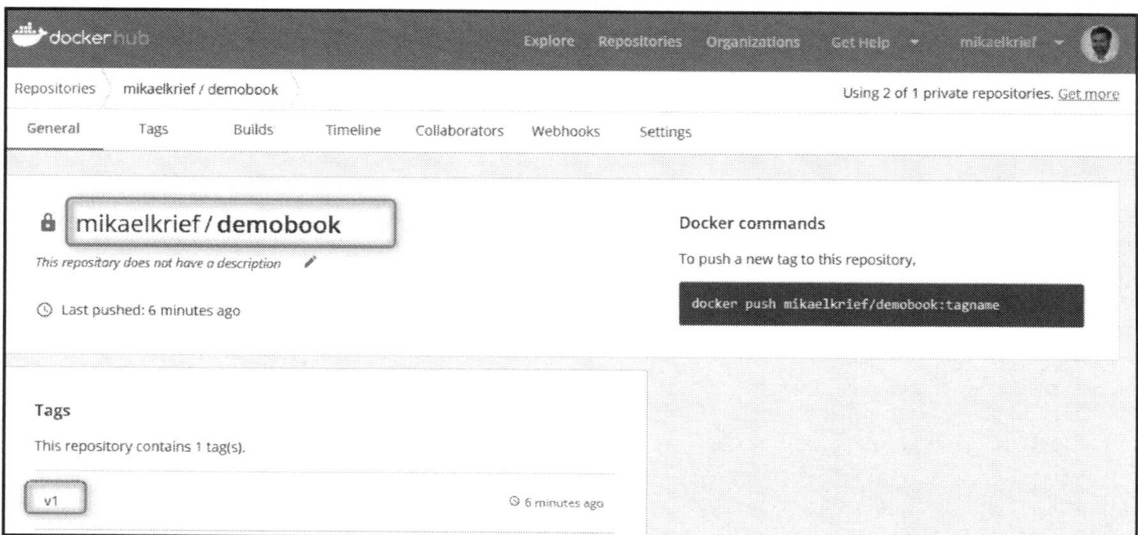

By default, the image pushed to Docker Hub is in public mode – everybody can view it in the explorer and use it.

So, we can access this image in Docker Hub in the Docker Hub search engine, as shown in the following screenshot:

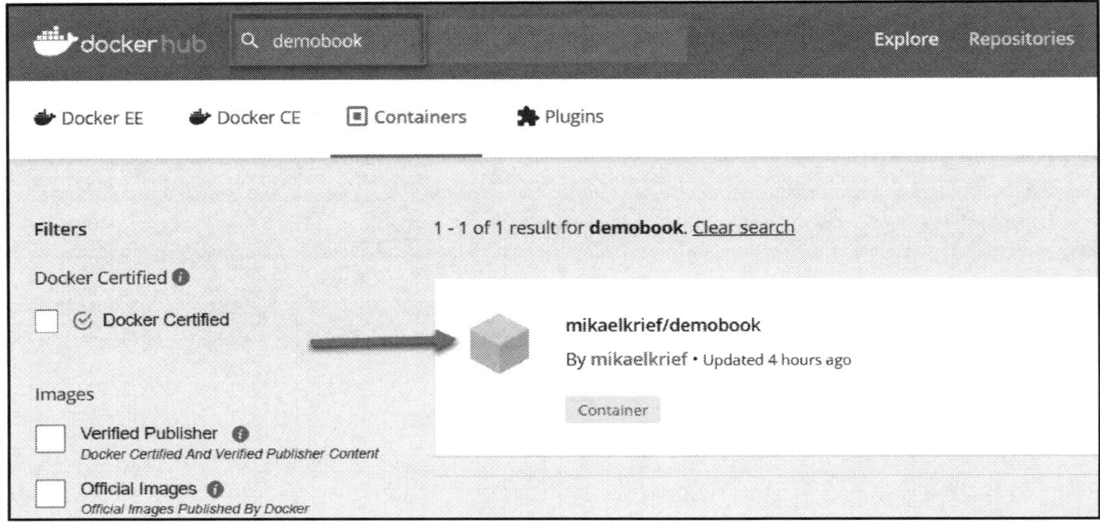

To make this image private – that is, you must be authenticated to be able to use it – you must go to the **Settings** of the image and click on the **Make private** button, as shown in the following screenshot:

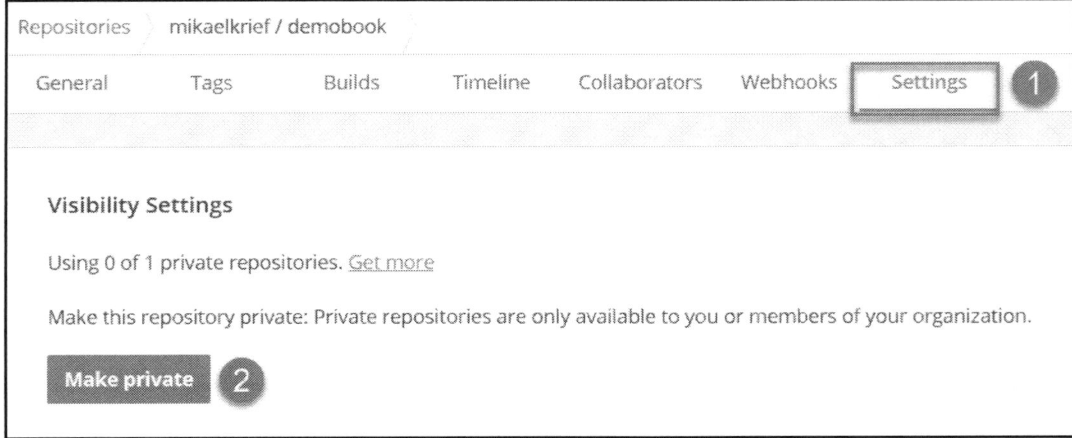

In this section, we looked at the steps and Docker commands for logging in to Docker Hub via the command line, then we looked at the `tag` and `push` commands for uploading Docker image to Docker Hub.

In the next section, we will see how to deploy this image with a CI/CD pipeline in a managed cloud container service – ACI and Terraform.

Deploying a container to ACI with a CI/CD pipeline

One of the reasons Docker has quickly become attractive to developers and operations teams is that the deployment of Docker images and containers has made CI and CD pipelines for enterprise applications easier.

To automate the deployment of our application, we will create a CI/CD pipeline that deploys the Docker image that contains our application in ACI.

ACI is a managed service from Azure that allows you to deploy containers very easily, without having to worry about the hardware architecture.

To learn more about ACI, head to the official page: `https://azure.microsoft.com/en-us/services/container-instances/`.

In addition, we will use Terraform for Infrastructure as a Code, which we discussed in `Chapter 2`, *Provisioning Cloud Infrastructure with Terraform*, using the Azure ACI resource and its integration with the Docker image.

We will, therefore, divide this section into two parts:

- The Terraform code of the Azure ACI and its integration with our Docker image
- An example of a CI/CD pipeline in Azure Pipelines, which allows you to execute the Terraform code

To start, we will write the Terraform code that allows you to provision an ACI resource in Azure.

The Terraform code for ACI

To provision an ACI resource with Terraform, we navigate to a new terraform-aci directory and create a Terraform file, main.tf.

In this code, we will provide Terraform code for a resource group and ACI resource using the azurerm_container_group Terraform object.

The documentation of the Terraform ACI resource is available here: https://www.terraform.io/docs/providers/azurerm/r/container_group.html.

This main.tf file contains the following Terraform code:

The following code creates the resource group:

```
resource "azurerm_resource_group" "acidemobook" {
 name = "demoBook"
 location = "westus2"
}
```

We add the Terraform code for the variable declarations:

```
variable "imageversion" {
 description ="Tag of the image to deploy"
}
variable "dockerhub-username" {
 description ="Tag of the image to deploy"
}
```

And we add the Terraform code for the ACI with the azurerm_container_group resource block:

```
resource "azurerm_container_group" "aci-myapp" {
    name = "aci-agent"
    location = "West Europe"
    resource_group_name = azurerm_resource_group.acidemobook.name
    os_type = "linux"
    container {
         name = "myappdemo"
         image = "docker.io/mikaelkrief/${var.dockerhub-
username}:${var.imageversion}"
         cpu = "0.5" memory = "1.5"
         ports {
          port = 80
          protocol = "TCP"
```

```
            }
        }
    }
```

In this code, we do the following:

- We declare `imageversion` and `dockerhub-username` variables, which will be instantiated during the CI/CD pipeline and include the username and the tag of the image to be deployed.
- We use the `azurerm_container_group` resource from Terraform to manage the ACI. In its `image` property, we indicate the information of the image to be deployed; that is, its full name in Docker Hub as well as its tag, which in our example is deported in the `imageversion` variable.

Finally, in order to protect the tfstate file, we can use the Terraform remote backend by using an Azure blob storage, as we discussed in the *Protecting tfstate in the remote backend* section of `Chapter 2`, *Provisioning Cloud Infrastructure with Terraform*.

The complete source code of this Terraform code is available here: `https://github.com/PacktPublishing/Learning_DevOps/tree/master/CHAP07/terraform-aci`.

Now that we have the Terraform code that allows us to create an Azure ACI that will execute a container of our image, we will create a CI/CD pipeline that will automatically deploy the container of the application.

Creating a CI/CD pipeline for the container

To create a CI/CD pipeline that will build our image and execute the Terraform code, we can use all the tools that we discussed in detail in `Chapter 6`, *Continuous Integration and Continuous Delivery*.

In this chapter, to visualize the pipeline, we will use **Azure Pipelines**, which is one of the previously detailed tools. It is advisable to carefully read the *Using Azure Pipelines* section of `Chapter 6`, *Continuous Integration and Continuous Delivery*. For this reason, we will not detail all the stages of the pipeline, but only those relevant to our container subject.

To implement the CI/CD pipeline in Azure Pipelines, we will proceed with these steps:

1. We'll create a new build definition whose **Source code** will point to the fork of the GitHub repository (`https://github.com/PacktPublishing/Learning_DevOps`), and select the root folder of this repository, as shown in the following screenshot:

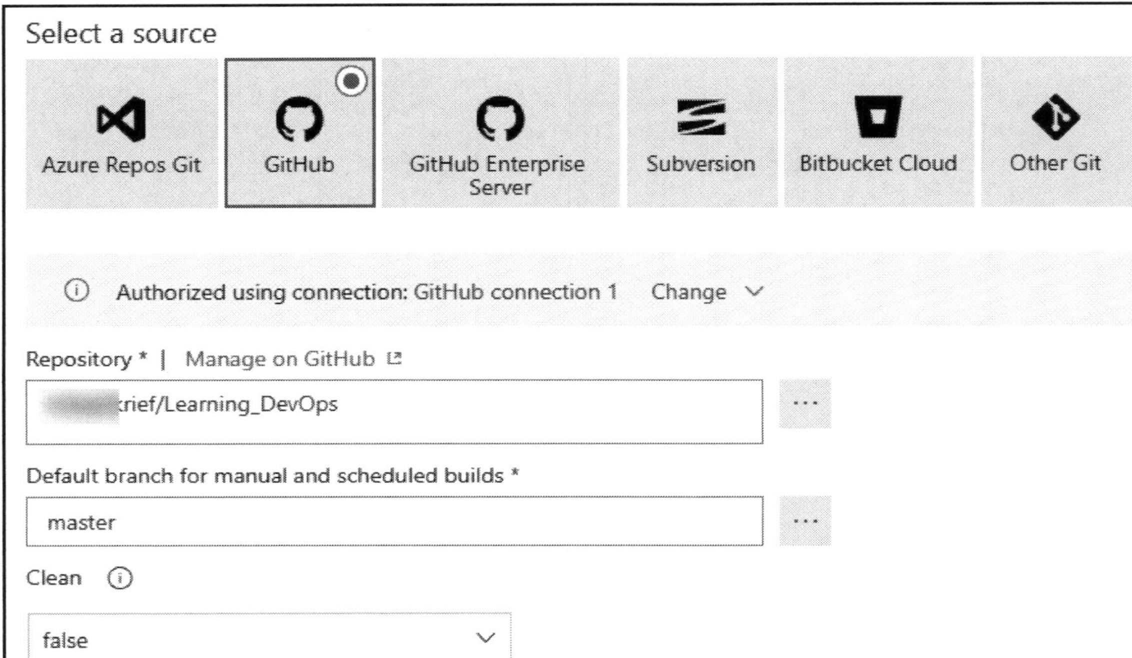

For more information about the fork in GitHub, read `Chapter 14`, *DevOps for Open Source Projects*.

You are free to use any source control version available in Azure Pipelines.

2. Then, on the **Variables** tab, we will define the variables that will be used in the pipeline. The following screenshot shows the information on the **Variables** tab:

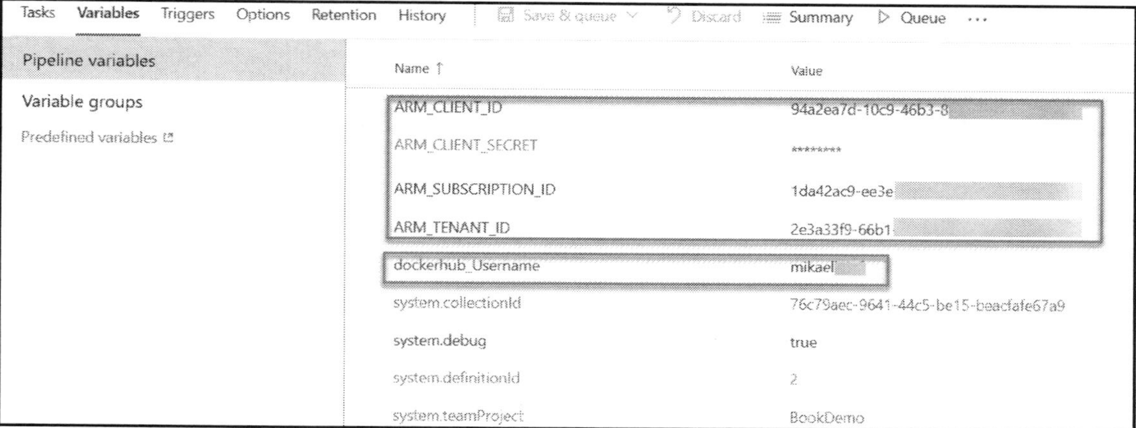

We defined the four pieces of Terraform connection information for Azure and the username of the Docker Hub.

3. Then, on the **Tasks** tab, we must take the following steps:
    1. Run the `docker build` command on the Dockerfile.
    2. Push the image to Docker Hub.
    3. Run the Terraform code to update the ACI with the new version of the updated image.

The following screenshot shows the configuration of the tasks:

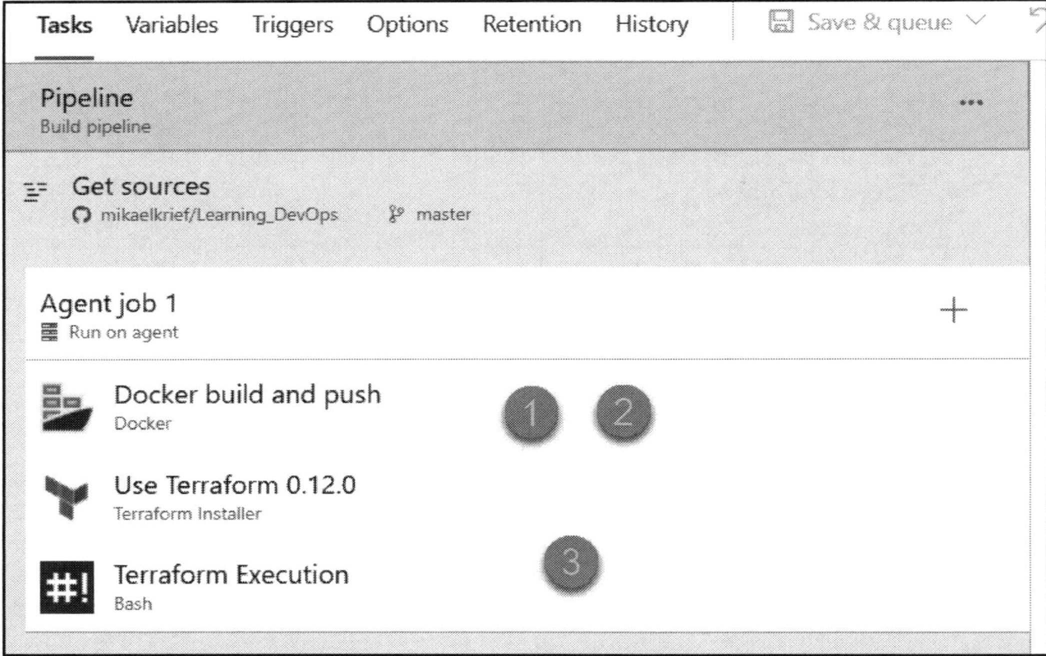

We configure these tasks with these steps:

4. The first task, **Docker build and push**, allows you to build the Docker image and push it to Docker Hub. Its configuration is quite simple:

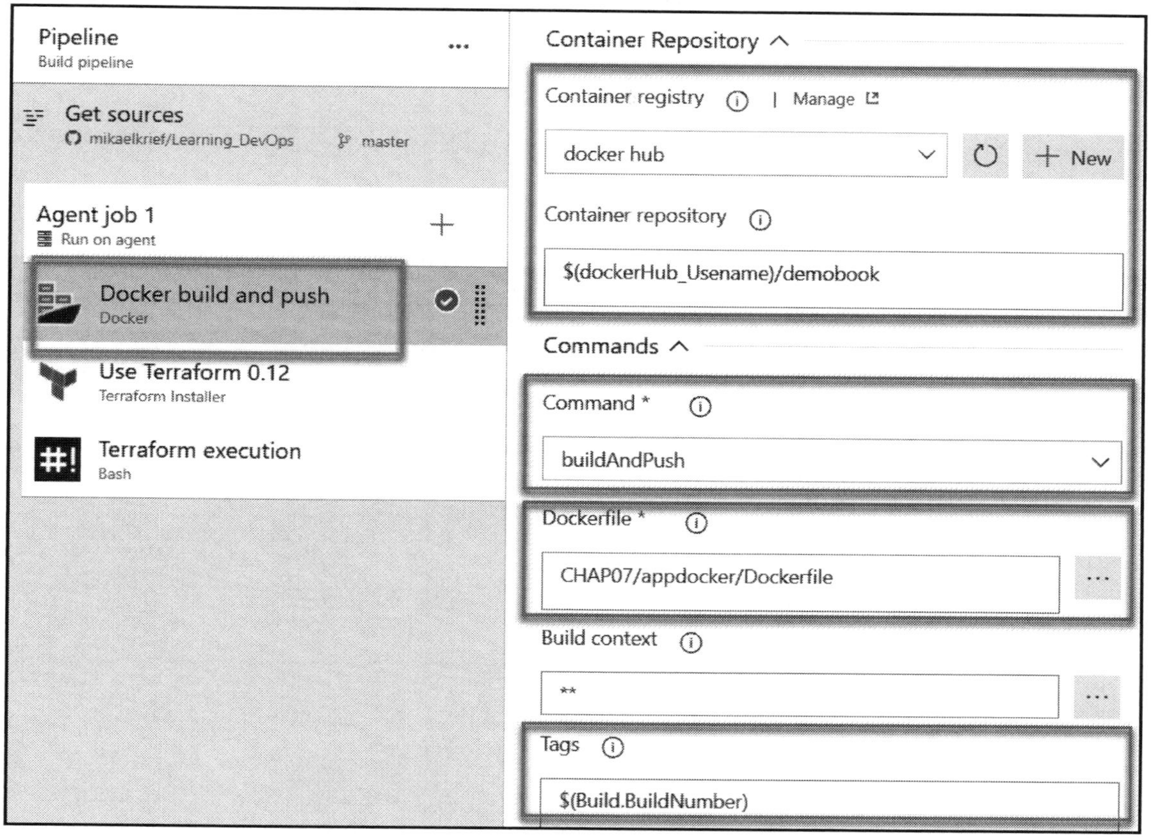

Its required parameters are:

- The connection to Docker Hub
- The tag of the image that will be pushed to Docker Hub

5. The second task, **Terraform Installer**, allows you to download Terraform on the pipeline agent by specifying the version of Terraform that you want.

This task is available in the marketplace at `https://marketplace.visualstudio.com/items?itemName=charleszipp.azure-pipelines-tasks-terraformtargetId=76c79aec-9641-44c5-be15-beacfafe67a9`.

The following screenshot shows its configuration, which is very simple:

6. The last task, **Bash**, allows you to execute a Bash script, and this screenshot shows its configuration:

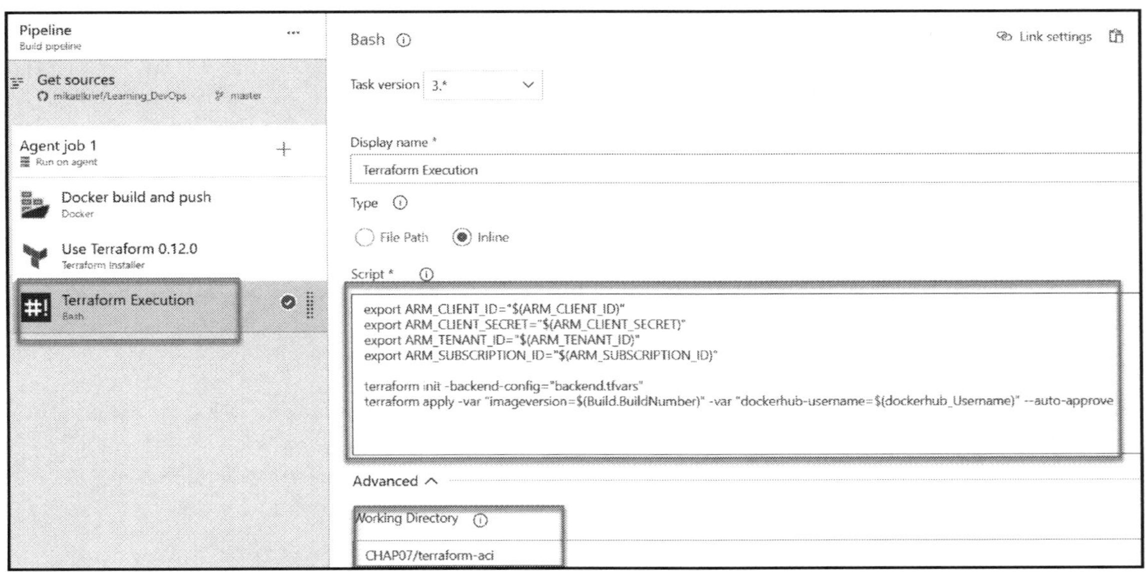

The configured script is as follows:

```
export ARM_CLIENT_ID="$(ARM_CLIENT_ID)"
export ARM_CLIENT_SECRET="$(ARM_CLIENT_SECRET)"
export ARM_TENANT_ID="$(ARM_TENANT_ID)"
export ARM_SUBSCRIPTION_ID="$(ARM_SUBSCRIPTION_ID)"
terraform init -backend-config="backend.tfvars"
terraform apply -var "imageversion=$(Build.BuildNumber)" -var
"dockerhub-username=$(dockerhub_Username)" --auto-approve
```

This script performs three actions, which are done in order:

1. Exports the environment variables required for Terraform.
2. Executes the `terraform init` command.
3. Executes `terraform apply` to apply the changes, with the two `-var` parameters, which are our Docker Hub username as well as the tag to apply. These parameters allow the execution of a container with the new image that has just been pushed to Docker Hub.

7. Then, to configure the build agent to use in the **Agent job** options, we use the Azure Pipelines agent hosted Ubuntu 16.04, shown in the following screenshot:

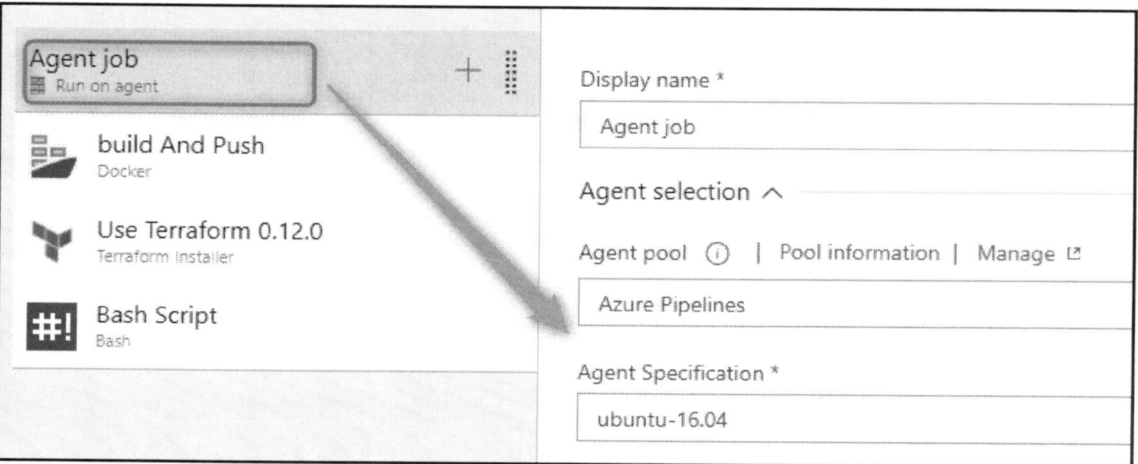

8. Finally, the last configuration is the trigger configuration on the **Triggers** tab, to enable the continuous integration with the trigger of this build at each commit, as shown in this screenshot:

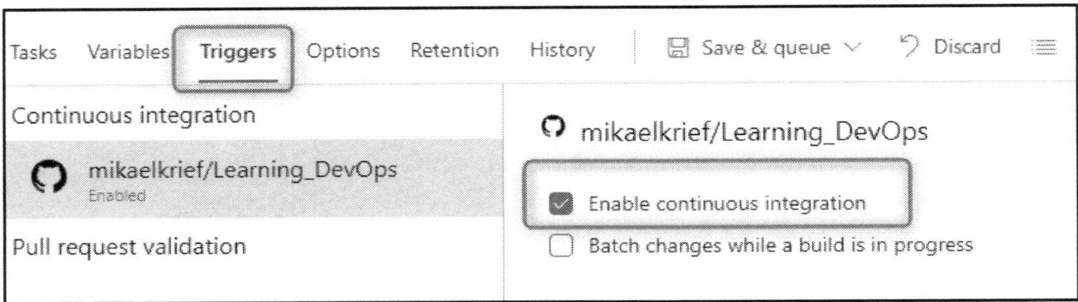

That is the configuration of the CI/CD pipeline in Azure Pipelines.

We trigger this build and at the end of its execution, we notice a new version of the Docker image which corresponds to the number of the build that pushed the Docker image into the Docker Hub:

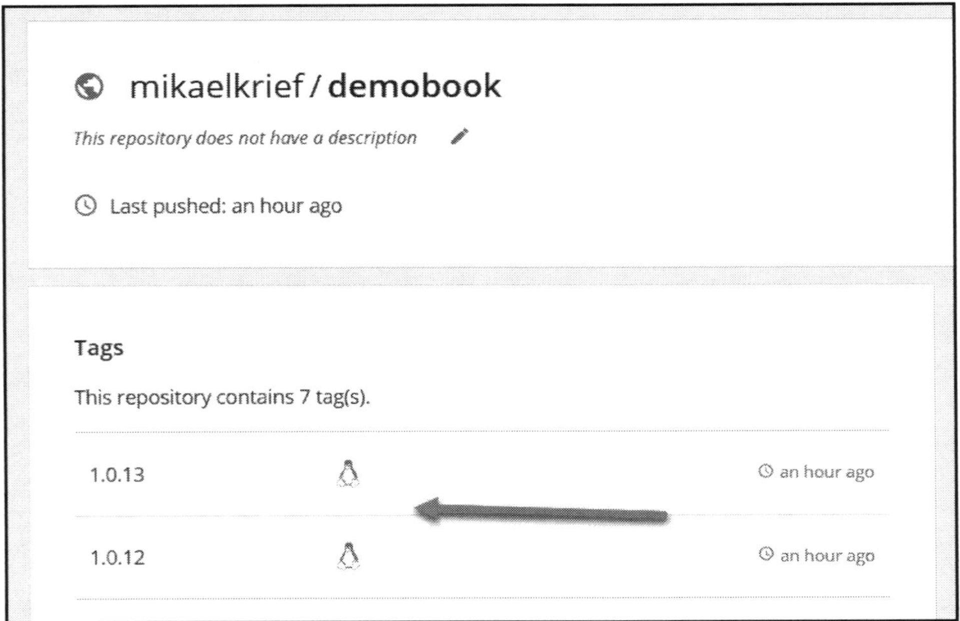

In the Azure portal, we have our ACI, `aci-app`, with our container, `mydemoapp`, as you can see in the following screenshot:

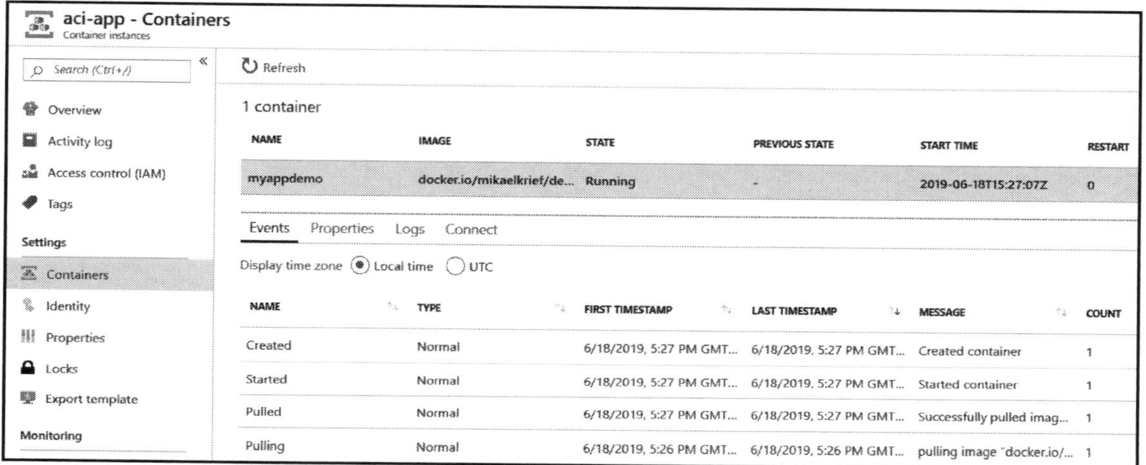

Notice that the container is running well.

Now, to access our application, we need to retrieve the public FQDN URL of the container provided in the Azure portal:

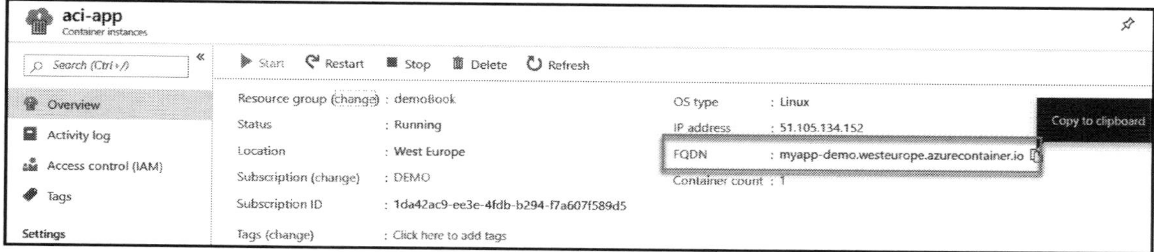

We open a web browser with this URL:

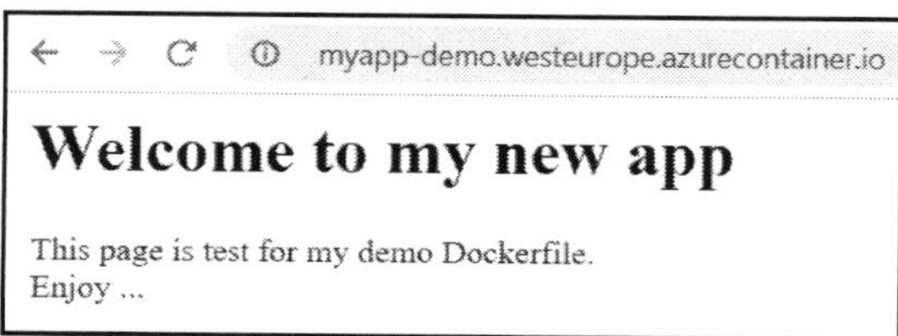

Our web application is displayed correctly.

The next time the application is updated, the CI/CD build is triggered, a new version of the image will be pushed into Docker Hub, and a new container will be loaded with this new version of the image.

In this section, we have looked at writing Terraform code to manage an ACI and the creation of a CI/CD pipeline in Azure Pipelines, which allows you to deploy the application's image in Docker Hub and then update the ACI with the new version of the image.

Summary

In this chapter, we presented Docker and its essential concepts. We discussed the necessary steps to create a Docker Hub account, then we installed Docker locally with Docker Desktop.

We created a Dockerfile that details the composition of a Docker image of a web application, and we also looked at the principal instructions that it is composed of – `FROM`, `COPY`, and `RUN`.

We executed the `docker build` and `run` commands to build an image from our Dockerfile and execute it locally, then we pushed it to Docker Hub using the `push` command.

Finally, to automate this entire DevOps mechanism, we implemented and executed a CI/CD pipeline in Azure Pipelines to deploy our container in an ACI that was provisioned with Terraform.

In the next chapter, we will continue with the subject of containers and we will look at the use of Kubernetes, which is a tool to manage containers on a large scale. We will use the **Azure Kubernetes Service** (**AKS**) and Azure Pipelines to deploy an application in Kubernetes with a CI/CD pipeline.

Questions

1. What is Docker Hub?
2. What is the basic element that allows you to create a Docker image?
3. In a Dockerfile, what is the instruction that defines the base image to use?
4. Which Docker command allows you to create a Docker image?
5. Which Docker command allows you to instantiate a new container?
6. Which Docker command allows you to publish an image in Docker Hub?

Further reading

If you want to know more about Docker, here are some great books:

- Docker Cookbook: `https://www.packtpub.com/virtualization-and-cloud/docker-cookbook-second-edition`
- Beginning DevOps with Docker: `https://www.packtpub.com/virtualization-and-cloud/beginning-devops-docker`

8
Managing Containers Effectively with Kubernetes

In the previous chapter, we learned in detail about containers with Docker, about the construction of a Docker image, and about the instantiation of a new container on the local machine. Finally, we set up a CI/CD pipeline that builds an image, deploys it in the Docker Hub, and executes its container in **Azure Container Instance** (**ACI**).

All this works well, and does not pose too many problems when working with a few containers. But, in so-called **microservice applications**, that is, applications that are composed of several services (each of them is a container), we will need to manage and orchestrate these containers.

There are two major container orchestration tools on the market, which are Docker Swarm and Kubernetes.

For some time now, Kubernetes, also known as **K8S**, has proved to be a true leader in the field of container management, and is, therefore, becoming a *must* for the containerization of applications.

In this chapter, we will learn how to install K8S on a local machine, as well as an example of how to deploy an application in K8S, both in a standard way and with Helm. Then, we will talk about **Azure Kubernetes Service** (**AKS**), as an example of a Kubernetes cluster, and finally, we will learn how to create a CI/CD pipeline with Azure Pipelines that allows us to deploy in a K8S cluster.

This chapter will cover the following topics:

- Installing Kubernetes
- First example of Kubernetes application deployment
- Using HELM as a package manager
- Using AKS
- Creating a CI/CD pipeline for Kubernetes with Azure Pipelines

Technical requirements

This chapter is a continuation of the previous chapter on Docker, so, to understand it properly, it is necessary to have read it and to have installed Docker Desktop (for Windows OS).

In the CI/CD part of this chapter, you will need to retrieve the source code that was provided in the previous chapter on Docker, which is available at `https://github.com/PacktPublishing/Learning_DevOps/tree/master/CHAP07/appdocker`.

The entire source code for this chapter is available at `https://github.com/mikaelkrief/Learning_DevOps/tree/master/CHAP08/k8sdeploy`.

Check out the following video to see the Code in Action:
`http://bit.ly/2MK9U8A`

Installing Kubernetes

Before installing Kubernetes, we need to have an overview of its architecture and main components, because Kubernetes is not a simple tool, but it is a cluster, that is, it consists of a **master** server and other slave servers called **nodes**.

I suggest you explore the architecture of Kubernetes in a simplified way.

Kubernetes architecture overview

Kubernetes is a platform that is made up of several components that assemble together and extend on demand, in order to enable a better scalability of applications. The architecture of Kubernetes, which is a client-server type, can be represented simply, as shown in the following diagram:

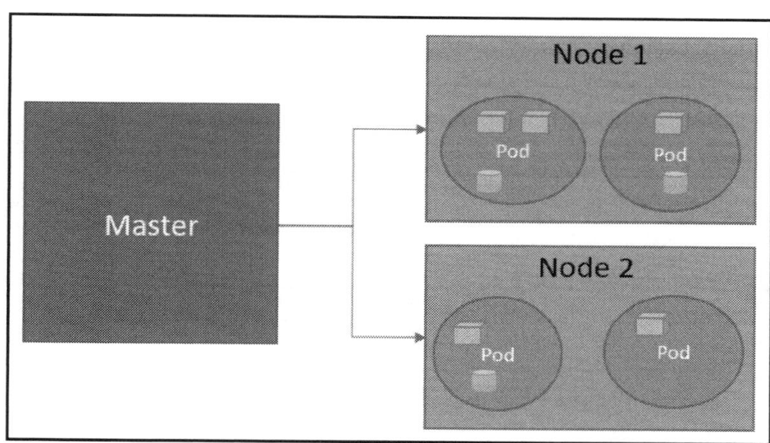

In the previous diagram, we can see that the cluster is made up of a master component and nodes (also called **worker** nodes), which represent the slave servers.

In each of these nodes, there are **pods**, which are virtual elements that will contain containers and volumes.

Put simply, we can create one pod per application, and it will contain all the containers of the application. For example, one pod can contain a web server container, a database container, and a volume that will contain persistent files for images and database files.

Finally, **kubectl** is the client tool that allows us to interact with a Kubernetes cluster. With this, we have the main requirements that allow us to work with Kubernetes, so let's look at how we can install it on a local machine.

Installing Kubernetes on a local machine

When developing a containerized application that is to be hosted on Kubernetes, it is very important to be able to run the application (with its containers) on your local machine, before deploying it on remote Kubernetes production clusters.

In order to install a Kubernetes cluster locally, there are several solutions, which are as follows:

The first solution is to use **Docker Desktop** by performing the following steps:

1. If we have already installed Docker Desktop, which we learned in `Chapter 7`, *Containerizing Your Application with Docker*, we can activate the **Enable Kubernetes** option in **Settings** in **Kubernetes** tab, as shown in the following screenshot:

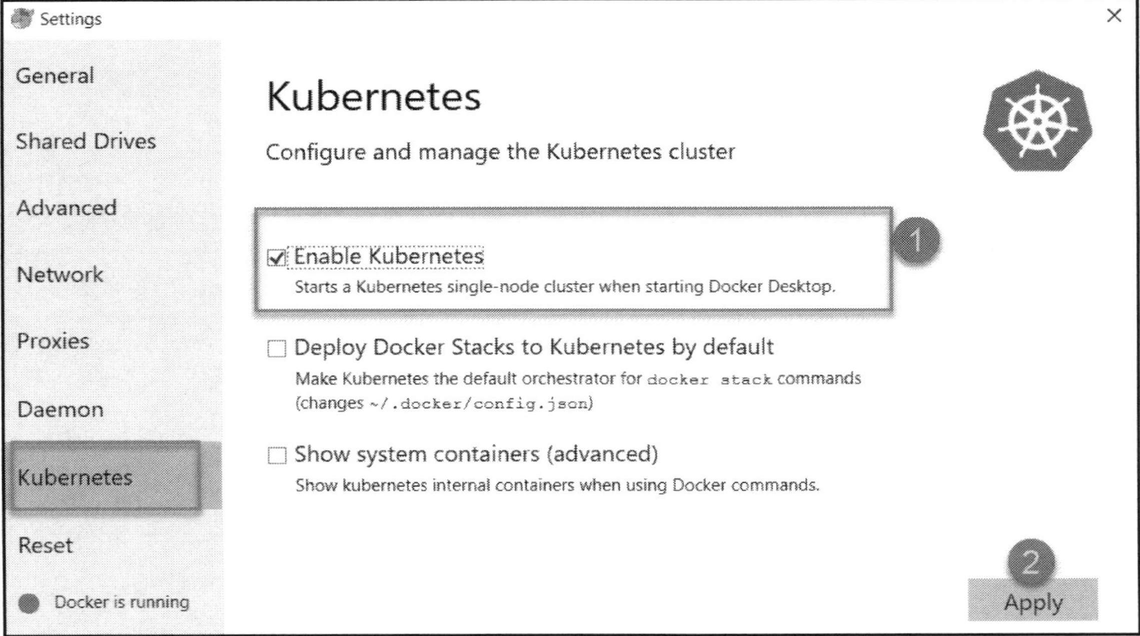

2. After clicking on the **Apply** button, Docker Desktop will install a mini Kubernetes cluster, and the kubectl client tool, on the local machine.

The second way of installing Kubernetes locally is to install **Minikube**, which also installs a simplified Kubernetes cluster locally. Here is the official documentation that you can read: `https://kubernetes.io/docs/setup/learning-environment/minikube/`. This is very detailed.

Following the local installation of Kubernetes, we will check its installation by executing the following command in a Terminal:

```
kubectl version --short
```

The following screenshot shows the results for the preceding command:

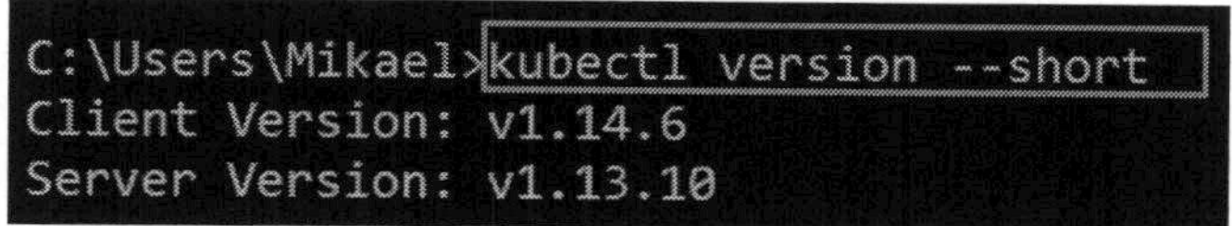

All of the operations that we carry out on our Kubernetes cluster will be done with kubectl commands.

After installing our Kubernetes cluster, we'll need another element, which is the Kubernetes dashboard. This is a web application that allows us to view the status, as well as all the components, of our cluster.

In the next section, we'll discuss how to install and test the Kubernetes dashboard.

Installing the Kubernetes dashboard

In order to install the Kubernetes dashboard, which is a pre-packaged containerized web application that will be deployed in our cluster, we will run the following command in a Terminal:

```
kubectl apply -f
https://raw.githubusercontent.com/kubernetes/dashboard/master/aio/deploy/re
commended.yaml
```

Its execution is shown in the following screenshot:

```
>kubectl apply -f https://raw.githubusercontent.com/kubernetes/dashboard/v2.0.0-beta1/aio/deploy/recommended.yaml
namespace "kubernetes-dashboard" created
serviceaccount "kubernetes-dashboard" created
service "kubernetes-dashboard" created
secret "kubernetes-dashboard-certs" created
secret "kubernetes-dashboard-csrf" created
secret "kubernetes-dashboard-key-holder" created
configmap "kubernetes-dashboard-settings" created
role.rbac.authorization.k8s.io "kubernetes-dashboard" created
clusterrole.rbac.authorization.k8s.io "kubernetes-dashboard" created
rolebinding.rbac.authorization.k8s.io "kubernetes-dashboard" created
clusterrolebinding.rbac.authorization.k8s.io "kubernetes-dashboard" created
deployment.apps "kubernetes-dashboard" created
service "dashboard-metrics-scraper" created
deployment.apps "kubernetes-metrics-scraper" created
```

From the preceding screenshot, we can see that different artifacts are created, which are as follows: secrets, two web applications, RBAC roles, permissions, and services.

Note that the URL mentioned in the parameters of the command that installs the dashboard may change depending on the versions of the dashboard. To find out the last valid URL to date, consult the official documentation by visiting `https://kubernetes.io/docs/tasks/access-application-cluster/web-ui-dashboard/`.

Now that we have installed the dashboard, we will connect to this Kubernetes dashboard and configure it.

To open the dashboard and connect to it from our local machine, we must first create a proxy between the Kubernetes cluster and our machine by performing the following steps:

1. To create the proxy, we execute the `kubectl proxy` command in a Terminal, and the detail of the execution is shown in the following screenshot:

We can see that the proxy is open on the localhost address (`127.0.0.1`) with the `8001` port.

2. Then, in a web browser, open the following URL, `http://localhost:8001/api/v1/namespaces/kubernetes-dashboard/services/https:kubernetes-dashboard:/proxy/#/login`, which is a local URL (localhost and `8001`) that is created by the proxy, and that points to the Kubernetes dashboard application that we have installed.

The following screenshot is used to select the Kubernetes configuration file, or enter the authentication token:

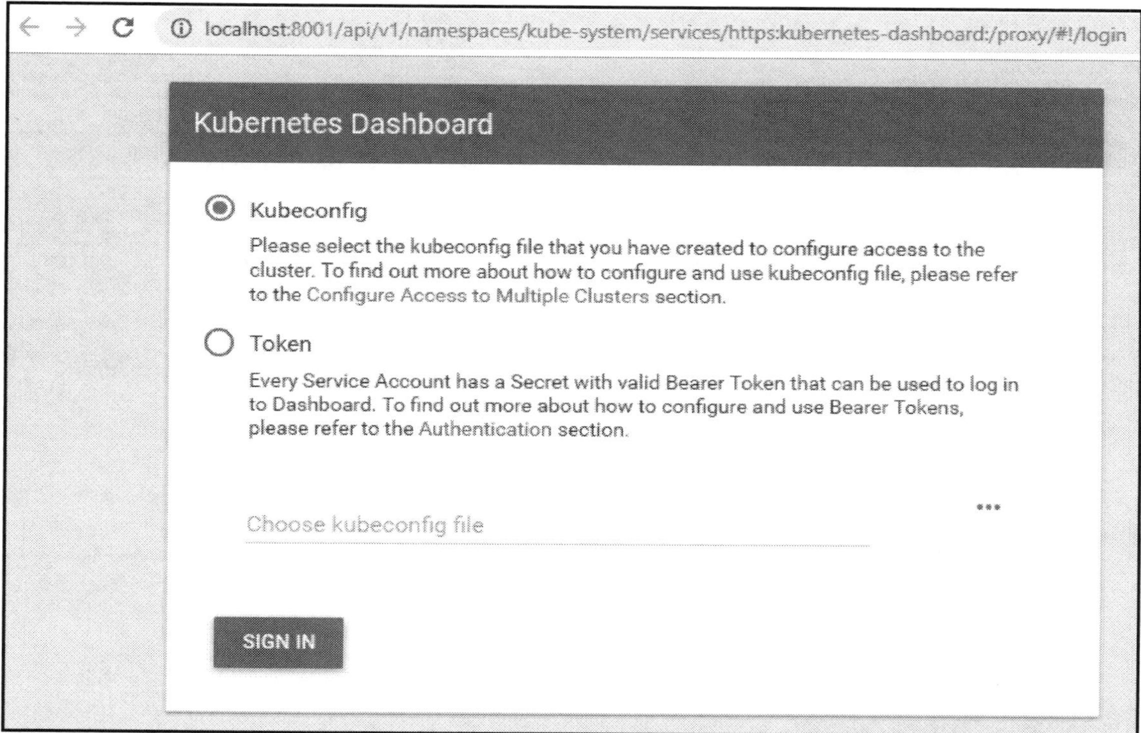

3. To create a new user authentication token, we will execute the following script in a PowerShell Terminal:

```
$TOKEN=((kubectl -n kube-system describe secret default | Select-
String "token:") -split " +")[1]
kubectl config set-credentials docker-for-desktop --
token="${TOKEN}"
```

The execution of this script creates a new token inside the local config file.

4. Finally, in the dashboard, we will select the `config` file, which is located in the `C:\Users\<user name>.kube\` folder, as shown in the following screenshot:

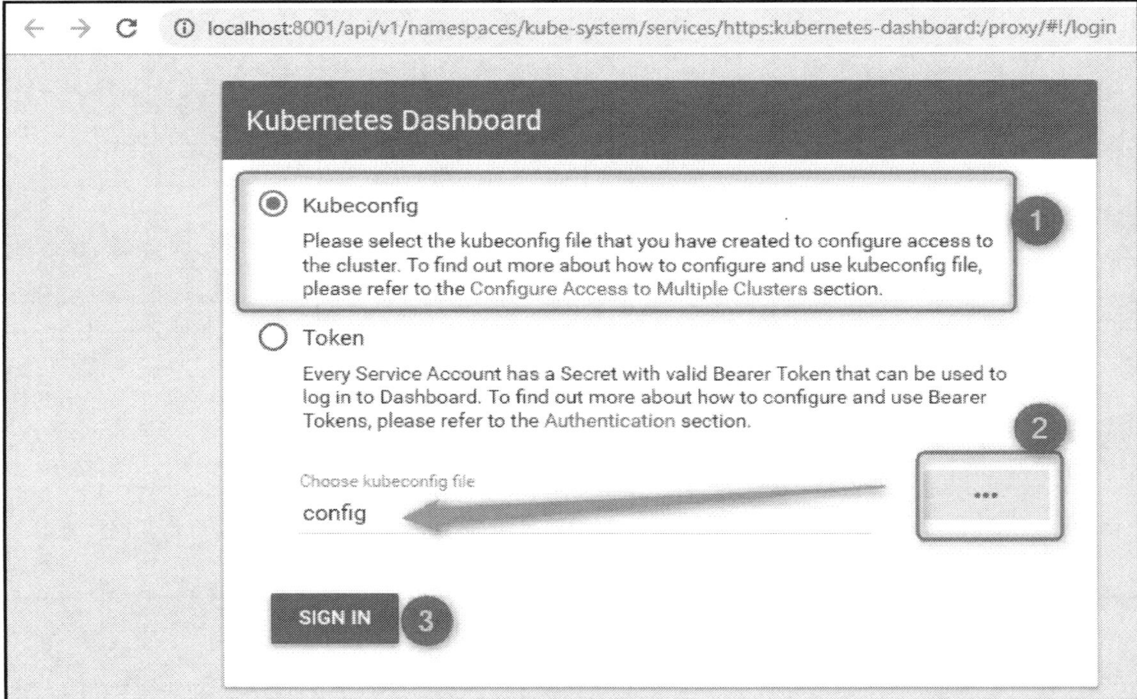

5. After clicking on the **SIGN IN** button, the dashboard is displayed as follows:

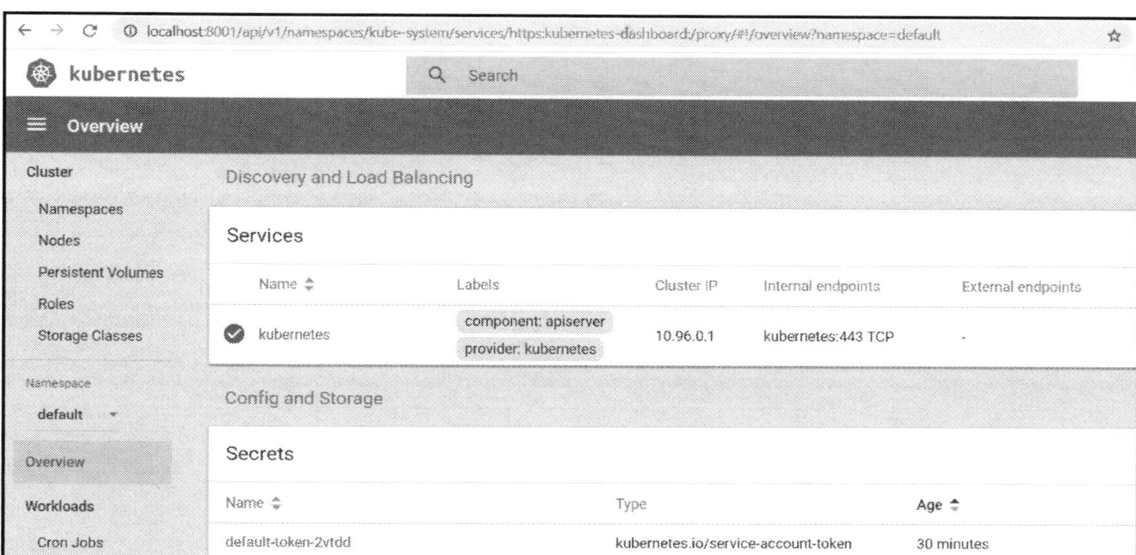

We have just seen how to install a Kubernetes cluster on a local machine, and then we installed and configured the Kubernetes web dashboard in this cluster. We will now deploy our first application in the local Kubernetes cluster using the YAML specification files and kubectl commands.

First example of Kubernetes application deployment

After installing our Kubernetes cluster, we will deploy an application in it. First of all, it is important to know that when we deploy an application in Kubernetes, we create a new instance of the Docker image in a cluster pod, so we need to have a Docker image that contains the application.

For our example, we will use the Docker image that contains a web application that we have pushed in the Docker Hub in `Chapter 7`, *Containerizing Your Application with Docker*.

To deploy this instance of the Docker image, we will create a new `k8sdeploy` folder, and, inside it, we will create a Kubernetes deployment YAML specification file (`myapp-deployment.yml`) with the following content:

```
---
apiVersion: apps/v1
kind: Deployment
metadata:
 name: webapp
spec:
 selector:
    matchLabels:
      app: webapp
 replicas: 2
 template:
    metadata:
     labels:
       app: webapp
   spec:
      containers:
      - name: demobookk8s
        image: mikaelkrief/demobook:latest
        ports:
        - containerPort: 80
```

In this code, we describe our deployment as follows:

- The `apiVersion` property is the version of `api` that should be used.
- In the `Kind` property, we indicate that the specification type is deployment.
- The `replicas` property indicates the number of pods that Kubernetes will create in the cluster; here, we choose two instances.

In this example, we chose two replicas, which can, at the very least, distribute the traffic charge of the application (if there is a high volume of load, we can put in more replicas), while also ensuring the proper functioning of the application. Therefore, if one of the two pods has a problem, the other, which is an identical replica, will ensure the proper functioning of the application.

Then, in the `containers` section, we indicate the image (from the Docker Hub) with `name` and `tag`. Finally, the `ports` property indicates the port that the container will use within the cluster.

This source code is also available at https://github.com/PacktPublishing/Learning_DevOps/blob/master/CHAP08/k8sdeploy/myapp-deployment.yml.

To deploy our application, we go to our Terminal, and execute one of the essential kubectl commands (`kubectl apply`) as follows:

```
kubectl apply -f myapp-deployment.yml
```

The `-f` parameter corresponds to the YAML specification file.

This command applies the deployment that is described in the YAML specification file on the Kubernetes cluster.

Following the execution of this command, we will check the status of this deployment, by displaying the list of pods in the cluster. To do this in the Terminal, we execute the `kubectl get pods` command, which returns the list of cluster pods. The following screenshot shows the execution of the deployment and displays the information in the pods, which we use to check the deployment:

```
[        \Learning_DevOps\CHAP08\k8sdeploy>kubectl apply -f myapp-deployment.yml
deployment.apps "webapp" created

[        \Learning_DevOps\CHAP08\k8sdeploy>kubectl get pods
NAME                         READY     STATUS    RESTARTS   AGE
webapp-69d88f7d99-jrd7w      1/1       Running   0          7s
webapp-69d88f7d99-t55k5      1/1       Running   0          7s
```

What we can see in the preceding screenshot is that the second command displays our two pods, with the name (`webapp`) specified in the YAML file, followed by a unique ID, and that they are in **Running** status.

We can also visualize the status of our cluster on the Kubernetes web dashboard, the `webapp` deployment with the Docker image that has been used, and the two pods that have been created. For more details, we can click on the different links of the elements.

Our application has been successfully deployed in our Kubernetes cluster, but, for the moment, it is only accessible inside the cluster, and for it to be usable, we need to expose it outside the cluster.

In order to access the web application outside the cluster, we must add a service type and a `NodePort` category element to our cluster. To add this service type and `NodePort`, in the same way as for deployment, we will create a second YAML file (`myapp-service.yml`) of the service specification in the same `k8sdeploy` directory, which has the following code:

```
---
apiVersion: v1
kind: Service
metadata:
 name: webapp
 labels:
 app: webapp
spec:
 type: NodePort
 ports:
 - port: 80
   targetPort: 80
   nodePort: 31000
selector:
 app: webapp
```

In this code, we specify the kind, `Service`, as well as the type of service, `NodePort`.

Then, in the `ports` section, we specify the port translation: the `80` port, which is exposed internally, and the `31000` port, which is exposed externally to the cluster.

The source code of this file is also available at `https://github.com/PacktPublishing/Learning_DevOps/blob/master/CHAP08/k8sdeploy/myapp-service.yml`.

To create this service on the cluster, we execute the `kubectl apply` command, but this time with our `myapp-service.yaml` file as a parameter, as follows:

```
kubectl apply -f myapp-service.yml
```

The execution of the command creates the service within the cluster, and, to test our application, we open a web browser with the `http://localhost:31000` URL, and our page is displayed as follows:

Our application is now deployed on a Kubernetes cluster, and it can be accessed from outside the cluster.

In this section, we have learned that the deployment of an application, as well as the creation of objects in Kubernetes, is achieved using specification files in YAML format, and several kubectl command lines.

The next step is to use HELM packages to simplify the management of the YAML specification files.

Using HELM as a package manager

As previously discussed, all the actions that we carry out on the Kubernetes cluster are done via the **kubectl** tool and the **YAML specification files**.

In a company that deploys several microservice applications on a K8S cluster, we often notice a large number of these YAML specification files, and this poses a maintenance problem. In order to solve this maintenance problem, we can use **HELM**, which is the package manager for Kubernetes.

For more information on package managers, you can also read the *Using a package manager* section in `Chapter 6`, *Continuous Integration and Continuous Delivery*.

HELM is, therefore, a repository that will allow the sharing of packages called **charts**, and that contain ready-to-use Kubernetes specification file templates.

To learn more about HELM and to access its documentation, visit `https://helm.sh/`.

So, we'll see how to install Helm on our local K8S cluster, and later we'll go through the installation of an application with HELM.

HELM is composed of two parts: a **client tool**, which allows us to list the packages of a repository, and to indicate the package(s) to be installed; and another server tool called **Tiller**, which is in the Kubernetes cluster, and receives information from the client tool and installs the package charts.

The following steps show us how to install Helm, and how to use it to deploy an application:

1. **Install the Helm client**: To install the HELM client, please refer to the installation documentation at `https://helm.sh/docs/using_helm/#installing-the-helm-client`, which details the installation procedure according to the different OSes.

   In Windows, for example, we can install it via the **Chocolatey** package manager, with the execution of the following command:

   ```
   choco install kubernetes-helm -y
   ```

   To check its installation, execute the `helm --help` command, as shown in the following screenshot:

   ```
   PS C:\Users\Mikael> helm --help
   The Kubernetes package manager

   To begin working with Helm, run the 'helm init' command:

           $ helm init

   This will install Tiller to your running Kubernetes cluster.
   It will also set up any necessary local configuration.

   Common actions from this point include:

   - helm search:    search for charts
   - helm fetch:     download a chart to your local directory to view
   - helm install:   upload the chart to Kubernetes
   - helm list:      list releases of charts
   ```

2. **Install the Tiller**: To install the Helm server component on our Kubernetes cluster, execute the following command:

```
helm init
```

The following screenshot shows the execution of the preceding command:

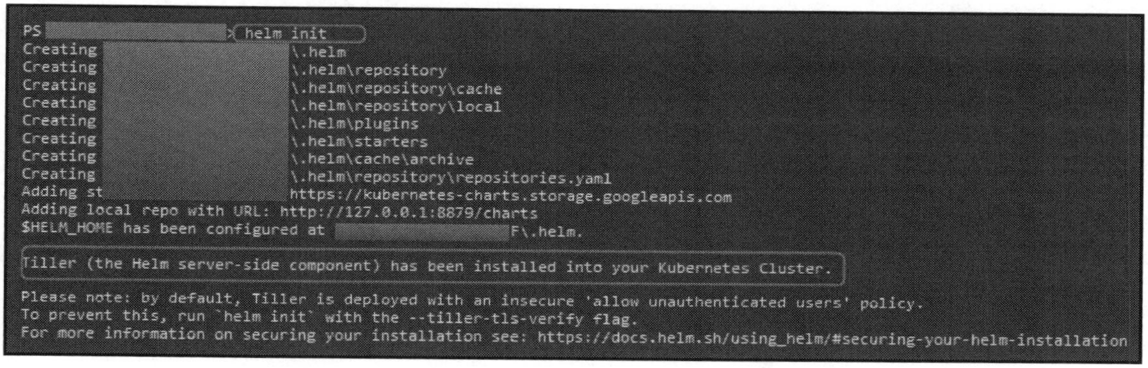

```
PS > helm init
Creating \.helm
Creating \.helm\repository
Creating \.helm\repository\cache
Creating \.helm\repository\local
Creating \.helm\plugins
Creating \.helm\starters
Creating \.helm\cache\archive
Creating \.helm\repository\repositories.yaml
Adding st https://kubernetes-charts.storage.googleapis.com
Adding local repo with URL: http://127.0.0.1:8879/charts
$HELM_HOME has been configured at F\.helm.

Tiller (the Helm server-side component) has been installed into your Kubernetes Cluster.

Please note: by default, Tiller is deployed with an insecure 'allow unauthenticated users' policy.
To prevent this, run `helm init` with the --tiller-tls-verify flag.
For more information on securing your installation see: https://docs.helm.sh/using_helm/#securing-your-helm-installation
```

The execution of the command tells us that the Tiller is properly installed.

3. **Search charts**: The packages that are contained in a HELM repository are called charts. Charts are composed of files that are templates of Kubernetes specification files for an application.

With the charts, it's possible to deploy an application in Kubernetes without having to write any YAML specification files. So, to deploy an application, we will use its corresponding chart, and we will pass some configuration variables of this application.

Once HELM is installed, we will install a chart that is in the HELM public repository, but first, to display the list of public charts, we run the following command:

```
helm search stable/
```

The `stable/` parameter is the name of Helm's public repository.

Here is an extract from the result, which includes more than a hundred charts:

```
Windows PowerShell
PS C:\Users\Mikael> helm search stable/
NAME                                CHART VERSION   APP VERSION     DESCRIPTION
stable/acs-engine-autoscaler        2.2.2           2.1.1           DEPRECATED Scales worker nodes within agent pools
stable/aerospike                    0.2.8           v4.5.0.5        A Helm chart for Aerospike in Kubernetes
stable/airflow                      3.0.1           1.10.2          Airflow is a platform to programmatically author, schedul...
stable/ambassador                   2.8.2           0.72.0          A Helm chart for Datawire Ambassador
stable/anchore-engine               1.1.1           0.4.0           Anchore container analysis and policy evaluation engine s...
stable/apm-server                   2.1.3           7.0.0           The server receives data from the Elastic APM agents and ...
stable/ark                          4.2.2           0.10.2          DEPRECATED A Helm chart for ark
stable/artifactory                  7.3.1           6.1.0           DEPRECATED Universal Repository Manager supporting all ma...
stable/artifactory-ha               0.4.1           6.2.0           DEPRECATED Universal Repository Manager supporting all ma...
stable/atlantis                     3.5.3           v0.7.1          A Helm chart for Atlantis https://www.runatlantis.io
stable/auditbeat                    1.1.0           6.7.0           A lightweight shipper to audit the activities of users an...
stable/aws-cluster-autoscaler       0.3.3                           Scales worker nodes within autoscaling groups.
stable/aws-iam-authenticator        0.1.0           1.0             A Helm chart for aws-iam-authenticator
stable/bitcoind                     0.2.2           0.17.1          Bitcoin is an innovative payment network and a new kind o...
stable/bookstack                    1.1.0           0.25.2          BookStack is a simple, self-hosted, easy-to-use platform ...
stable/buildkite                    0.2.4           3               DEPRECATED Agent for Buildkite
```

Of course, it is possible to create our private or corporate Helm repository with tools such as Nexus, Artifactory, or even Azure Container Registry. Let's now install an application with Helm.

4. **Deploy an application with Helm**: To illustrate the use of Helm, we will deploy a WordPress application in our Kubernetes cluster by using a Helm chart.

   In order to do this, execute the `helm install` command as follows:

   ```
   helm install stable/wordpress --name mywp
   ```

   Helm installs a WordPress instance called `mywp`, and all of the Kubernetes components, on the local Kubernetes cluster.

   We can also display the list of Helm packages that are installed on the cluster by executing the following command:

   ```
   helm ls
   ```

   And, if we want to remove a package and all of its components, for example, to remove the application installed with this package, we execute the `helm delete` command:

   ```
   helm delete mywp --purge
   ```

   The `purge` parameter indicates that everything has been deleted from this application.

To learn more about Helm, visit `https://helm.sh/docs/related/`, which has many interesting resources.

In this section, we have seen an overview of the installation and use of Helm, which is the package manager for Kubernetes. Hitherto, we have worked with a local Kubernetes cluster. Now let's look at an example of a Kubernetes service that is managed in Azure with AKS.

Using AKS

A production Kubernetes cluster can often be complex to install and configure. This type of installation requires the availability of servers, human resources who have the requisite skills regarding the installation and management of a K8S cluster, and especially the implementation of an enhanced security policy to protect the applications.

To overcome these problems, cloud providers offer managed Kubernetes cluster services. This is the case with Amazon with EKS, Google with Kubernetes Engine, and finally, Azure with AKS. In this section, I propose an overview of AKS, while also highlighting the advantages of a managed Kubernetes cluster.

AKS is, therefore, an Azure service that allows us to create and manage a real Kubernetes cluster as a managed service.

The advantage of this managed Kubernetes cluster is that we don't have to worry about its hardware installation, and that the management of the master part is done entirely by Azure when the nodes are installed on VMs.

The use of this service is free; what is charged is the cost of the VMs on which the nodes are installed.

To learn more about the benefits offered by AKS, you can read the documentation at `https://docs.microsoft.com/en-us/azure/aks/intro-kubernetes`.

Let's now look at how to create an AKS service.

Creating an AKS service

The creation of an AKS cluster in Azure can be done in three different ways:

- **Manually, via the Azure portal**: The standard way to create an AKS service is to do so via the Azure portal, by creating a Kubernetes service, and then entering its basic Azure properties, that is, the type and number of nodes desired, as shown in the following screenshot:

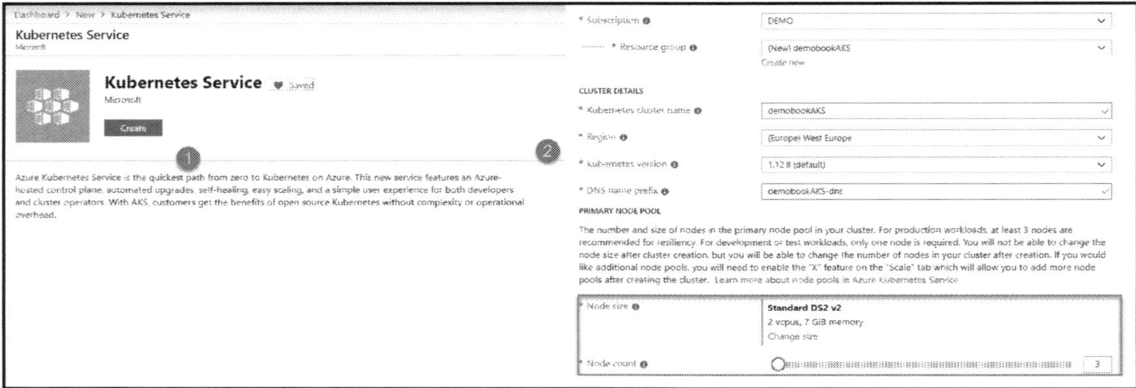

- **Creation via an Az CLI script**: You can also use an `az cli` script to automate the creation of the AKS cluster. The script is as follows:

```
#Create the Resource group
az group create --name Rg-AKS --location westeurope

#Create the AKS resource
az aks create --resource-group Rg-AKS --name demoBookAKS --node-
count 2 --generate-ssh-keys --enable-addons monitoring
```

  The `node-count` property indicates the number of nodes, and the `enable-addons` property enables us to monitor the AKS service.

- **Creation with Terraform**: It is also possible to create the AKS service with Terraform. The complete Terraform script is available in the Azure documentation at `https://docs.microsoft.com/en-us/azure/terraform/terraform-create-k8s-cluster-with-tf-and-aks`, and, to learn more about using Terraform, you can read `Chapter 2`, of this book (*Provisioning Cloud Infrastructure with Terraform*).

Now that the AKS cluster has been created, we will be able to configure kubectl in order to connect to it.

Configuring kubectl for AKS

To configure kubectl when connecting to the AKS service, we will use the `az cli` tool by executing the following commands in a Terminal:

```
az login
#If you have several Azure subscriptions
az account set --subscription <subscription Id>

az aks get-credentials --resource-group Rg-AKS --name demoBookAKS
```

This last command takes the resource group as the parameter, and as the name of the created AKS cluster. The role of this command is to automatically configure kubectl for connection to the AKS cluster, as shown in the following screenshot:

```
C:              l>az aks get-credentials --resource-group Rg-AKS --name demoBookAKS
Merged "demoBookAKS" as current context in              \.kube\config
```

To test the connection to AKS, we can execute the following command, `kubectl get nodes`, which displays the number of nodes that are configured when creating the AKS cluster, as shown in the following screenshot:

```
              >kubectl get nodes
NAME                        STATUS   ROLES   AGE   VERSION
aks-nodepool1-41966373-0    Ready    agent   8m    v1.12.8
aks-nodepool1-41966373-1    Ready    agent   8m    v1.12.8
```

All of the operations that we have seen in the *First example of Kubernetes application deployment* section of this chapter are identical, whether deploying an application with AKS, or with kubectl.

After having seen the steps that are taken to create an AKS service in Azure, we will now provide an overview of its advantages.

Advantages of AKS

AKS is a Kubernetes service that is managed in Azure. This has the advantage of being integrated with Azure, some of which are listed as follows:

- **Ready to use**: In AKS, the Kubernetes web dashboard is natively installed, and the documentation at `https://docs.microsoft.com/en-us/azure/aks/kubernetes-dashboard` explains how to access it.
- **Integrated monitoring services**: AKS also has all of Azure's integrated monitoring services, including container monitoring, cluster performance management, and log management, as shown in the following screenshot:

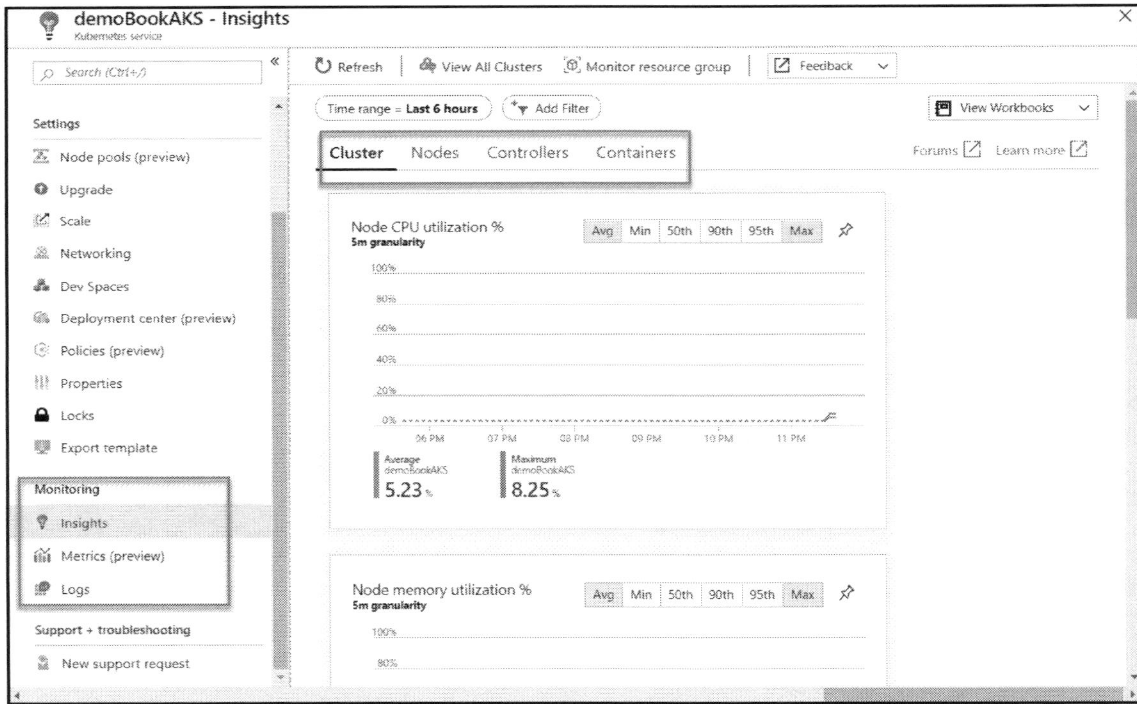

- **Very easy to scale**: AKS allows the quick and direct scaling of the number of nodes of a cluster via the portal, or via scripts.

As we can see in the following screenshot, we choose the number of nodes that we want in the Azure portal, and the change is effective immediately:

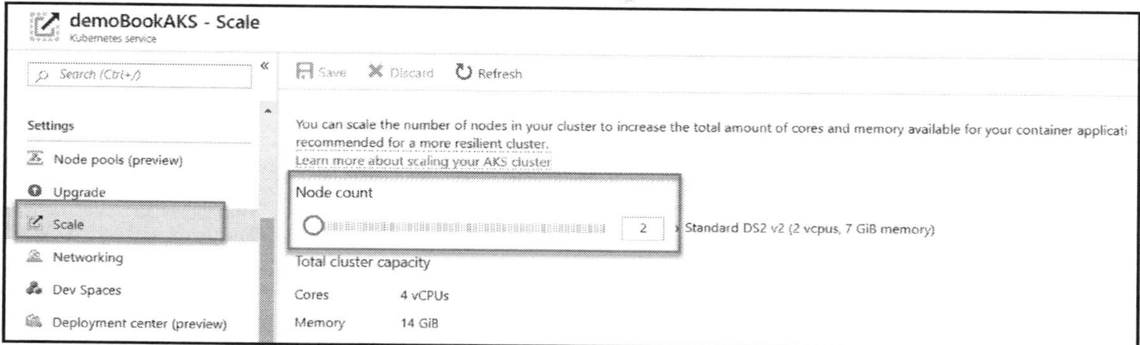

If we have an Azure subscription and we want to use Kubernetes, it's intuitive and quick to install. AKS has a number of advantages, such as integrated monitoring and scaling in the Azure portal. Using the kubectl tool does not require any changes compared to a local Kubernetes.

In this section, we have discussed AKS, which is a managed Kubernetes service in Azure. Then, we created an AKS instance and configured kubectl in order to connect to it. Finally, we listed its advantages, which are mainly integrated monitoring and fast scalability.

In the next section, we will see how to deploy an application in AKS Kubernetes by using a CI/CD pipeline with Azure Pipelines.

Creating a CI/CD pipeline for Kubernetes with Azure Pipelines

So far, we have seen how to use kubectl to deploy a containerized application in a local K8S cluster, or in a remote cluster with AKS.

Now, we will see how to create a complete CI/CD pipeline for Kubernetes, from the creation of a new Docker image pushed in the Docker Hub, to its deployment in an AKS cluster.

To build this pipeline, we'll use the **Azure Pipelines** service that is in Azure DevOps, which we have previously discussed in `Chapter 6`, *Continuous Integration and Continuous Delivery,* and `Chapter 7`, *Containerizing Your Application with Docker*.

This continuous integration pipeline will be composed of the following:

- A build that will be in charge of building and promoting a new Docker image in the Docker Hub.
- A release that will use our YAML deployment specification file to deploy the latest version of the image in an AKS cluster.

By way of a reminder, the source code of the Docker image is available at `https://github.com/PacktPublishing/Learning_DevOps/tree/master/CHAP07/appdocker`, and the YAML specification files are available at `https://github.com/PacktPublishing/Learning_DevOps/tree/master/CHAP08/k8sdeploy`.

To be able to use this code, we will either have to make a fork of this repository, or copy these sources into another repository of your choice.

Let's start our CI/CD pipeline with the build definition that will push our Docker image in the Docker Hub.

The build and push of the image in the Docker Hub

In Azure DevOps, we will create a new build definition that will be in **Classic design editor** mode, and that will point to the source code that contains the following Docker file: `https://github.com/PacktPublishing/Learning_DevOps/tree/master/CHAP07/appdocker/Dockerfile`.

In this build definition, we will configure the **Tasks** tab with two steps, in this order:

- The build and push of the Docker image, as we discussed in the *Deploying the container in ACI with CI/CD pipeline* section of `Chapter 7`, *Containerizing Your Application with Docker*.
- The publication of the build artifacts, which are the K8S YAML specification files that will be deployed during the release.

The sequences of the tasks that configure the build pipeline are demonstrated in the following screenshot:

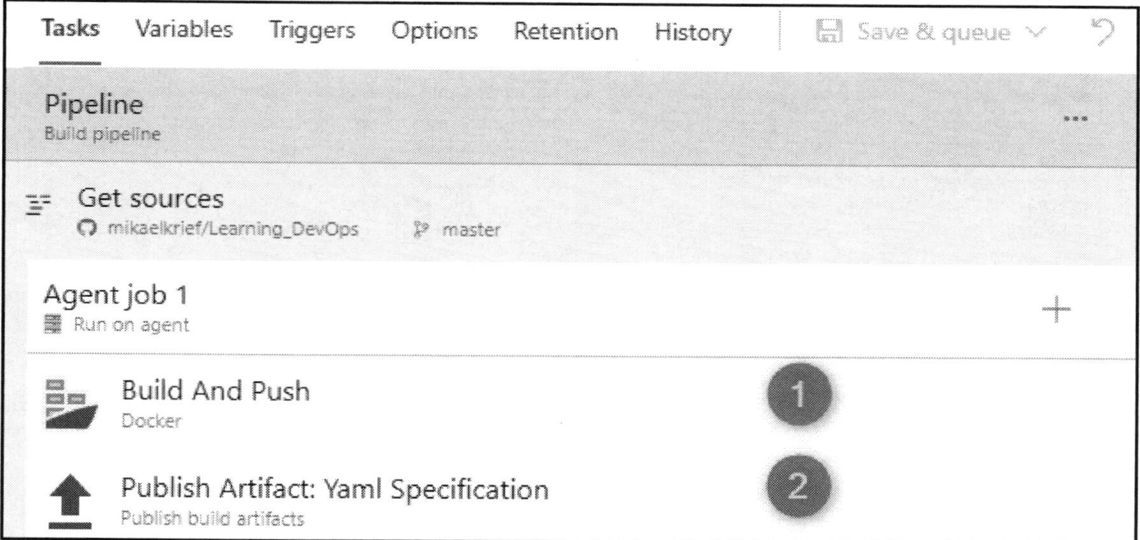

Let's look in detail at the configuration steps of this build pipeline:

1. The configuration of the task that **builds and pushes** the Docker image, as follows:

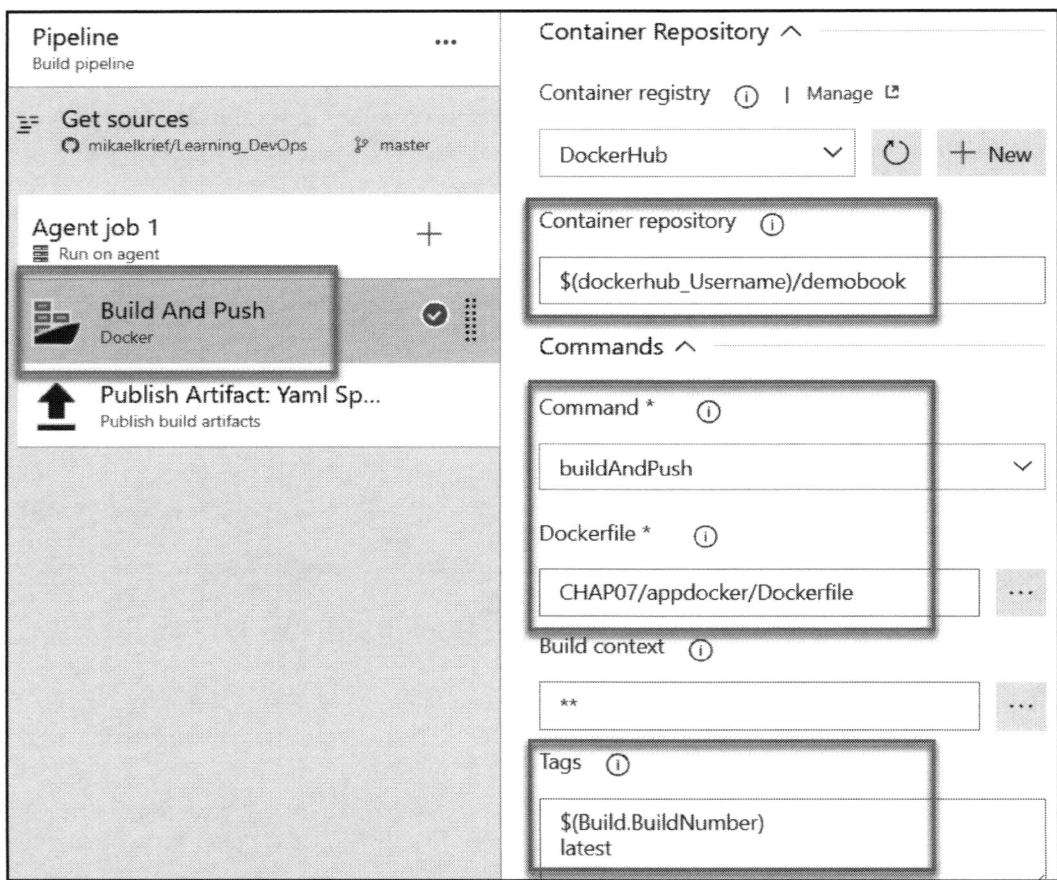

2. The configuration of the task that **publishes artifacts** of the Kubernetes YAML files as release artifacts, as follows:

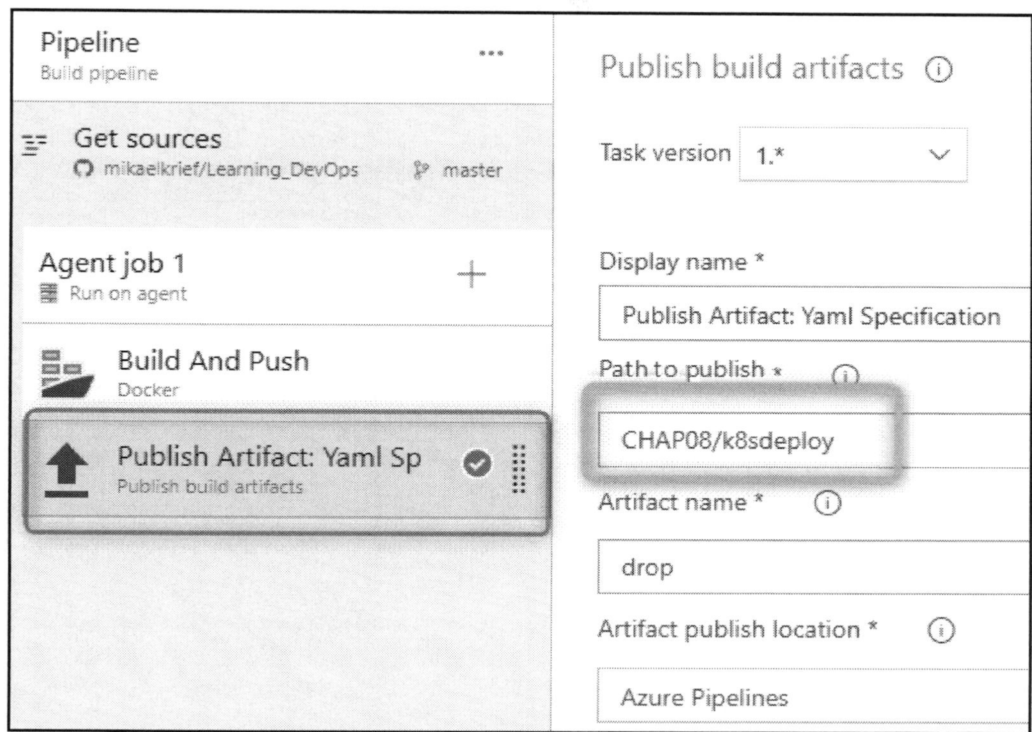

3. In the **Variables** tab, a variable is added that contains the Docker Hub username, as shown here:

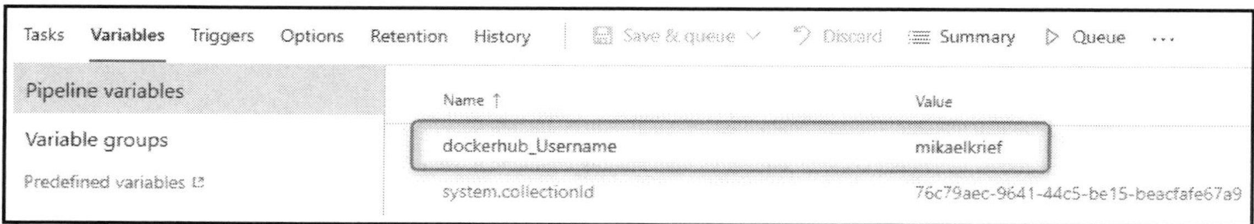

4. In the **Triggers** tab, continuous integration is enabled, as shown in the following screenshot:

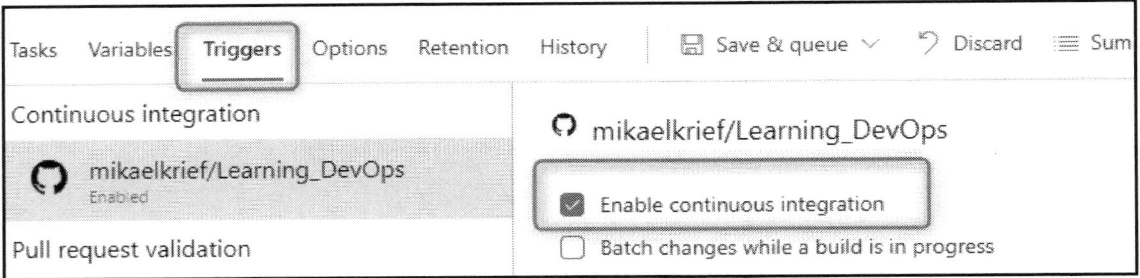

5. In the **Options** tab, we indicate the build number with the `2.0.patch` pattern:

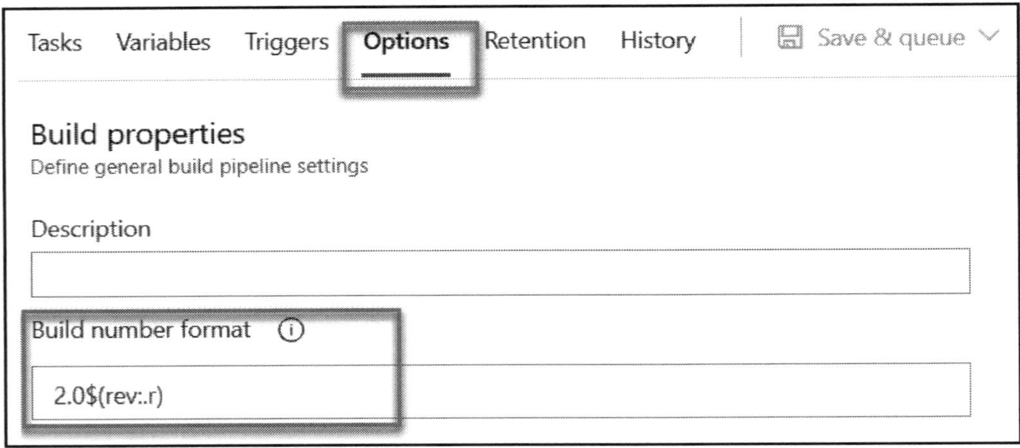

This build number will be the tag of the Docker image that is uploaded into the Docker Hub. Once the configuration is finished, we save the build definition and execute it. If the builds were successfully executed, we will notice the following:

- Build artifacts that contain the YAML specification for Kubernetes files:

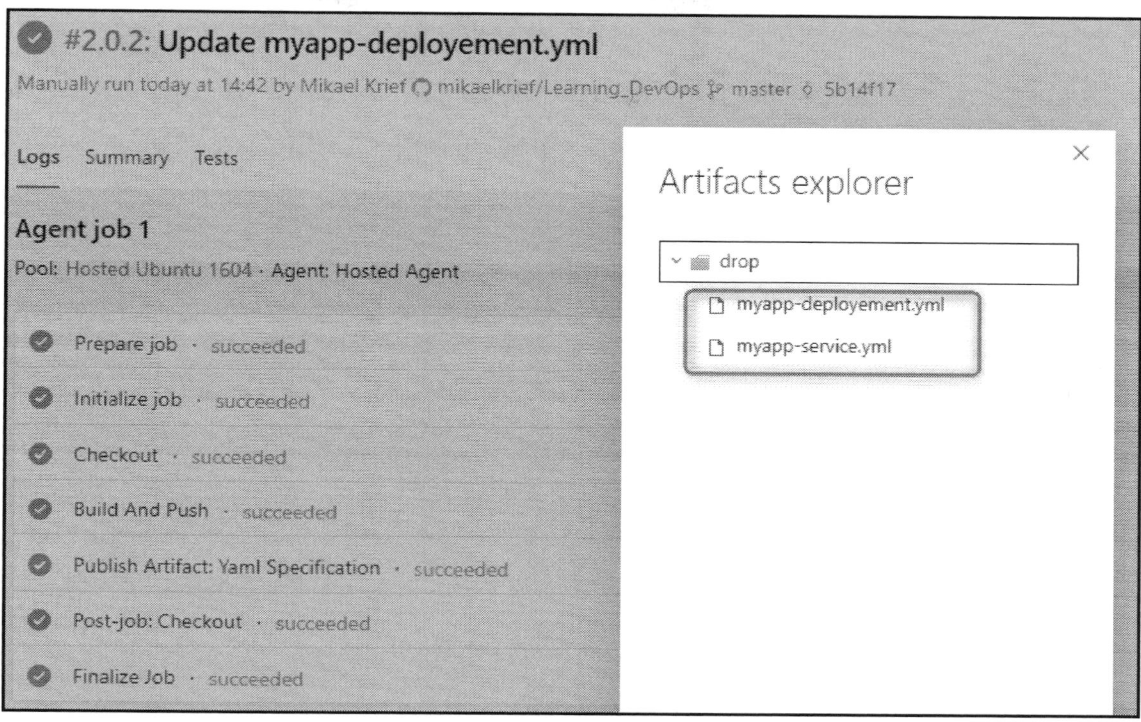

- In the Docker Hub, a new tag on the image that corresponds to the build number, as well as the latest tag, as shown in the following screenshot:

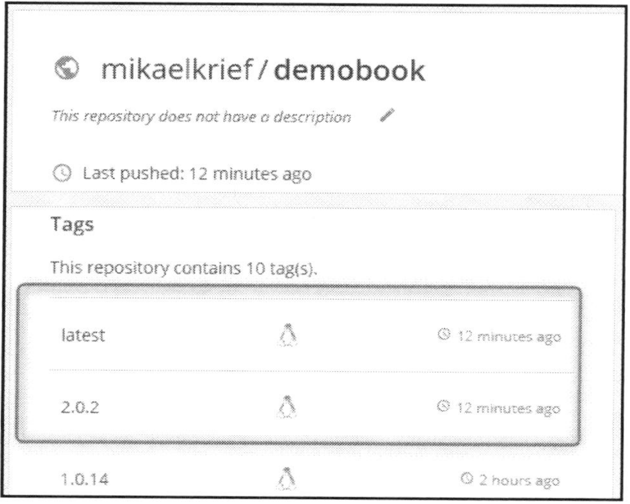

So, in this section, we have created and executed a build in Azure Pipelines that creates and pushes a Docker image into the Docker Hub, and then publishes the YAML files as artifacts for release.

We will now create the release definition that will automatically deploy our application in AKS.

Automatic deployment of the application in Kubernetes

We will now create a new definition of release that automatically deploys our application in the AKS cluster that we created in the previous *Using AKS* section. For this deployment, in Azure Pipelines, we create a new release by performing the following steps:

1. Regarding the choice of template for the release, select the **Empty template**.
2. Create a stage called **AKS**, and inside add a task that allows the `kubectl` commands (this task is present by default in the Azure DevOps tasks catalog), as shown in the following screenshot:

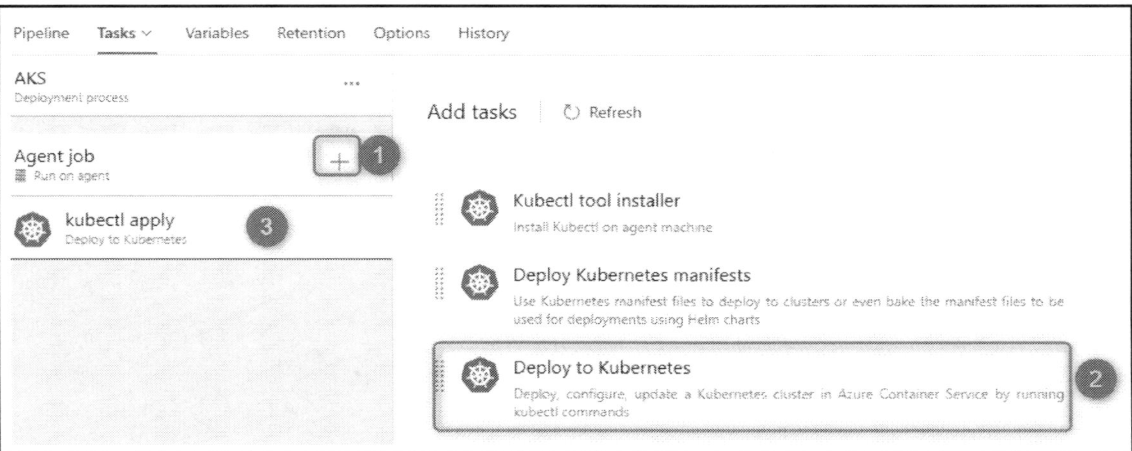

3. Add the **Deploy to Kubernetes** task to the Azure Pipelines tasks catalog with the following configuration:

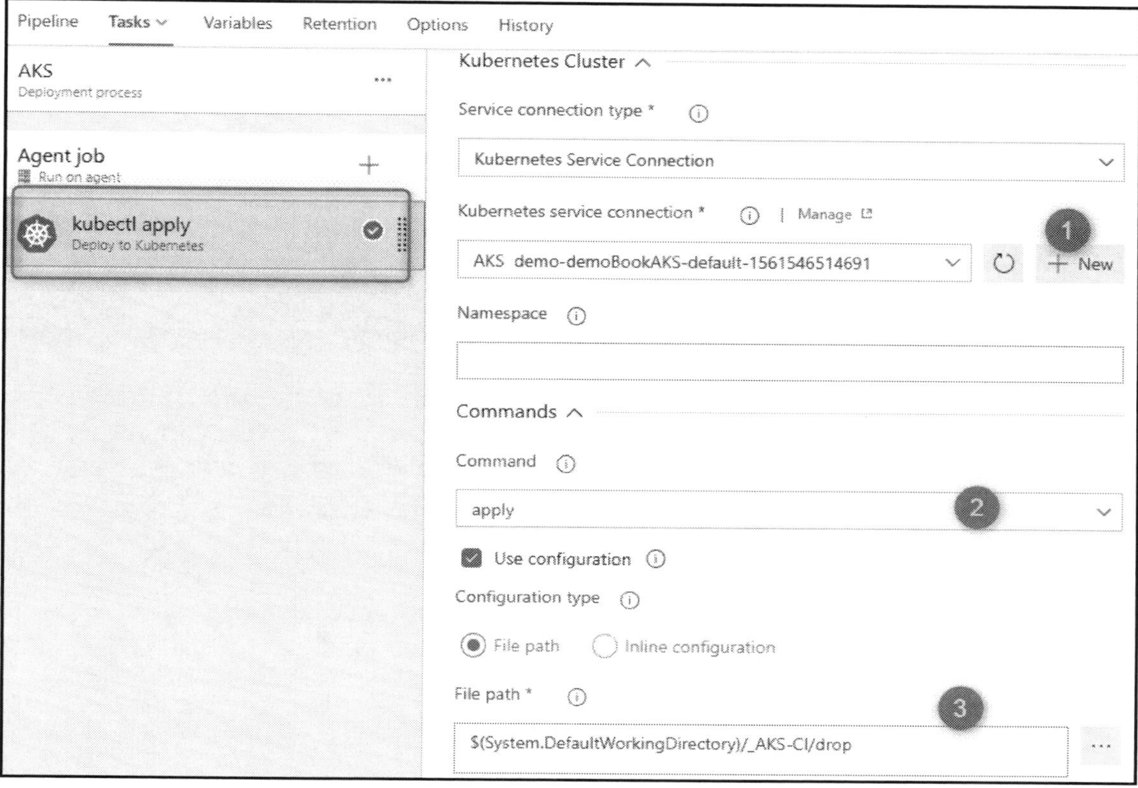

The settings for the **Deploy to Kubernetes** task are as follows:

- We choose the endpoint of the Kubernetes cluster—the **New** button allows us to add a new endpoint configuration of a cluster.
- Then, we choose the `apply` command to be executed by kubectl—here, we will execute an application.
- Finally, we choose the directory, coming from the artifacts, which contains the YAML specification files.

4. We save the release definition by clicking on the **Save** button.
5. Finally, we click on the **Create a new release** button, which triggers a deployment in our AKS cluster.

At the end of the release execution, it is possible to check that the application has been deployed by executing the command in a Terminal as follows:

```
kubectl get pods,services
```

This command displays the list of pods and services that are present in our AKS Kubernetes cluster, and the result of this command is shown in the following screenshot:

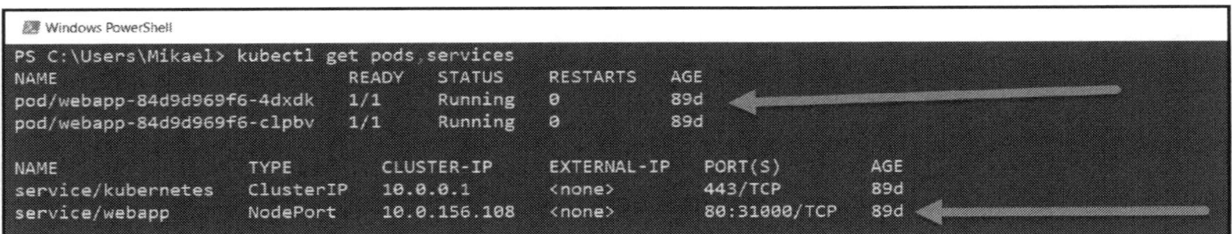

We can see our two deployed web applications pods and the `NodePort` service that exposes our applications outside the cluster. Then, we open a web browser with the `http://localhost:31000` URL, and our application is displayed correctly:

We have created a complete CI/CD pipeline that deploys an application in a Kubernetes cluster. If our application (HTML file) is modified, the build will create and push a new version of the image (in the latest tag), and then the release will apply the deployment on the Kubernetes cluster.

In this section, we have created an end-to-end DevOps CI/CD pipeline in order to deploy an application in a Kubernetes cluster (AKS for our example) with Azure Pipelines.

Summary

In this chapter, we have seen an advanced use of containers with the use of Kubernetes, which is a container manager.

We discussed the different options for installing a small cluster on a local machine using Docker Desktop. Then, using the YAML specification file and the `kubectl` command, we realized the deployment of a Docker image in our Kubernetes cluster in order to run a web application.

We installed and configured Helm, which is the package manager of Kubernetes. Then, we applied it in practice with an example of a chart deployment in Kubernetes.

We also had an overview of AKS, which is a Kubernetes service, managed by Azure, with its creation and configuration.

Finally, we finished this chapter with an example of implementation on a CI/CD pipeline with Azure Pipelines, which deploys a containerized application in a Kubernetes cluster.

The next chapter begins a new part of this book, which deals with application testing, and we will start with API testing with Postman.

Questions

1. What is the role of Kubernetes?
2. Where is the configuration of the objects that are written in Kubernetes?
3. What is the name of the Kubernetes customer tool?
4. Which command allows us to apply a deployment in Kubernetes?
5. What is HELM?
6. What is AKS?

Further reading

If you want to know more about Kubernetes, take a look at the following resources:

- The DevOps 2.3 Toolkit: `https://www.packtpub.com/business/devops-23-toolkit`
- Hands-On Kubernetes on Azure: `https://www.packtpub.com/virtualization-and-cloud/hands-kubernetes-azure`

4
Section 4: Testing Your Application

This section explains some ways to test APIs with Postman. Then, we talk about static code analysis with SonarQube and performance tests with Azure DevOps.

We will have the following chapters in this section:

- `Chapter 9`, *Testing APIs with Postman*
- `Chapter 10`, *Static Code Analysis with SonarQube*
- `Chapter 11`, *Security and Performance Tests*

9
Testing APIs with Postman

In the previous chapters, we talked about DevOps culture and **Infrastructure as Code** (**IaC**) with Terraform, Ansible, and Packer. Then, we saw how to use a source code manager with Git, and the implementation of a CI/CD pipeline with Jenkins and Azure Pipelines. Finally, we showed the containerization of applications with Docker, and their deployment in a Kubernetes cluster.

If you are a developer, you should realize that you use APIs every day, either for client-side use (where you consume the API) or as a provider of the API.

An API, as well as an application, must be testable, that is, it must be possible to test the different methods of this API in order to verify that it responds without error, and that the response of the API is equal to the expected result.

In addition, the proper functioning of an API is much more critical to an application, because this API is potentially consumed by several client applications, and if it does not work, it will have an impact on all of these applications.

In this chapter, we will learn how to test an API with a specialized tool called **Postman**. We will explore the use of collections and variables, then we will write Postman tests, and finally we will see how to automate the execution of Postman tests with Newman in a CI/CD pipeline.

This chapter covers the following topics:

- Creating a Postman collection
- Using environments and variables
- Writing Postman tests
- Executing tests locally
- Understanding Newman concept
- Preparing Postman collections for Newman
- Running the Newman command line
- Integration of Newman in the CI/CD pipeline process

Technical requirements

In this chapter, we will use **Newman**, which is a Node.js package. Therefore, we need to install Node.js and npm on our computer beforehand, which we can download at `https://nodejs.org/en/`.

For the demo APIs that are used in this chapter, we will use an example that is provided on the internet: `https://jsonplaceholder.typicode.com/`.

The GitHub repository, which contains the complete code source of this chapter, can be found at `https://github.com/PacktPublishing/Learning_DevOps/tree/master/CHAP09`.

Check out the following video to see the Code in Action:
`http://bit.ly/2JkUiX7`

Creating a Postman collection with requests

Postman is a free client tool in a graphical format that can be installed on any type of OS. Its role is to test APIs through requests, which we will organize into collections. It also allows us to dynamize API tests through the use of variables and the implementation of environments. Postman is famous for its ease of use, but also for the advanced features that it offers.

In this section, we will learn how to create and install a Postman account, then we will create a collection that will serve as a folder to organize our requests, and finally we will create a request that will test a demo API.

Before we use Postman, we will need to create a Postman account by going to `https://www.getpostman.com/` and clicking on the **SIGN IN** button. Then, in the form, click on the **Create Account** link, as shown in the following screenshot:

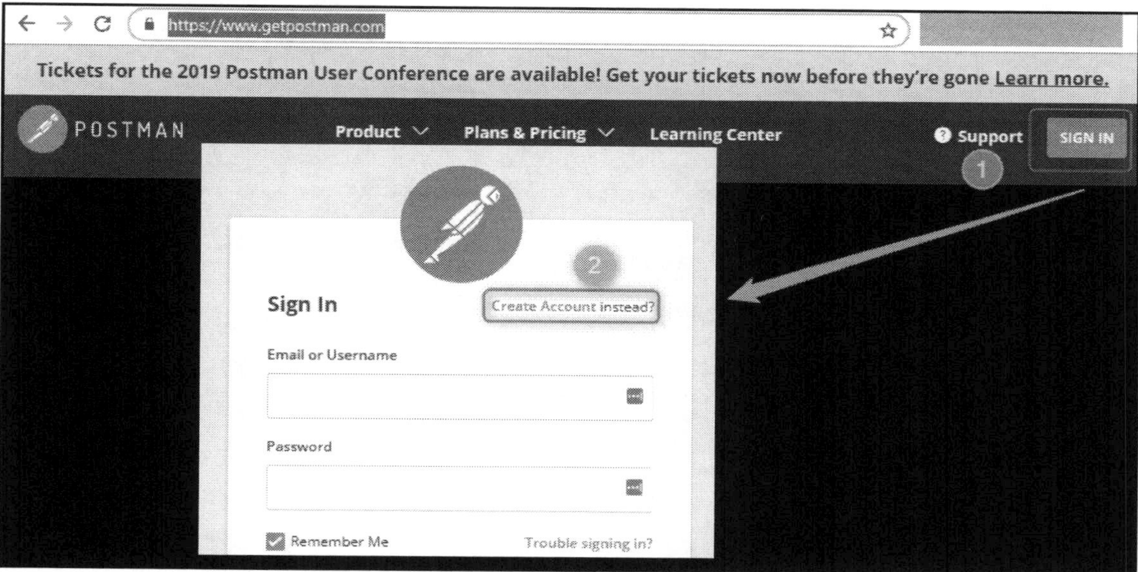

Then, you can either create a Postman account for yourself, by filling out the form, or you can create an account using your Google account.

This account will be used to synchronize Postman data between your machine and your Postman account. This way, the data will be accessible on all of your workstations.

After creating a Postman account, we will look at how to download and install it on a local machine.

Installation of Postman

Once the Postman account has been created, those who are using Windows can download Postman from `https://www.getpostman.com/downloads/`, and choose the version to install. For those who want to install it on Linux or macOS, just click on the link of your OS. The following screenshot shows the download links according to your OS:

Once Postman is downloaded, we need to install it by clicking on the download file for Windows, or for other OSes, follow the installation documentation at `https://learning.getpostman.com/docs/postman/launching_postman/installation_and_updates/`.

We have just seen that the installation of Postman is very simple; the next step is to create a collection in which we will create a request.

The API that we will test in this chapter is a demo API, which is provided freely on this site: `https://jsonplaceholder.typicode.com/`.

Creating a collection

In Postman, any request that we test must be added into a directory called `Collection`, which provides storage for requests and allows for better organization.

We will, therefore, create a `DemoBook` collection that will contain the requests to the demo API, and for this, we will perform the following tasks:

1. In Postman, in the left-hand panel, we click on the **Collections** | **New Collection** button.
2. Once the form opens, we will enter the name as `DemoBook`, and we validate it by clicking on the **Create** button. These steps for creating a new collection are illustrated in the following screenshot:

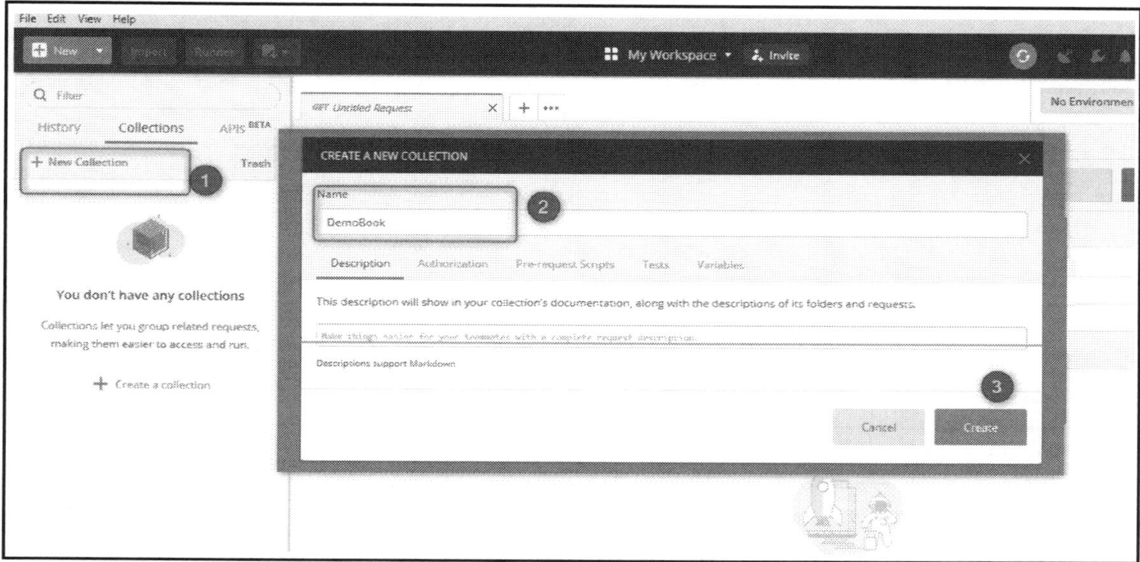

So, we have a `Demobook` collection that appears in the left-hand panel of Postman.

This collection is also synchronized with our Postman web account, and we can access it at `https://web.postman.co/me/collections`.

This collection will allow us to organize the requests of our API tests, and it is also possible to modify its properties in order to apply a certain configuration to all the requests that will be included in this collection.

These properties include request authentication, tests to be performed before and after requests, and common variables to all requests in this collection.

To modify the settings and properties of this collection, the following actions are performed:

1. Click on the **...** button of the context menu of the collection.
2. Choose the **Edit** option, and the edit form appears, in which we can change all the settings that will apply to the requests in this collection.

The following screenshot shows the steps that are taken to modify the properties of a collection:

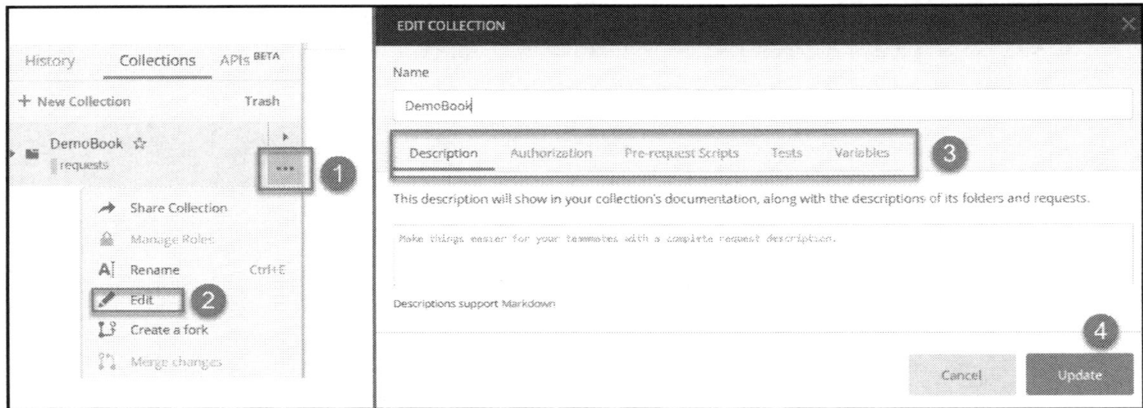

So, we have discussed the procedure that is followed in order to create a collection that is the first Postman artifact, and this will allow us to organize our API test requests.

We will now create a request that will call and test the proper functioning of our demo API.

Creating our first request

In Postman, the object that contains the properties of the API to be tested is called a **request**. This request contains the configuration of the API itself, but it also contains the tests that are to be performed to check it's functioning properly.

The main parameters of a request are as follows:

- The URL of the API
- Its method: Get/POST/DELETE/PATCH

- Its authentication properties
- Its querystring keys and its body request
- The tests that are to be performed before or after the execution of the API

The creation of a request is done in two steps: its creation in the collection and its configuration.

1. **The creation of the request**: To create the request of our API, here are the steps that need to be followed:
    1. We go to the context menu of the `DemoBook` collection and click on the **Add Request** option:

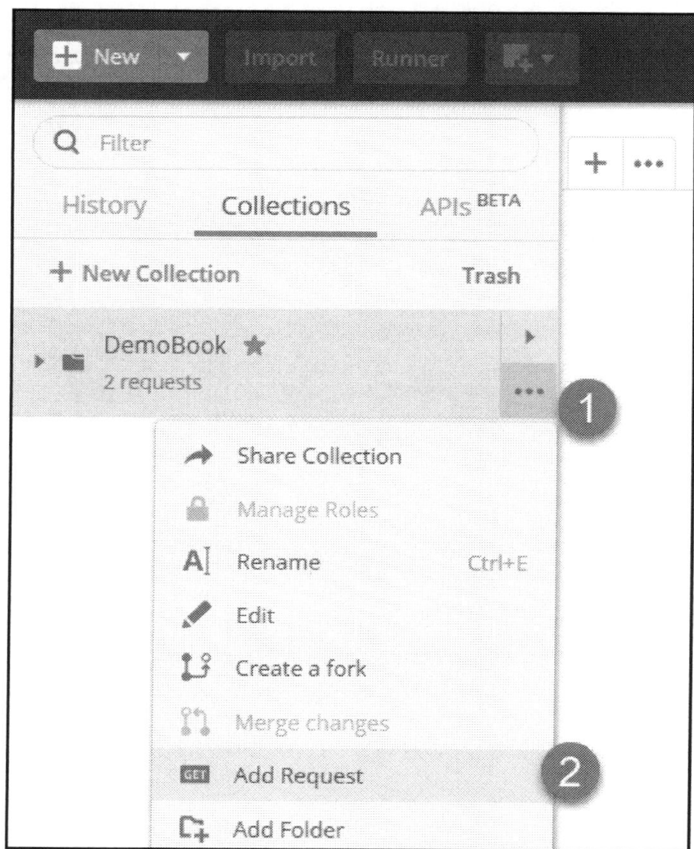

2. Then, in the form, enter the name of the request, `Get all posts`. Finally, we validate the form by clicking on the **Save to DemoBook** button, as shown in the following screenshot:

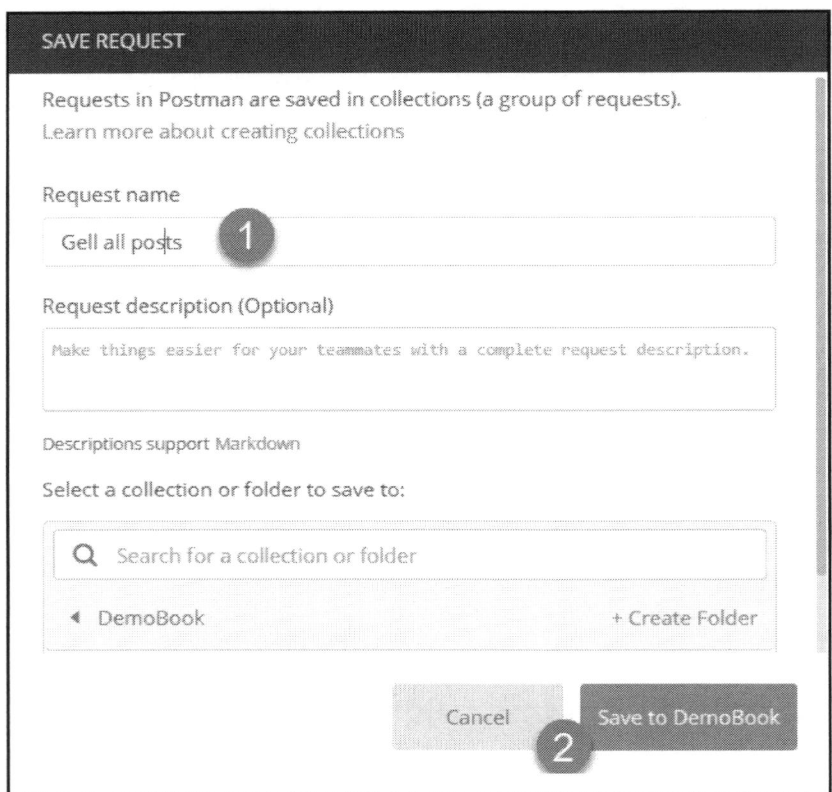

2. **The configuration of the request**: After creating the request, we will configure it by entering the URL of the API to be tested, which is `https://jsonplaceholder.typicode.com/posts`, in the **GET** method. After entering the URL, we saving the request configuration by clicking on **Save** button.

The following screenshot shows the parameters of this request with its URL and method:

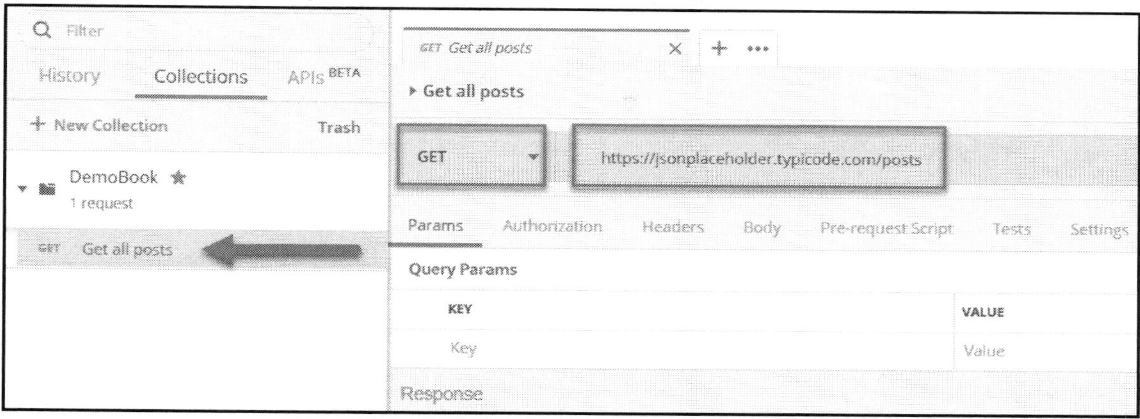

Finally, to complete the tests, and to add more content to our lab, we will add a second request to our collection, which we will call `Get a single post`. It will test another method of the API, and it will also ensure that we configure it with the `https://jsonplaceholder.typicode.com/posts/<ID of post>` URL.

The following screenshot shows the requests of our collection:

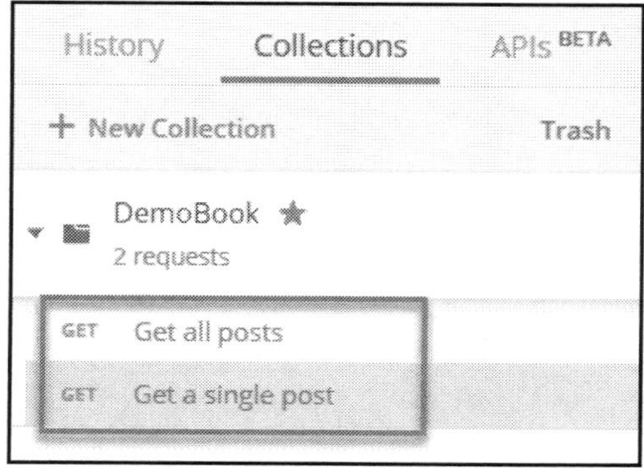

Note that the Postman documentation for collection creation can be found at `https://learning.getpostman.com/docs/postman/collections/creating_collections/`.

In this section, we have learned how to create a collection in Postman, as well as how to create requests and their configurations.

In the next section, we will learn how to dynamize our requests with the use of environments and variables.

Using environments and variables to dynamize requests

When we want to test an API, we need to test it on several environments for better results. For example, we will test it on our local machine and development environment, and then also on the QA environment. To optimize test implementation times and to avoid having a duplicate request in Postman, we will inject variables into this same request in order to make it testable in all environments.

So, in the following steps, we will improve our requests by creating an environment and two variables; then, we will modify our requests in order to use these variables:

1. In Postman, we will start by creating an **environment** that we call `Local`, as shown in the following screenshot:

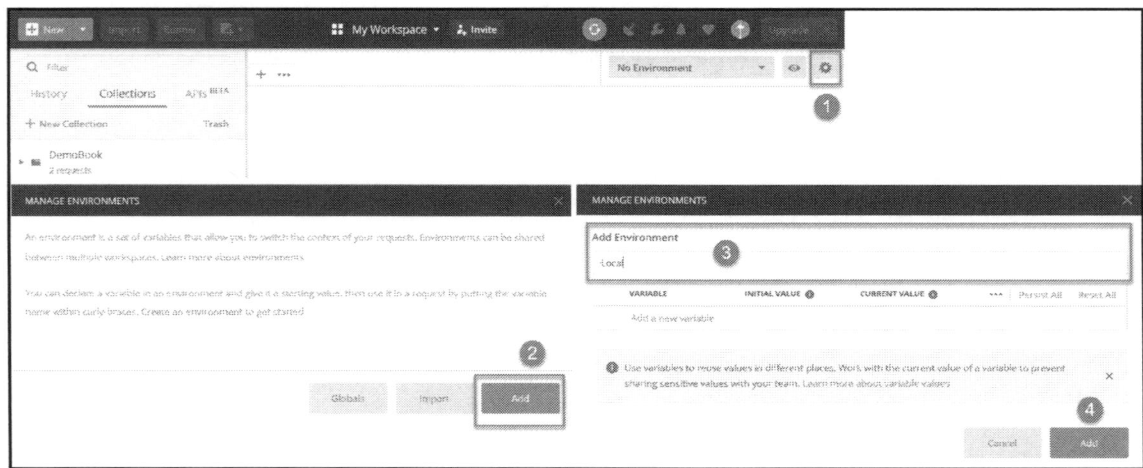

2. Then, in this `Local` environment, we will insert a **variable** named `PostID`, which will contain the value to pass in the URL of the request. This following screenshot shows the creation of the `PostID` variable:

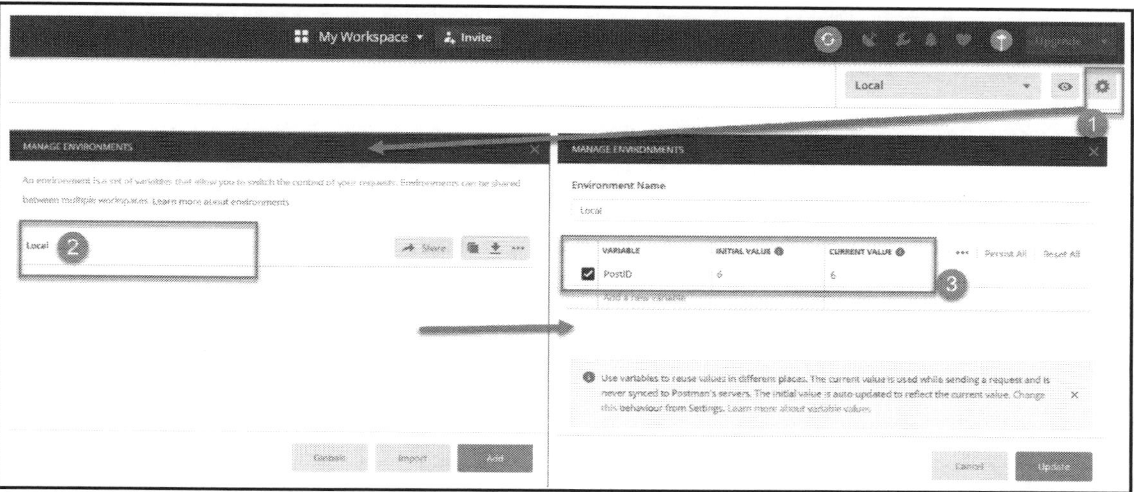

3. Thus, for the `Local` environment, the value of the `PostID` variable is `6`. To have a different value for other environments, it is necessary to create other environments using the same steps that we have just seen, and then adding the same variables (with the same name) and their corresponding values. This is, for example, shows the variable screen for the `QA` environment:

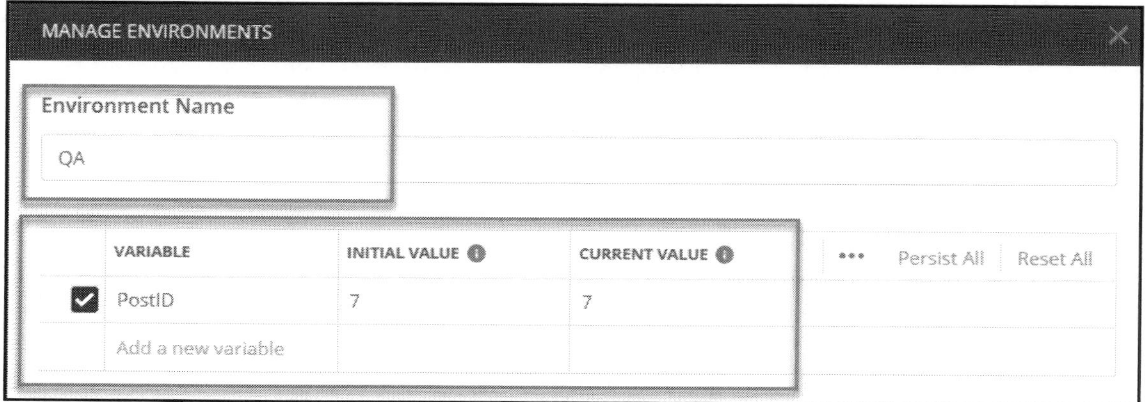

4. Finally, we will modify the request in order to use the variable that we have just declared. In Postman, the usage of a variable is done using the `{{variable name }}` pattern. So first, we select the desired environment from the dropdown in the top-right corner. Then, in the request, we will replace the post's ID at the end of the URL with `{{PostID}}`, as shown in the following screenshot:

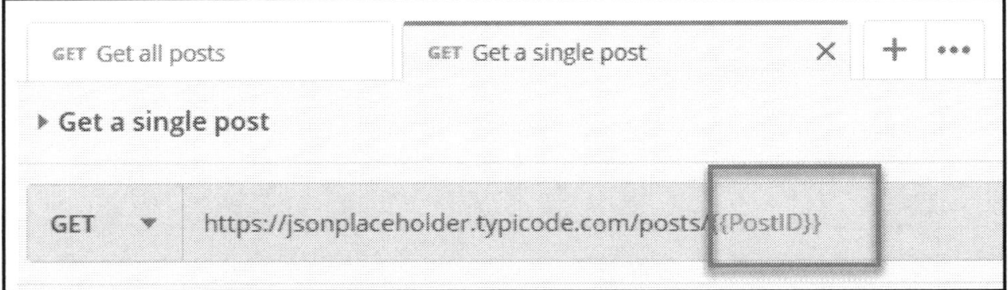

Note that the Postman documentation of environments and variables is available at `https://learning.getpostman.com/docs/postman/environments_and_globals/intro_to_environments_and_globals/`.

In this section, we created a Postman request that will allow us to test an API. Then we made its execution more flexible by creating an environment in Postman that contains variables that are also used in Postman's requests.

In the next section, we will write Postman tests to verify the API result.

Writing Postman tests

Testing an API is not only about checking that its call returns a return code of 200, that is, that the API responds well, but also that its return result corresponds to what is expected, or that its execution time is not too long.

For example, consider an API that returns a response in JSON format with several properties. In the tests of this API, it will be necessary to verify that the result returned is a JSON text that contains the expected properties, and even more so to verify the values of these properties

In Postman, it is possible to write tests that will ensure that the response of the request corresponds to the expected result in terms of return or execution time using the JavaScript language.

Postman tests are written in the **Tests** tab of the request:

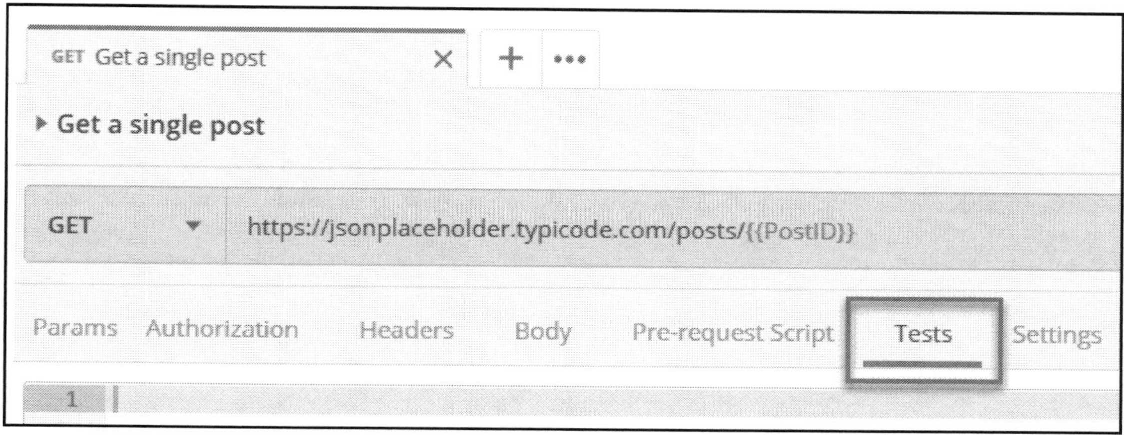

To test our request, we will write several tests, which are as follows:

- That the return code of the request is 200
- That the response time of the request is less than 400 ms
- That the answer is not an empty JSON
- That the return JSON response contains the `userId` property, which is equal to 1

To perform these tests, we will write the code for the following in the **Tests** tab.

The following code illustrates the return code of the request:

```
pm.test("Status code is 200", function () {
 pm.response.to.have.status(200);
});
```

The following code illustrates a response time of less than 400 ms:

```
pm.test("Response time is less than 400ms", function () {
 pm.expect(pm.response.responseTime).to.be.below(400);
});
```

The following code illustrates that the response in JSON format is not empty:

```
pm.test("Json response is not empty", function (){
 pm.expect(pm.response).to.be.json;
});
```

The following code illustrates that, in the JSON response the `userId` property is equal to `1`:

```
pm.test("Json response userId eq 1", function (){
 var jsonRes = pm.response.json();
 pm.expect(jsonRes.userId).to.eq(1);
});
```

And so, finally, the **Tests** tab of our request, which tests our API, contains all this code, as shown in the following screenshot:

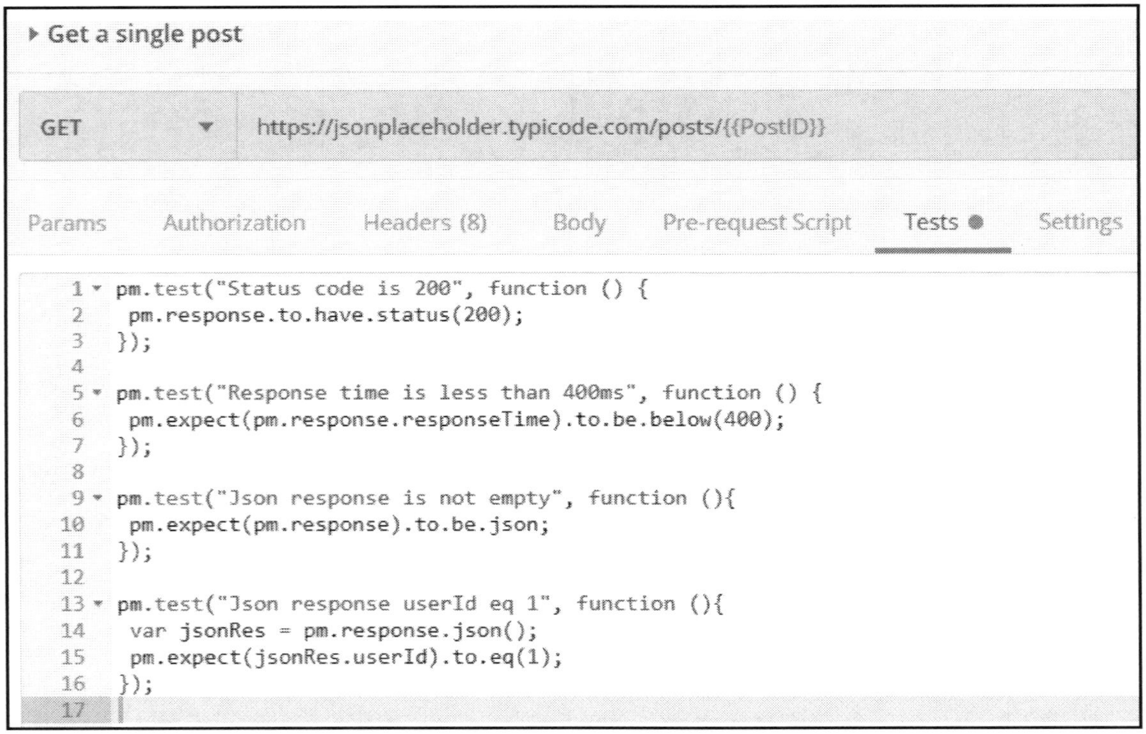

▸ Get a single post

GET ▾ https://jsonplaceholder.typicode.com/posts/{{PostID}}

Params Authorization Headers (8) Body Pre-request Script Tests ● Settings

```
pm.test("Status code is 200", function () {
 pm.response.to.have.status(200);
});

pm.test("Response time is less than 400ms", function () {
 pm.expect(pm.response.responseTime).to.be.below(400);
});

pm.test("Json response is not empty", function (){
 pm.expect(pm.response).to.be.json;
});

pm.test("Json response userId eq 1", function (){
 var jsonRes = pm.response.json();
 pm.expect(jsonRes.userId).to.eq(1);
});
```

We have completed our Postman request with test writing, which will check the proper functioning of the API according to its feedback code, performance, and response content.

For more information about the Postman tests and script, you can read the documentation at `https://learning.getpostman.com/docs/postman/scripts/intro_to_scripts`.

In this section, we have just seen how, in Postman, we can write API tests to check the proper functioning of our API. We will now run our Postman request locally in order to test our API.

Executing Postman request tests locally

So far, in Postman, we have created a collection, in which there are two requests that contain the parameters and tests of our APIs that are to be tested. To test the proper functioning of the APIs with their parameters and tests, we must now execute our requests that are in Postman. Note that it will only be at the end of this execution that we will know whether our APIs correspond to our expectations.

To execute a Postman request, we will perform the following actions:

1. You must first choose the desired environment.
2. Click on the **Send** button of the request, as shown in the following screenshot:

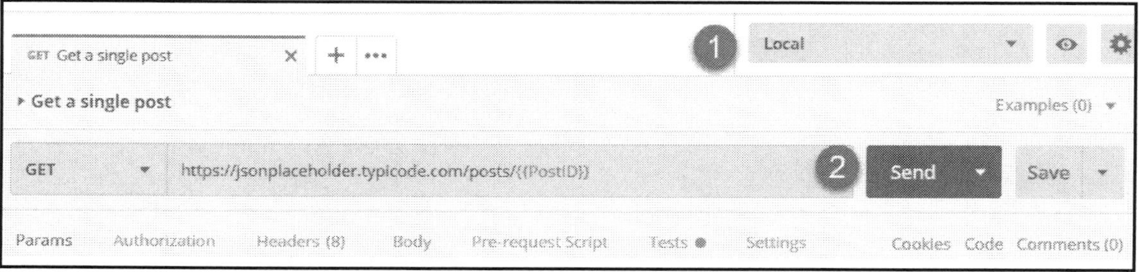

3. In the **Body** tab, we can then view the content of the query response, and if we want to display it in **JSON** format, we can choose the display format. The following screenshot shows the response of the request displayed in JSON format:

4. The **Test Results** tab displays the results of the execution of the tests that we previously wrote, and in our case, the four tests have been executed correctly—they are all green, as shown in the following screenshot:

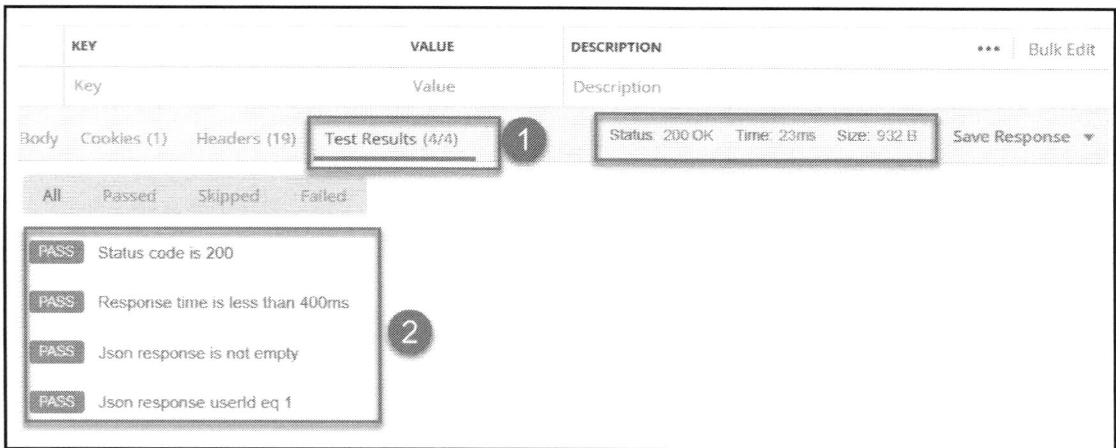

In the preceding screenshot, we can see that the return code of the Postman request is equal to 200, which corresponds to the request's successful execution return code, and its execution time of 23 ms, which is below the threshold (400 ms) that I set for myself as an example.

In the event that one of the tests fails, it will be displayed in red in order to clearly identify it. An example of a failed test is shown in the following screenshot:

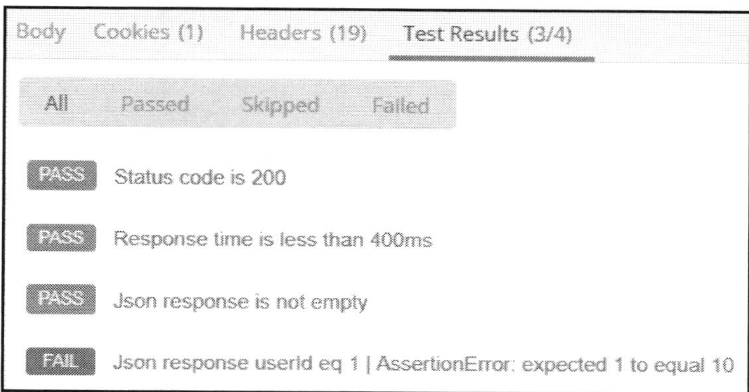

We have just seen the execution of a Postman request to test an API, but this execution is only for the current request. If we want to execute all Postman requests in a collection, we can use the **Postman Collection Runner**.

The Postman Collection Runner is a Postman feature that automatically executes all the requests in a collection in the order in which they have been organized.

You can learn more in the Collection documentation by visiting `https://learning.getpostman.com/docs/postman/collection_runs/starting_a_collection_run/`.

The following two screenshots show the **Runner** execution steps, in which we choose the collection to execute, the environment, and the number of iterations. To start its execution, we click on the **Run DemoBook** button:

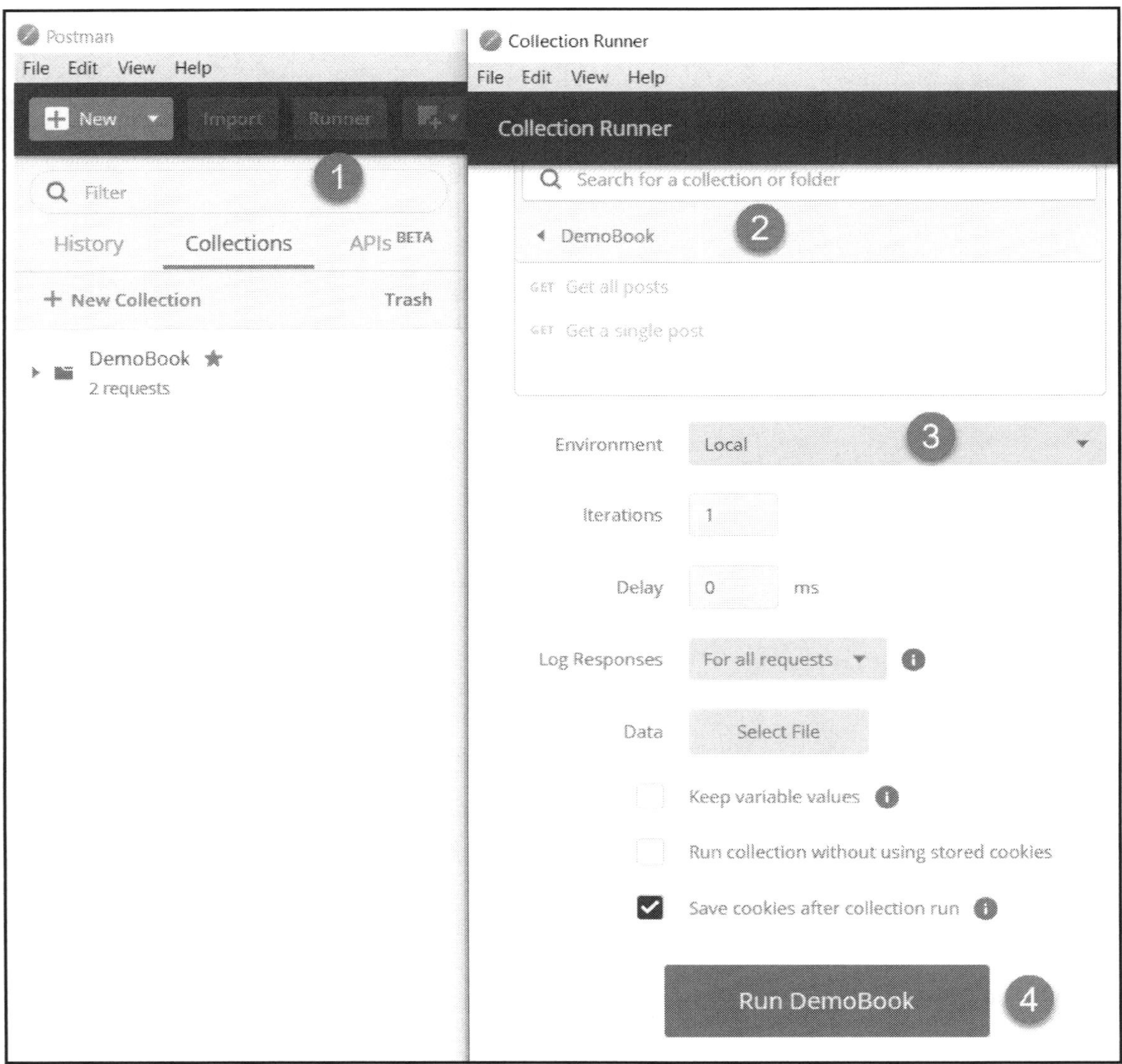

And so, in the **Collection Runner** screen, we can see the test execution result of all the requests in the collection, as shown in the following screenshot:

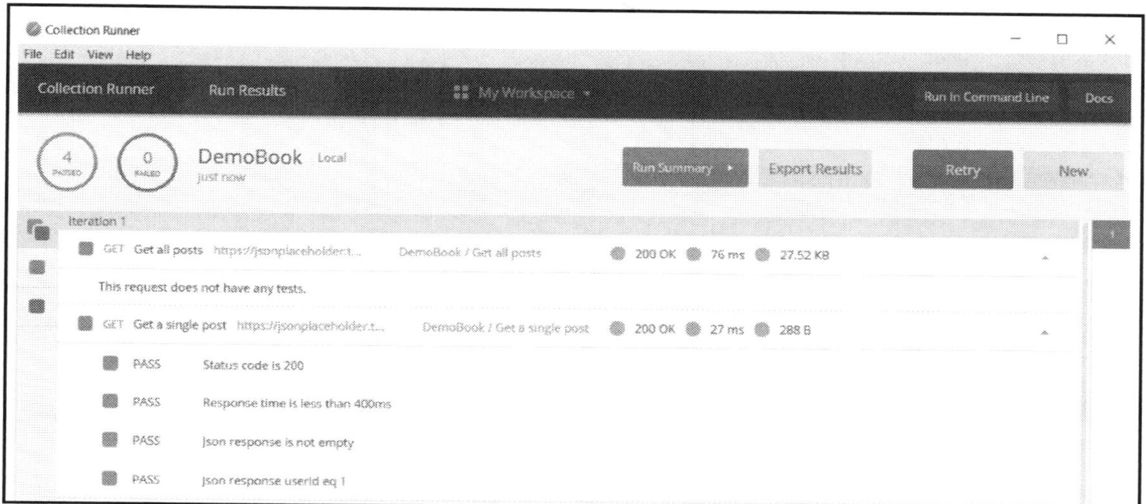

The documentation for the Collection Runner can be found at `https://learning.getpostman.com/docs/postman/collection_runs/intro_to_collection_runs`.

In this section, we have learned how to execute Postman requests in order to test an API in a unitary way, before executing all the requests in the collection using the Postman Collection Runner. In the following section, we will introduce Newman, which will allow us to automate the execution of Postman tests.

Understanding the Newman concept

So far in this chapter, we have talked about using Postman locally to test the APIs that we develop or consume. But, what is important in unit, acceptance, and integration tests is that they are automated in order that they are able to be executed within a CI/CD pipeline.

Postman, as such, is a graphical tool that does not automate itself, but there is another tool called **Newman** that automates tests that are written in Postman.

Newman is a free command-line tool that has the great advantage of automating tests that are already written in Postman. It allows us to integrate API test execution into CI/CD scripts or processes.

In addition, it offers the possibility of generating the results of the tests of reports of different formats (HTML, JUnit, and JSON).

Nevertheless, Newman does not allow us to do the following:

- To create or configure Postman requests; as we will see, requests that are executed by Newman will be exported from Postman.
- To execute only one request that is in a collection—it executes all the requests in a collection.

To learn more about Newman, you can visit the product page at `https://www.npmjs.com/package/newman`.

In order to use Newman, we will need—as stated in the *Technical requirements* section of this chapter—to have installed Node.js and npm, which are available at `https://nodejs.org/en/` (this installer installs both tools).

Then, to install Newman, we must execute the command in Terminal:

```
npm install -g newman
```

The following screenshot shows the execution of the command:

```
>npm install -g newman
\AppData\Roaming\npm\newman -> C:\Users\MikaelKRIEF\AppData\Roaming\npm\node_modules\newman\bin\newman.js
+ newman@4.5.1
added 152 packages from 191 contributors in 13.755s
```

This command installs the npm `newman` package and all its dependencies globally, that is, it is accessible on the entire local machine.

Once installed, we can test its installation by running the `newman --help` command, which displays the arguments and options to use, as shown in the following screenshot:

```
>newman --help
Usage: newman [options] [command]

Options:
  -v, --version                 output the version number
  -h, --help                    output usage information

Commands:
  run [options] <collection>  URL or path to a Postman Collection.

To get available options for a command:
  newman [command] -h
```

In this section, we introduced Newman by talking about its advantages, and we learned how to install it. In the next section, we will export Postman's collection and environment for use with Newman.

Preparing Postman collections for Newman

As we have just seen, Newman is Postman's client tool, and in order to work, it needs the configuration of the collections, requests, and environments that we have created in Postman.

That's why, before running Newman, we will have to export Postman's collection and environments, and this export will serve as Newman's arguments. So, let's start exporting the `DemoBook` collection that we created in Postman.

Exporting the collection

The export of a Postman collection consists of obtaining a JSON file that contains all the settings of this collection and the requests that are inside it.

It is from this JSON file that Newman will be able to run the same API tests as when we ran them from Postman.

To do this export, we perform the following tasks:

1. Go to the context menu of the collection that we want to export.
2. Choose the **Export** action.
3. Then, in the window that opens, uncheck the **Collection v2.1 (recommended)** checkbox.
4. Finally, validate by clicking on the **Export** button.

These steps are shown in the following screenshot:

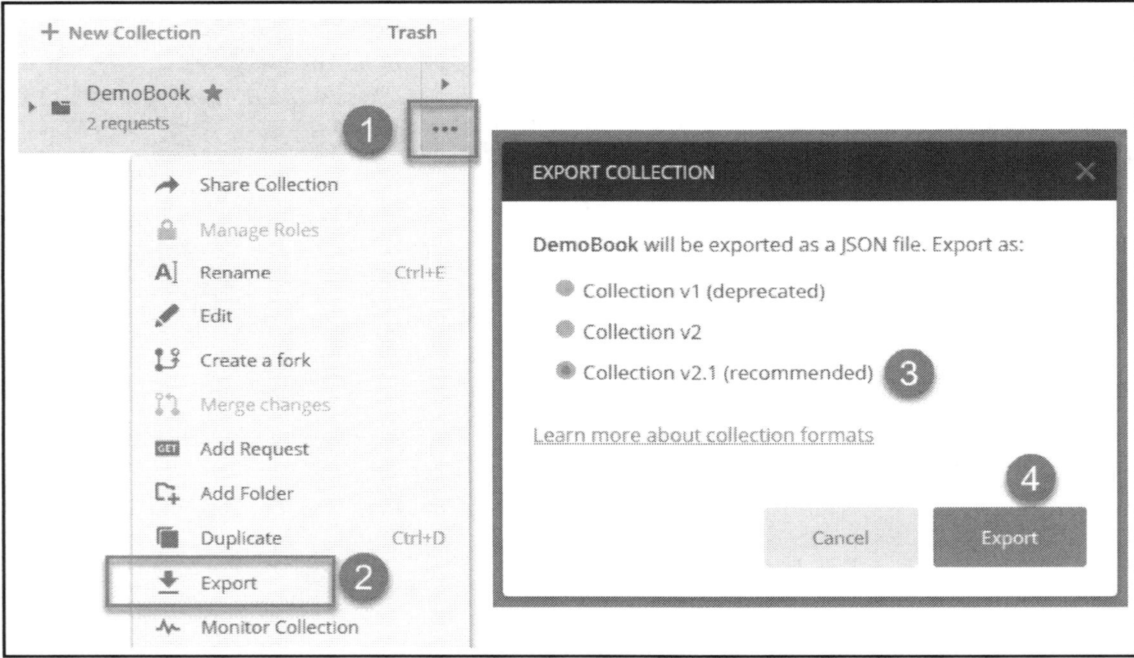

Clicking on the **Export** button exports the collection to a JSON format file, `DemoBook.postman_collection.json`, which we save in a folder that we create, which is dedicated to Newman.

After exporting the collection, we also need to export the environment and variable information, because the requests in our collection depend on it.

Exporting the environments

We could stop there for Newman's configuration, but the problem is that our Postman requests use variables that are configured in environments.

It is, therefore, for this reason that we will also have to export the information from each environment in JSON format, so that we can also pass it on as an argument to Newman.

To export the environments and their variables, we perform the following tasks:

1. We will open the **MANAGE ENVIRONMENTS** from settings in Postman.
2. Then, click on the download environment button.

These steps are shown in the following screenshot:

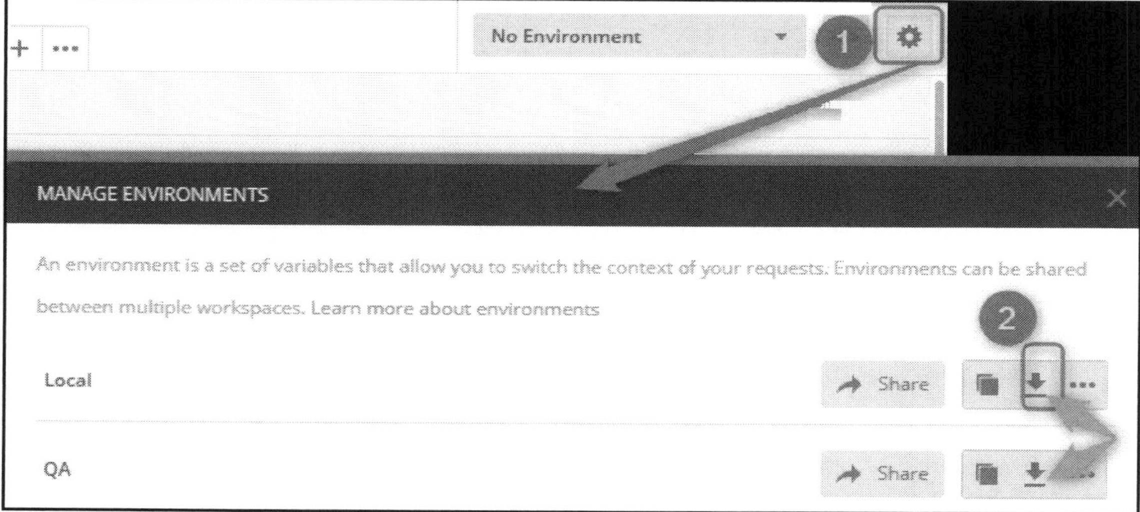

So, for each environment, we will export their configurations in a JSON file, which we save in the same folder where we exported the collection.

Finally, we have a folder on our machine that contains three Postman JSON files:

- One JSON file for the collection
- One JSON file for the `Local` environment
- One JSON file for the `QA` environment

The following screenshot shows the contents of the local folder that contains Postman's exports:

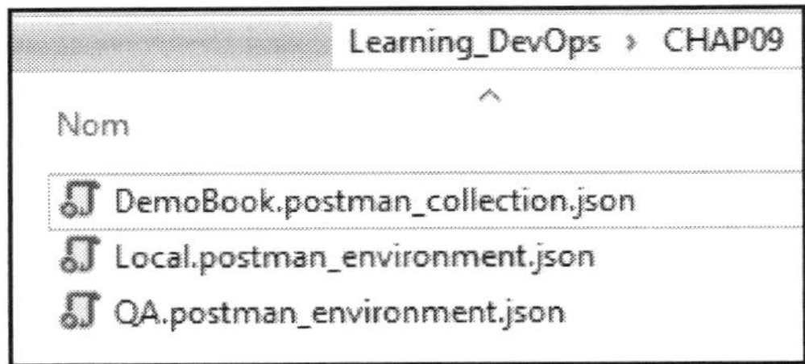

We have just seen the export of all the configurations of our Postman requests, including the collection and the environments, and we will now see the execution of the Newman command line.

Running the Newman command line

After exporting the Postman configuration that we saw earlier, we will run the Newman utility on our local machine.

To execute Newman, we go to Terminal, then to the folder where the JSON configuration files are located, and execute the following command:

```
newman run DemoBook.postman_collection.json -e
Local.postman_environment.json
```

The `newman run` command takes the JSON file of the collection that we exported as an argument, and a parameter, `-e`, which is the JSON file of the exported environment.

For more details about all the arguments of this command, read the documentation at `https://www.npmjs.com/package/newman#newman-options`.

Newman will execute the Postman requests from the collection we exported. It will also use the variables of the exported environment and will also perform the tests we wrote in the request.

The result of its execution, which is quite detailed, is shown in the following screenshot:

```
C:\Windows\System32\cmd.exe

C:\Users\Mikael\Documents\Postman>newman run DemoBook.postman_collection.json -e Local.postman_environment.json
newman

DemoBook

→ Get all posts
  GET https://jsonplaceholder.typicode.com/posts [200 OK, 27.64KB, 194ms]

→ Get a single post
  GET https://jsonplaceholder.typicode.com/posts/6 [200 OK, 932B, 27ms]
  √  Status code is 200
  √  Response time is less than 400ms
  √  Json response is not empty
  √  Json response userId eq 1

┌─────────────────────────┬──────────┬──────────┐
│                         │ executed │   failed │
├─────────────────────────┼──────────┼──────────┤
│              iterations │        1 │        0 │
├─────────────────────────┼──────────┼──────────┤
│                requests │        2 │        0 │
├─────────────────────────┼──────────┼──────────┤
│            test-scripts │        1 │        0 │
├─────────────────────────┼──────────┼──────────┤
│      prerequest-scripts │        0 │        0 │
├─────────────────────────┼──────────┼──────────┤
│              assertions │        4 │        0 │
├─────────────────────────┴──────────┴──────────┤
│ total run duration: 338ms                     │
├───────────────────────────────────────────────┤
│ total data received: 27.16KB (approx)         │
├───────────────────────────────────────────────┤
│ average response time: 110ms [min: 27ms, max: 194ms, s.d.: 83ms] │
└───────────────────────────────────────────────┘
```

And, here is also a screenshot that shows the result of its execution in case there is an error in the test:

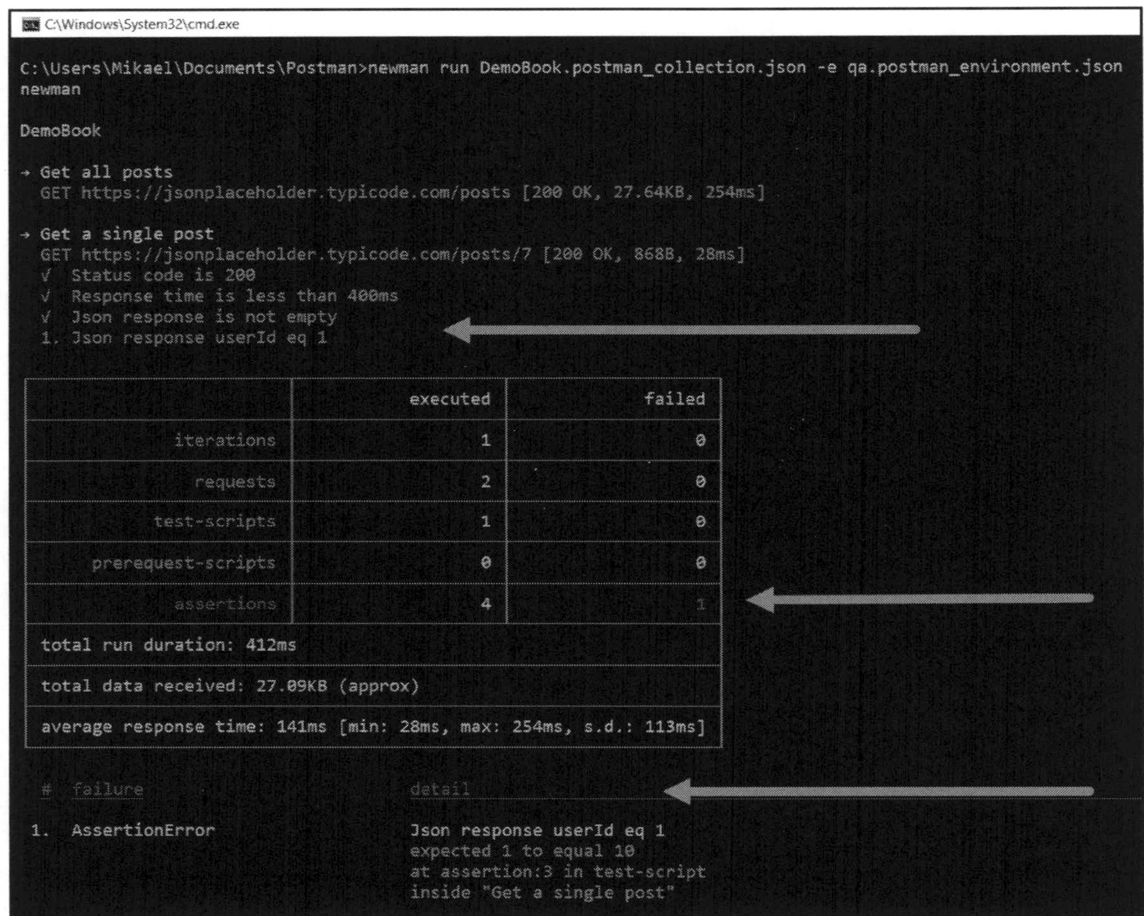

We can see the details of the test that shows an error and what is expected in the request. In this section, we have just learned how to run Newman on a local machine, and we will now learn now Newman is integrated into a CI/CD pipeline.

Integration of Newman in the CI/CD pipeline process

Newman is a tool that automates the execution of Postman requests from the command line, which will quickly allow us to integrate it into a CI/CD pipeline.

To simplify its integration into a pipeline, we go to the first step in the directory that contains the JSON files that were exported from Postman, and create an npm configuration file—`package.json`.

This will have the following content:

```
{
 "name": "postman",
 "version": "1.0.0",
 "description": "postmanrestapi",
 "scripts": {
     "testapilocal": "newman run DemoBook.postman_collection.json -e
Local.postman_environment.json -r junit,cli --reporter-junit-export result-
tests-local.xml",
     "testapiQA": "newman run DemoBook.postman_collection.json -e
QA.postman_environment.json -r junit,cli --reporter-junit-export result-
tests-qa.xml"
 },
 "devDependencies": {
     "newman": "^4.5.1"
 }
}
```

In the `scripts` section, we put the two scripts that will be executed with the command lines that we saw in the previous section, and we add to them the `-r` argument, which allows the output of the command with reporting in JUnit format, and in the `DevDependencies` section, we indicate that we need the Newman package.

That's it; we have all the files that are necessary for integrating Newman's execution into a CI/CD pipeline.

To show Newman's integration into a CI/CD pipeline, we will use **Azure Pipelines**—an Azure DevOps service that we have already seen in `Chapter 6`, *Continuous Integration and Continuous Delivery*, and `Chapter 7`, *Containerizing Your Application with Docker*, and which has the advantage of having a graphic representation of the pipeline.

As a prerequisite for the pipeline, the directory that contains the JSON files of the Postman export, as well as the `package.json` file, must be committed in a source control version.

In our case, we will use the GitHub repository, which contains the complete source code of this chapter: `https://github.com/PacktPublishing/Learning_DevOps/tree/master/CHAP09`.

Build and release configuration

In Azure Pipelines, we will create a build and a release configuration by following these actions:

1. We create a new **build definition** that copies the files that are needed to run Newman into the build artifacts, as shown in the following screenshot:

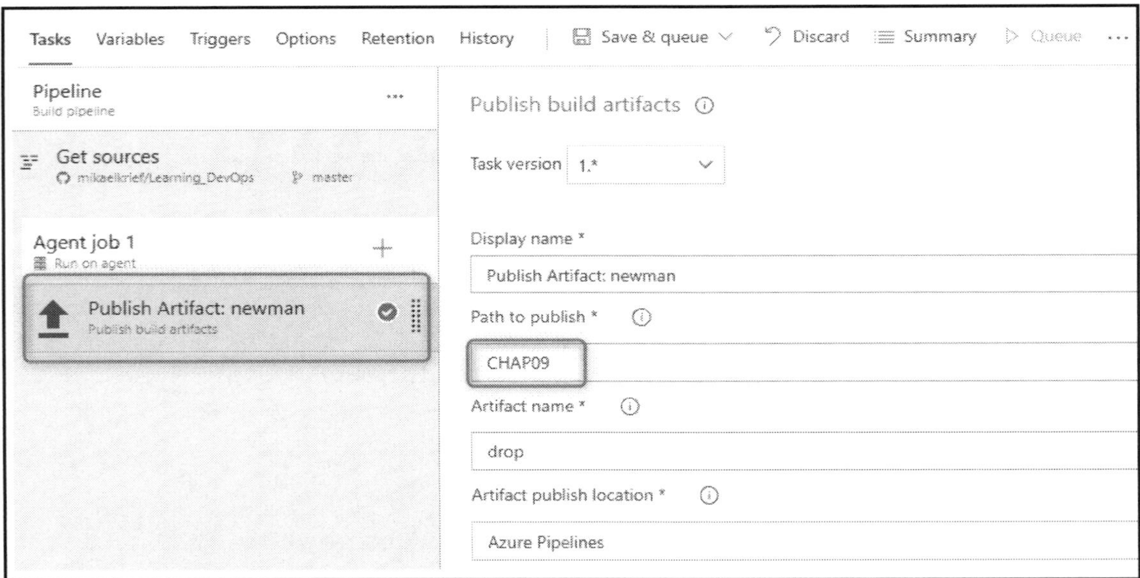

We also enable the continuous integration option in the **Triggers** tab. Then, to run this build, we save and queue this build definition.

2. Then, we create a new **release definition**, which will be in charge of running Newman for each environment. This release will get the artifacts of the build, and will be composed of two stages, `DEV` and `QA`, as shown in the following screenshot:

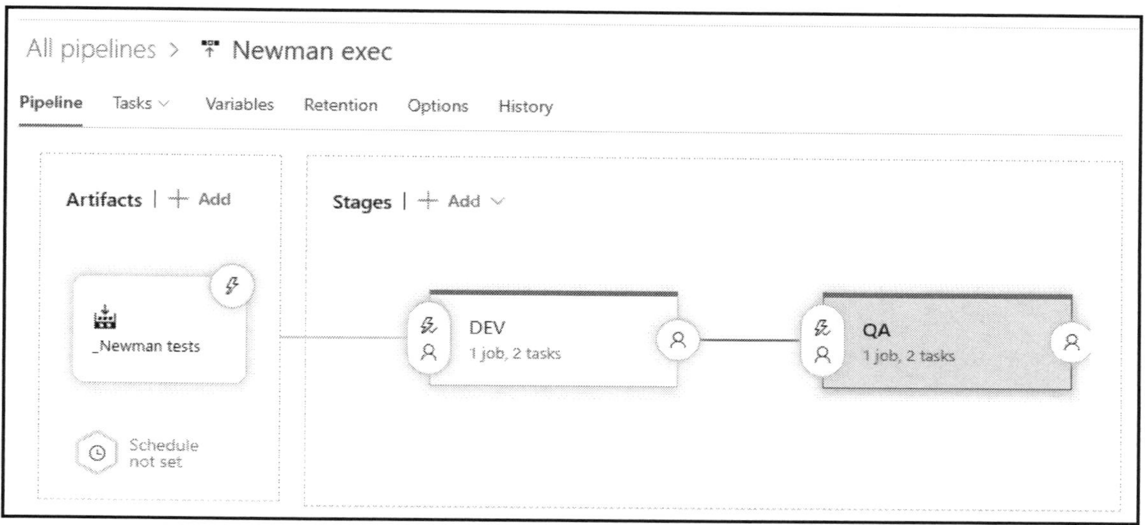

For each of these stages, we configure three tasks that go as follows, based on the `package.json` file:

1. Install Newman.
2. Run Newman.
3. Publish test results in Azure Pipelines.

The following screenshot shows the configuration of the tasks for each stage:

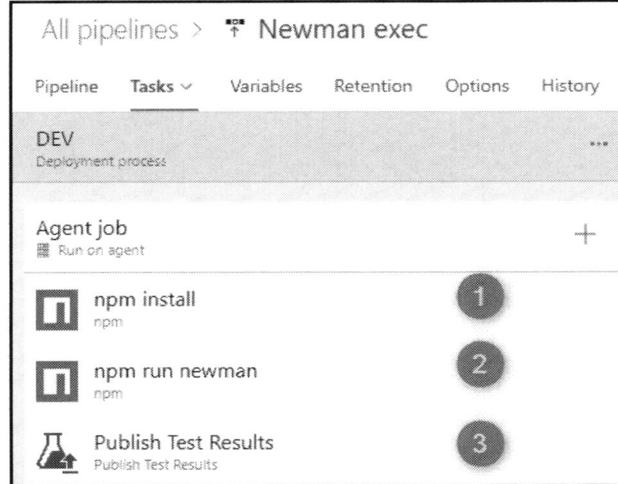

Let's look at the details of the parameters of these tasks in order:

Npm install

The parameters of the **npm install** task are as follows:

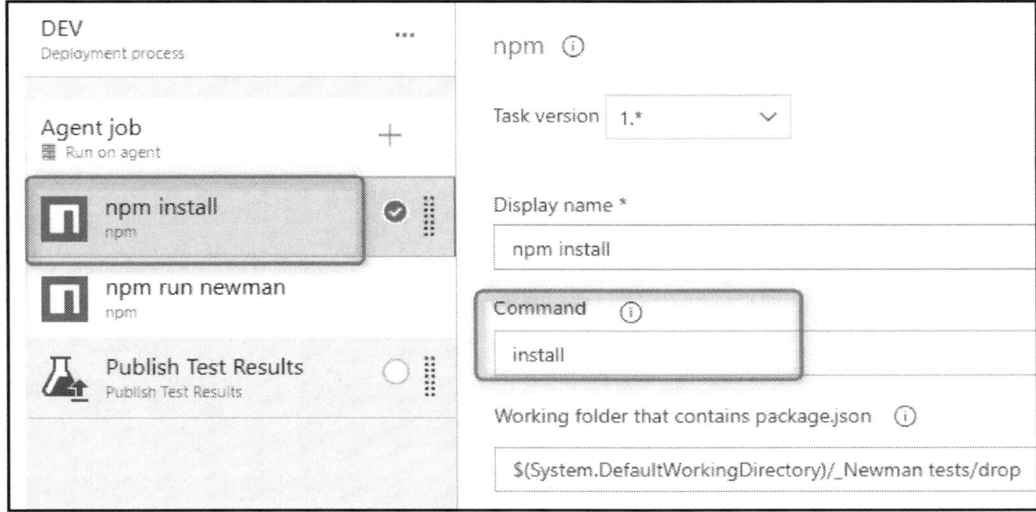

Here, the command that is to be executed in the directory containing the artifact files is `npm install`.

Npm run newman

The parameters of the **npm run newman** task are as follows:

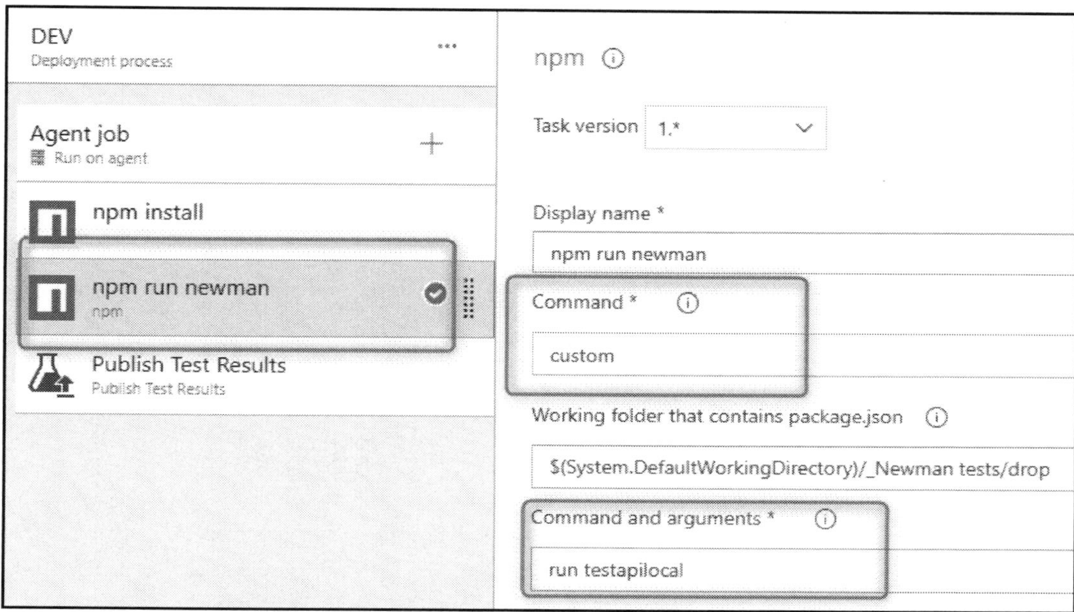

Here, the custom command that is to be executed is `npm run testapilocal` in the directory that contains the artifact files. This `testapilocal` command being defined in the `package.json` file in the script section (seen above) and which executes Newman's command line.

Publish test results

The parameters of the **Publish Test Results** task, which allows us to publish the results of the tests that are performed by Newman in Azure Pipelines, are as follows:

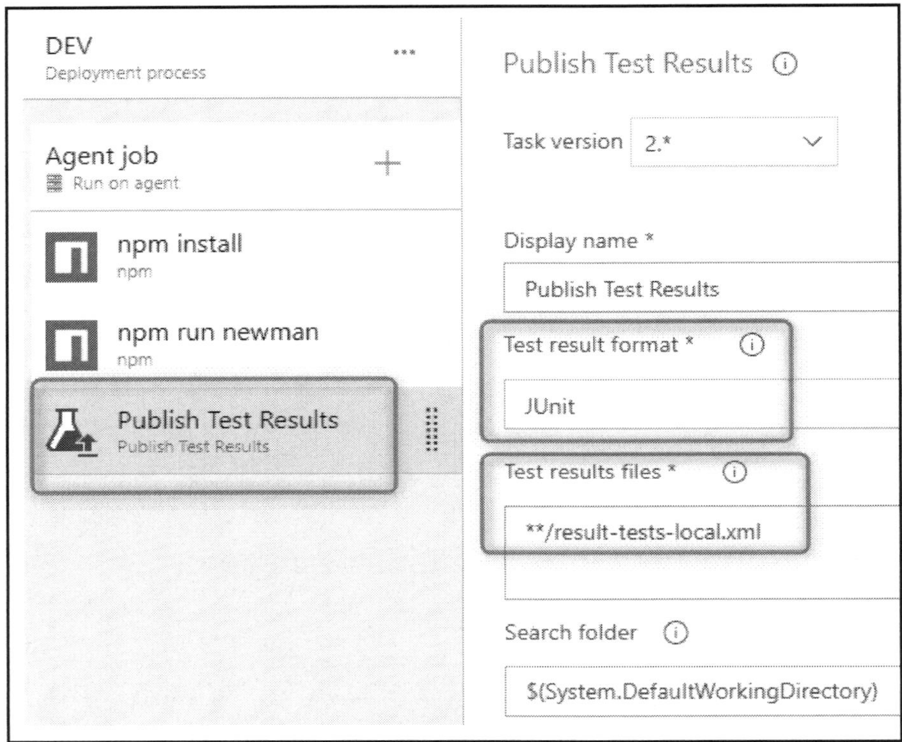

In the parameters of this task, we indicate the JUnit XML reporting files that are generated by Newman, and in the **Control Options** of this task, we select an option to execute the task, even if the **npm run newman** task fails.

The following screenshot shows the parameter of the **Controls Options** to run this task: **Even if a previous task failed, unless the deployment was canceled**:

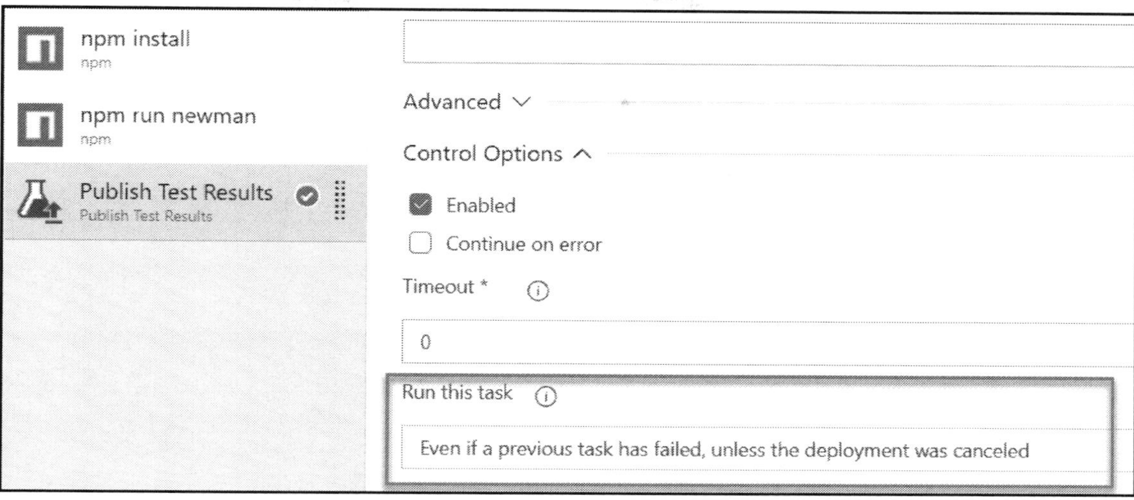

The configuration of the pipeline is complete—we will proceed to its execution.

The pipeline execution

Once the configuration of the release is finished, we can execute this release, and at the end, we can see the reporting of the Newman tests in the **Tests** tab.

The following screenshot shows the reporting of the Newman tests in Azure Pipelines:

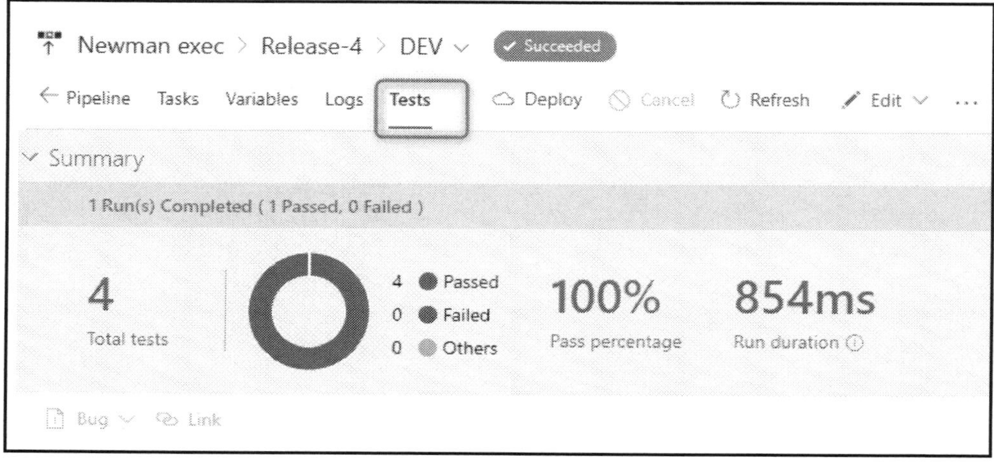

All Postman tests were successfully completed.

Here is a screenshot that shows the reporting of the tests in case one of the tests fails:

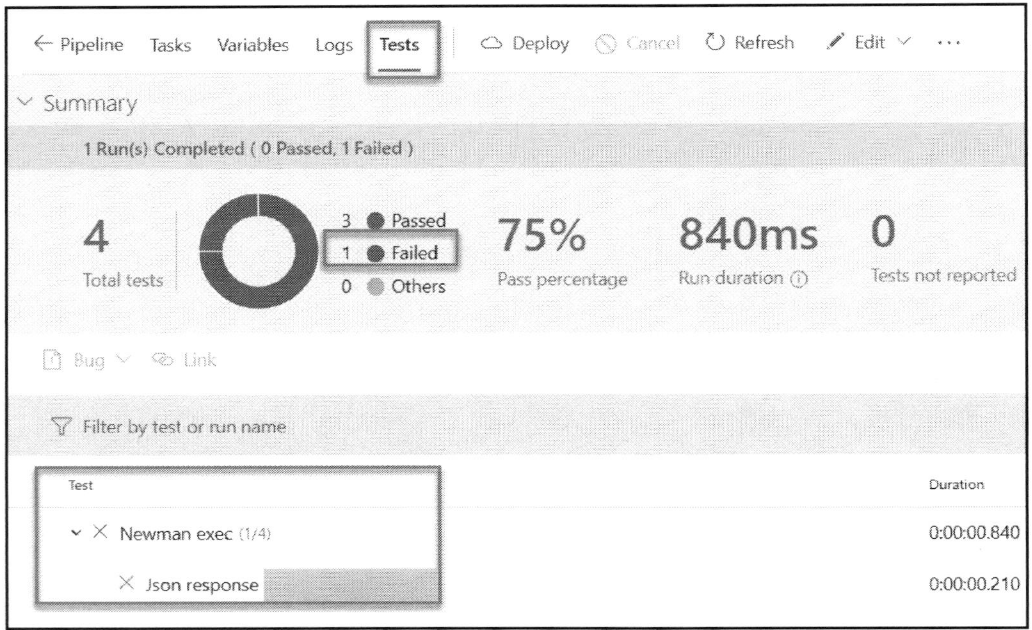

So, we integrated Newman's execution well, and thus we were able to automate the requests of our API, which we had configured in Postman, in a CI/CD pipeline.

For the integration of Newman executions in **Jenkins**, read the documentation at `https://learning.getpostman.com/docs/postman/collection_runs/integration_with_jenkins`, and for integration with **Travis CI**, documentation can be found at `https://learning.getpostman.com/docs/postman/collection_runs/integration_with_travis`.

In this section, we learned how to create and configure a CI/CD pipeline in Azure Pipelines, which performs Postman tests that have been exported for Newman.

Summary

In this chapter, we introduced Postman, which is an excellent tool for testing APIs. We created a Postman account and installed it locally.

Then, we created collections and environments, in which we created requests that contain the settings of our APIs that are to be tested.

We also talked about automating these tests using the Newman command-line tool, with the export of Postman collections and environments.

Finally, in the last part of this chapter, we created and executed a CI/CD pipeline in Azure DevOps that automates the execution of API tests in a DevOps process.

In the next chapter, we will stay on the subject of testing, and we will look at the analysis of static code with a well-known tool called **SonarQube**.

Questions

1. What is the goal of Postman?
2. What is the first element that needs to be created in Postman?
3. What is the name of the element that contains the configuration of the API that is to be tested?
4. Which tool in Postman allows us to execute all the requests of a collection?
5. Which tool allows us to integrate Postman API tests into a CI/CD pipeline?

Further reading

If you want to know more about Postman, here are some resources:

- Postman Learning Center: `https://learning.getpostman.com/`
- Videos and tutorials about Postman: `https://www.getpostman.com/resources/videos-tutorials/`

10
Static Code Analysis with SonarQube

In the previous chapter, we looked at how to test the functionality of an API with Postman, a free tool for testing APIs, and the integration and automation of these tests in a CI/CD pipeline using Newman.

Testing the functionality of an API or application is good practice when we wish to improve the quality of applications. In a company, the quality of an application must be considered by all its members because an application that brings business value to users increases the company's profits.

However, we often neglect to test the quality of the code because we think that what matters is how the application works and not how it is coded. This way of thinking is a big mistake because poorly written code can contain security vulnerabilities and can also cause performance problems. Moreover, the quality of the code has an impact on its maintenance and scalability because code that is too complex or poorly written is difficult to maintain and, therefore, will cost more for the company to fix.

In this chapter, we will focus on static code analysis with a well-known tool called **SonarQube**. We will provide a brief overview of it and go over how to install it. Then, we will use SonarLint to analyze the code locally. Finally, we will integrate SonarQube into a CI/CD pipeline on Azure Pipelines.

In this chapter, we will cover the following topics:

- Exploring SonarQube
- Installing SonarQube
- Real-time analysis with SonarLint
- Executing SonarQube in continuous integration

Technical requirements

To use SonarQube and SonarLint, we have to install **Java Runtime Environment** (**JRE**), which can be found at `https://www.oracle.com/technetwork/java/javase/downloads/jre8-downloads-2133155.html` (an Oracle account is required), on the server where we have SonarQube and on the local development environment where we have SonarLint.

To integrate SonarQube into an Azure DevOps pipeline, we must install the following extension on our Azure DevOps organization: `https://marketplace.visualstudio.com/items?itemName=SonarSource.sonarqube=SonarSource.sonararqube`.

The complete code source for this chapter is available at `https://github.com/PacktPublishing/Learning_DevOps/tree/master/CHAP10`.

Check out the following video to see the Code in Action:
`http://bit.ly/361msA3`

Exploring SonarQube

SonarQube is an open source tool from SonarSource (`https://www.sonarsource.com/`) that's written in Java. It allows us to perform static code analysis to verify the quality and security of an application's code.

SonarQube is designed for developer teams and provides them with a dashboard and reports that are customizable so that they can present the quality of the code in their applications.

It allows for the analysis of static code in a multitude of languages (over 25), such as PHP, Java, .NET, JavaScript, Python, and so on. The complete list can be found at `https://www.sonarqube.org/features/multi-languages/`. In addition, apart from code analysis with security issues, code smell, and code duplication, SonarQube also provides code coverage for unit tests.

Finally, SonarQube integrates very well into CI/CD pipelines so that it can automate code analysis during developer code commits. This reduces the risk of deploying an application that has security vulnerabilities or code complexity that is too high.

Now that we've provided an overview of SonarQube, we will look at its architecture and components. Finally, we will look at the different ways of installing it.

Installing SonarQube

SonarQube is an on-premises solution. In other words, it must be installed on servers or VMs. In addition, SonarQube consists of several components that will analyze the source code of applications, retrieve and store the data from this analysis, and provide reports on the quality and security of the code.

Before we install SonarQube, it is essential that we look at its architecture and components.

Overview of the SonarQube architecture

SonarQube is a client-server tool, which means that its architecture is composed of artifacts on the server side and also on the client side.

A simplified SonarQube architecture is shown in the following diagram:

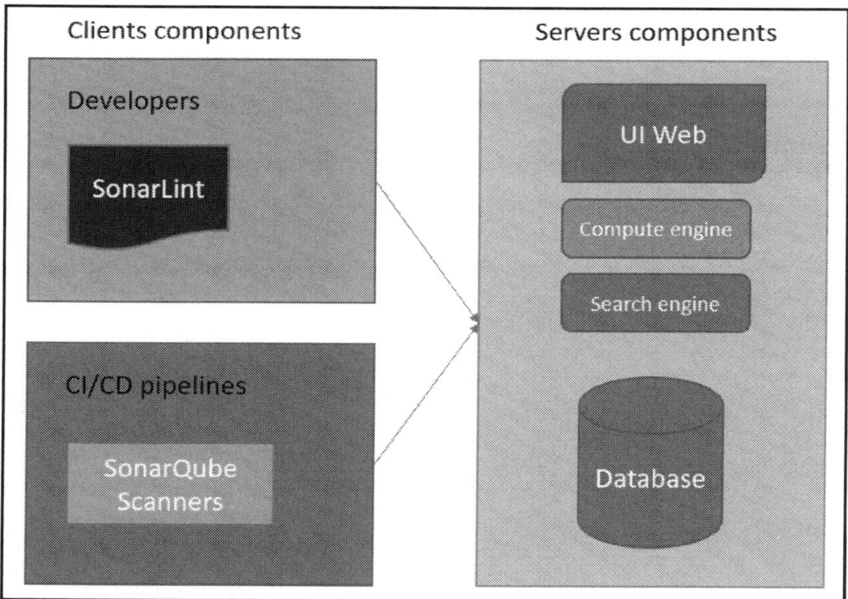

Let's look at the components that are shown in this preceding diagram. The components that make up SonarQube on the server side are as follows:

- A SQL Server, MySQL, Oracle, or PostgreSQL database that contains all the analysis data.
- A web application that displays dashboards.
- The compute engine, which is in charge of retrieving the analysis and processes. It puts them in the database.
- A search engine built with Elasticsearch.

The client-side components are as follows:

- The scanner, which scans the source code of the applications and sends the data to the compute engine.
- The scanner is usually installed on the build agents that are used to execute CI/CD pipelines.
- SonarLint is a tool that's installed on developers' workstations for real-time analysis. We will look at it in detail later in this chapter.

For more details on this architecture, we can consult the SonarQube architecture and integration documentation, which can be found at `https://docs.sonarqube.org/latest/architecture/architecture-integration/`.

Now that we have looked at its architecture and components, we will learn how to install it.

Installing SonarQube

SonarQube can be installed in different ways: either manually or by installing a Docker container from the Sonar image. Alternatively, if we have an Azure subscription, we can use a SonarQube VM from the Marketplace. Let's take a closer look at each of these options.

Manual installation of SonarQube

If we want to install the SonarQube server manually, we must take the prerequisites into account, which are that Java must be already installed on the server and that we need to check the hardware configuration shown at `https://docs.sonarqube.org/latest/requirements/requirements/`.

Then, we must manually install the server components in order, like so:

1. Install the database. This can be either MSSQL, Oracle, PostgreSQL, or MySQL.
2. Then, for the web application, download the Community Edition of SonarQube from `https://www.sonarqube.org/downloads/` and unzip the downloaded ZIP file.
3. In the `$SONARQUBE-HOME/conf/sonar.properties` file, configure the access to the database we installed in *step 1* and the storage path of Elasticsearch, as detailed in the following documentation: `https://docs.sonarqube.org/latest/setup/install-server/`.
4. Start the web server.

To find out about all the details regarding this installation according to the chosen database and our OS, we can consult the following documentation: `https://docs.sonarqube.org/latest/setup/install-server/`.

Installation via Docker

If we want to install SonarQube Community for tests or demonstration purposes, we can install it via the official Docker image that is available from Docker Hub at `https://hub.docker.com/_/sonarqube/`.

Be careful as this image uses a small integrated database that is not made for production.

Installation in Azure

If we have an Azure subscription, we can quickly access the entire SonarQube server using the SonarQube VM from the Azure Marketplace. Follow these steps to create a SonarQube VM in Azure:

1. In the Azure Marketplace, search and select the **SonarQube** image. The following screenshot shows the SonarQube page of the Marketplace:

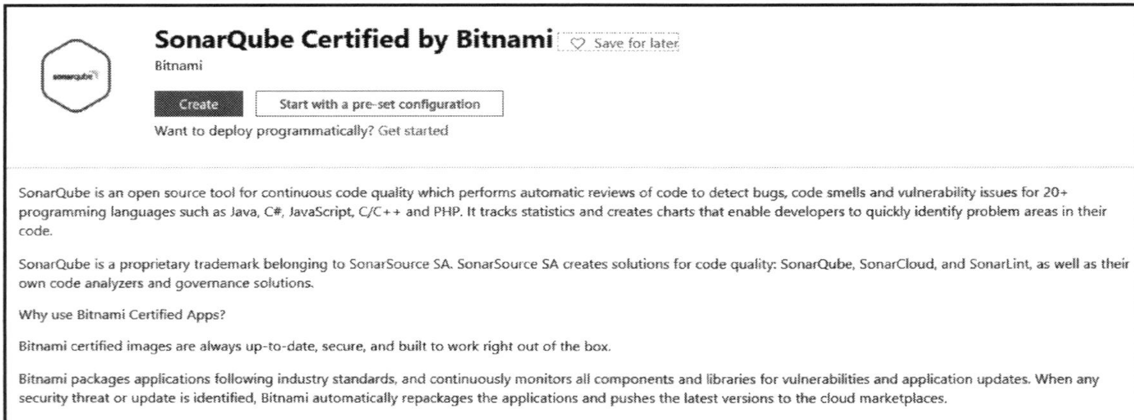

   Click on the **Create** button to get started.

2. In the VM form, on the **Basics** tab, select the **Resource group** and provide the **Virtual machine name** information, as shown in the following screenshot:

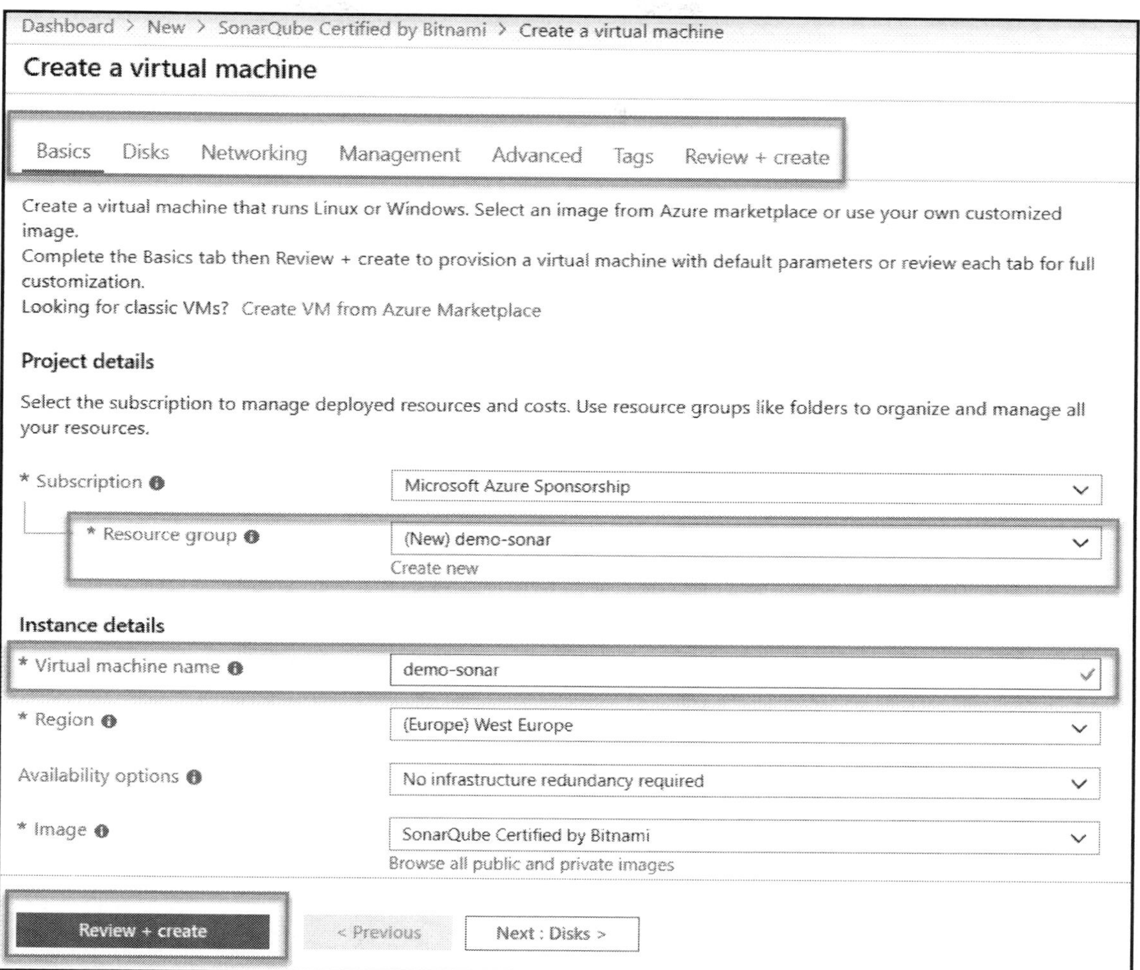

We can also change some optional VM options in the **Disks** and **Networking** tabs. Then, we validate these changes by clicking on the **Review + create** button.

3. At the end of the resource creation, in the Azure portal, we can view the status of the deployment, which, in this case, is successful:

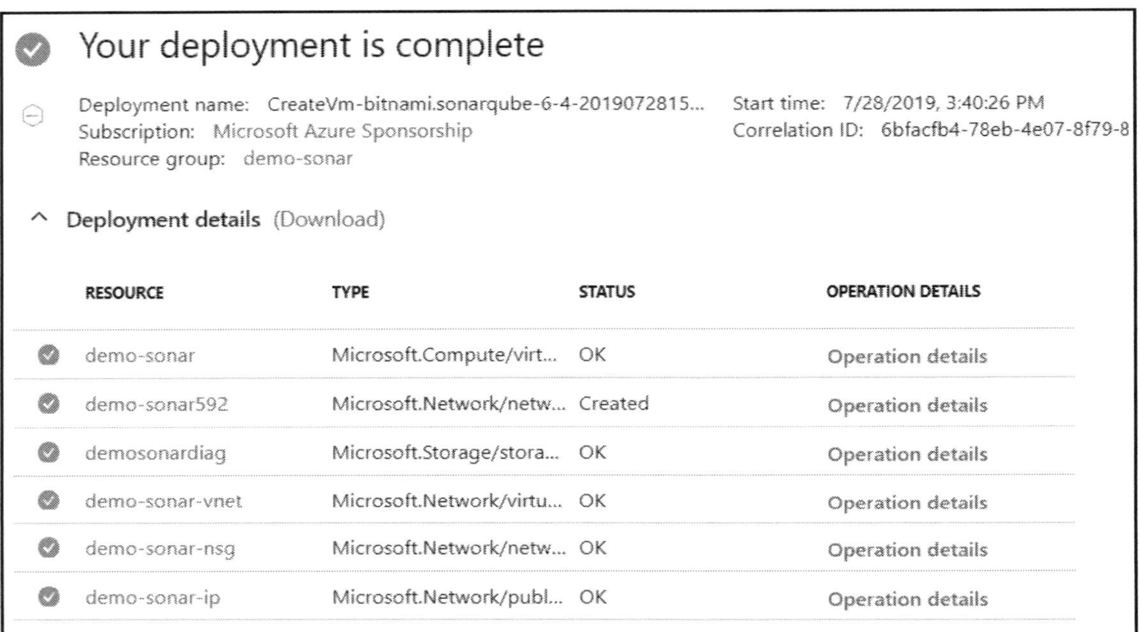

4. To access the installed SonarQube server, view the details of the VM and get the **Public IP address**, as shown in the following screenshot:

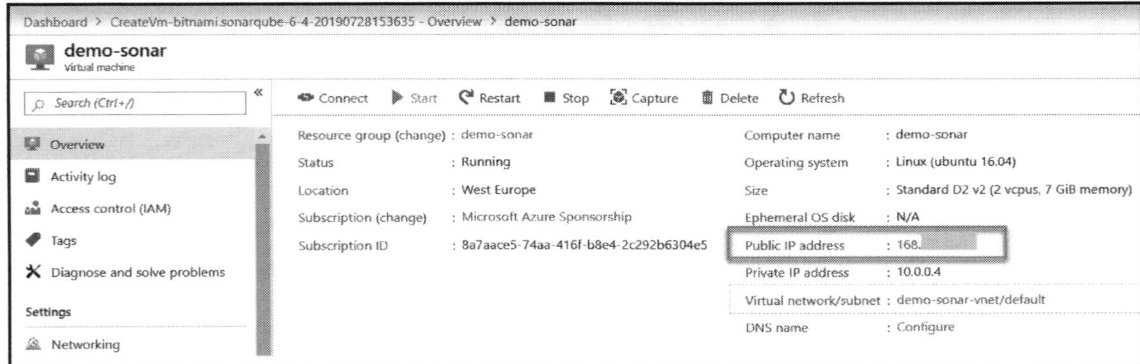

5. Open a web browser with this IP address as a URL. The SonarQube authentication page will be displayed:

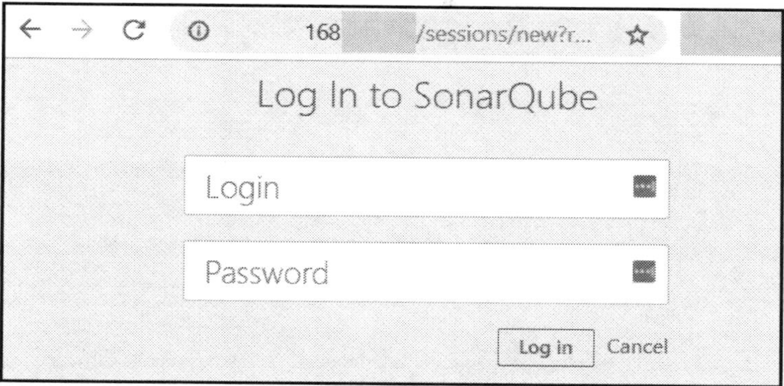

The default login is `admin` and is accessible via the VM boot diagnostics information, as indicated in the documentation (`https://docs.bitnami.com/azure/faq/get-started/find-credentials/`). The following screenshot shows how to perform password recovery via the **Boot diagnostics**:

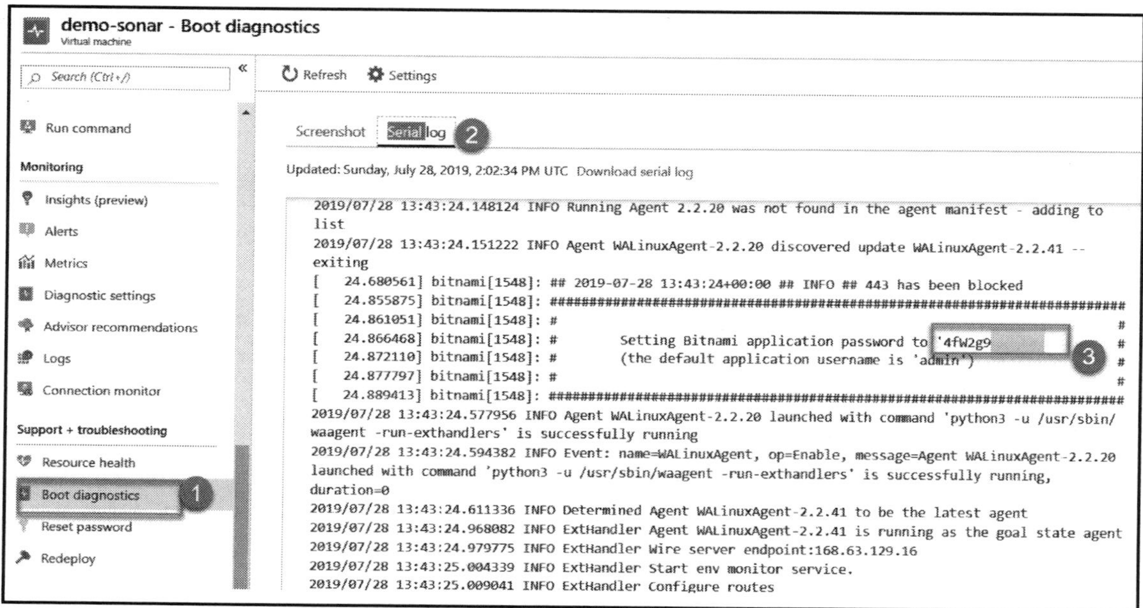

6. Once authenticated, we can access the SonarQube dashboard:

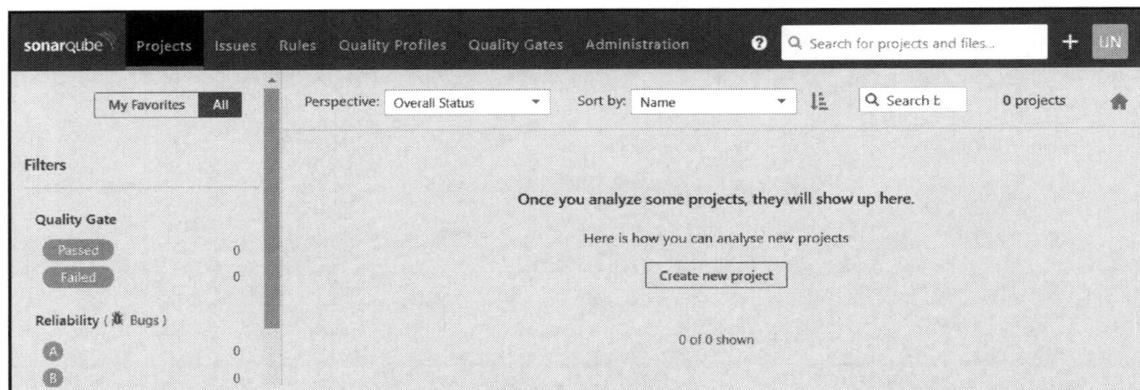

In this section, we have looked at the architecture of SonarQube, along with details about the client and server components. Then, we looked at the different installation methods that are available and the configuration of SonarQube. In the next section, we will look at how developers can perform real-time code analysis using SonarLint before they commit their code.

Real-time analysis with SonarLint

Developers who use SonarQube in a continuous integration context often face the problem of having to wait too long before they get the results of the SonarQube analysis. They must commit their code and wait for the end of the continuous integration pipeline before they get the results of the code analysis.

To address this problem and, therefore, improve the daily lives of developers, SonarSource – the editor of SonarQube – provides another tool, **SonarLint**, that allows real-time code analysis.

SonarLint is a free and open source tool (`https://www.sonarlint.org/`) that downloads differently depending on your development tool and development language. SonarLint is available for Eclipse, IntelliJ, Visual Studio, and Visual Studio Code IDEs.

In this book, we will look at an example of using SonarLint on an application written in TypeScript using the Visual Studio Code IDE. The prerequisite to using SonarLint is having the JRE installed on the local development computer. It can be downloaded from `https://www.oracle.com/technetwork/java/javase/downloads/jre8-downloads-2133155.html`.

To learn more about the concrete use of SonarLint, follow these steps:

1. In Visual Studio Code, install the SonarLint extension by going to the following page on the Azure Marketplace:

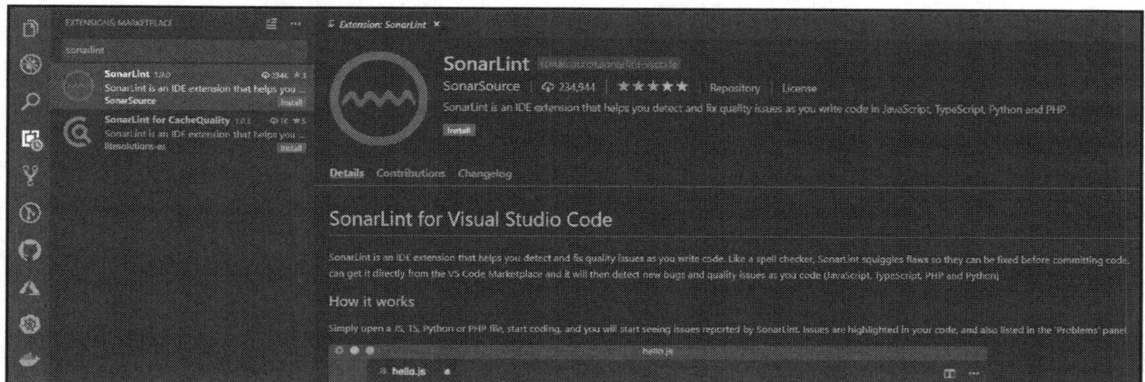

2. Then, in Visual Studio Code, in the **User** settings, configure the extension with the installation path of the JRE, as follows:

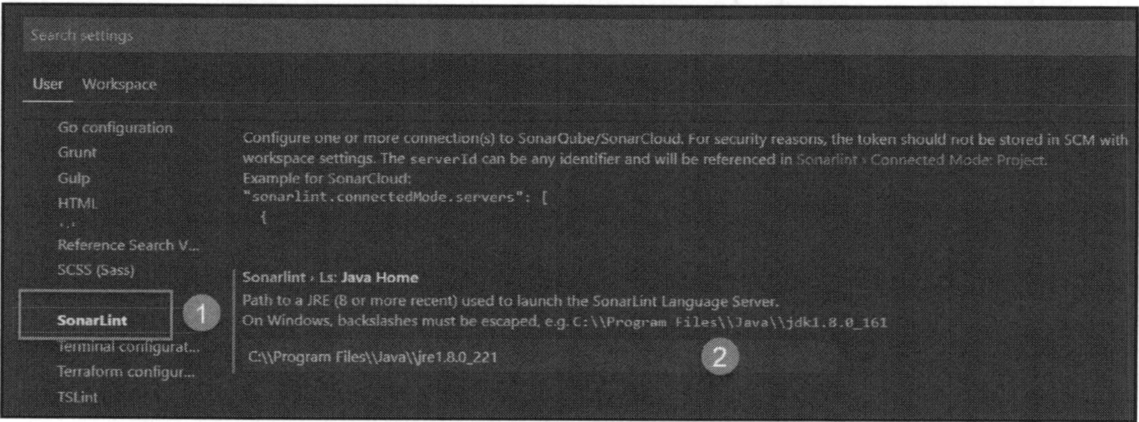

3. In our project, create a `tsApp` folder. Inside that folder, create an `app.ts` file that contains the code of our application. The source code is available here: `https://github.com/PacktPublishing/Learning_DevOps/blob/master/CHAP10/tsApp/app.ts`.

4. Note that in this SonarLint code, it states that the code is not correct, as shown in the following screenshot:

```
var x=4
if (x && x)

}
    var x: number

    Correct one of the identical sub-expressions on both sides of operator "&&"
    (typescript:S1764) sonarlint(typescript:S1764)

    Quick Fix...   Peek Problem
```

5. SonarLint allows us to learn more about this error by displaying detailed information regarding the error and how to fix it, as shown in the following screenshot:

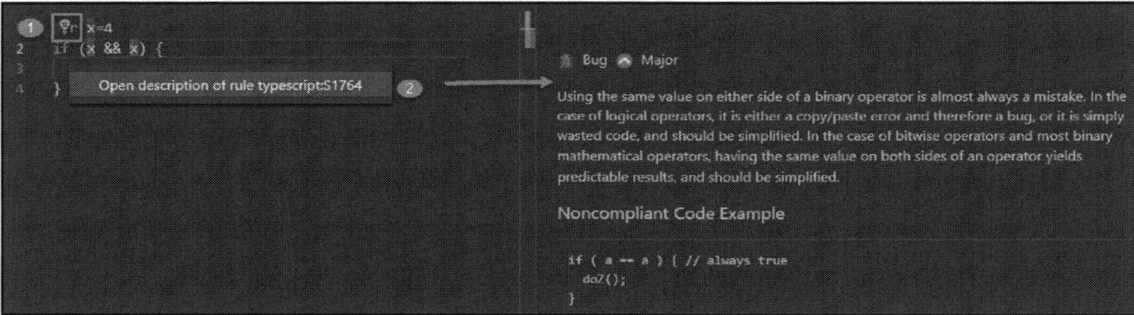

Thus, SonarLint and its integration with various IDEs allows us to detect static code errors in real time as soon as possible, that is, while the developer writes their code and before they commit it in source control version.

In this section, we learned how to install SonarLint in Visual Studio Code and how to use it to perform real-time code analysis.

In the next section, we will discuss how to integrate SonarQube analysis into a continuous integration process in Azure Pipelines.

Executing SonarQube in continuous integration

So far in this chapter, we have looked at how to install SonarQube and how developers use SonarLint on their local machines.

Now, we will look at how to perform code analysis during continuous integration to ensure that each time a code commit is made, we can check the application code that's provided by all team members.

In order to integrate SonarQube into a continuous integration process, we will need to perform the following actions:

1. Configure SonarQube by creating a new project.
2. Create and configure a continuous integration build in Azure Pipelines.

Let's start by examining the creation of a new project in SonarQube.

Configuring SonarQube

SonarQube's configuration consists of creating a new project and retrieving an identification token.

To create a new project, follow these steps:

1. Click on the **Create new project** link on the dashboard.
2. Then, in the form, enter a unique `demobook` key and a name for this project as `demo-book`.

3. To validate this, click on the **Set Up** button to create the project. These steps are shown in the following screenshot:

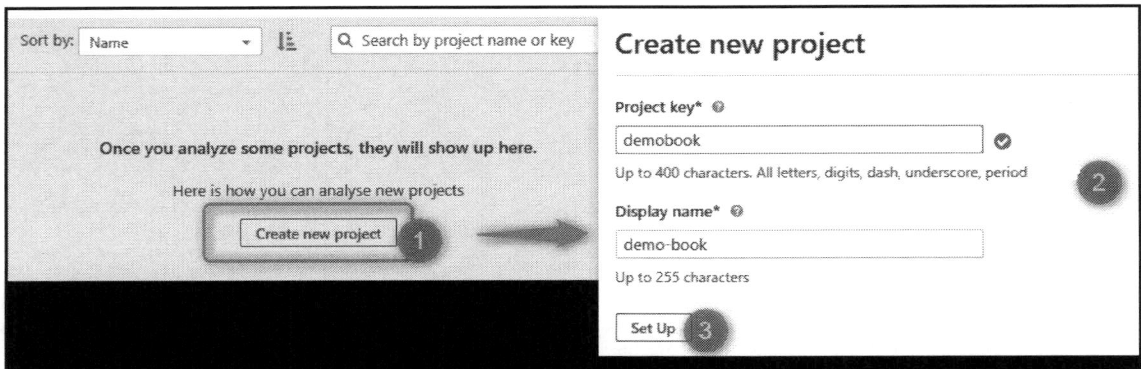

As soon as the project is created, the SonarQube assistant proposes that we create a token (unique key) that will be used for analysis.

To generate and create this token, follow these steps:

1. In the input, type a desired token name.
2. Then, validate it by clicking on the **Generate** button.
3. The unique key is then displayed on the screen. This key is our token, and we must keep it safe. The following screenshot shows the steps for generating the token:

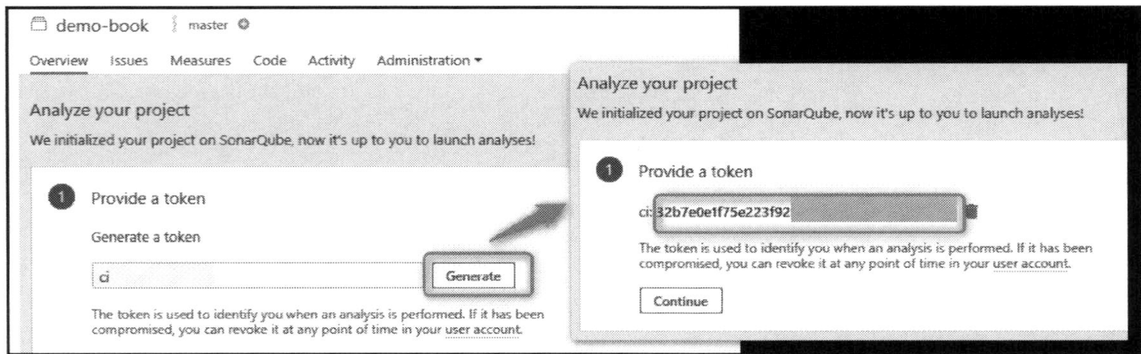

The configuration of SonarQube with our new project is complete. Now, we will configure our CI pipeline to perform the SonarQube analysis.

Creating a CI pipeline for SonarQube in Azure Pipelines

To illustrate the integration of a SonarQube analysis into a CI pipeline, we will use Azure Pipelines, which we looked at in detail in `Chapter 6`, *Continuous Integration and Continuous Delivery*.

The application that we'll use as an example in this section has been developed in Node.js, which is a simple calculator that contains some methods, including unit test methods.

Note that the purpose of this section is not to discuss the application code, but rather the pipeline. You can access the application source code at `https://github.com/PacktPublishing/Learning_DevOps/tree/master/CHAP10/AppDemo`.

To use SonarQube in Azure Pipelines, we must install the **SonarQube extension** in our Azure DevOps organization from the Visual Studio Marketplace, which is located at `https://marketplace.visualstudio.com/items?itemName=SonarSource.sonarqube`, as described in the *Technical requirements* section of this chapter.

The following screenshot shows the header and button to install the extension from the Visual Studio Marketplace:

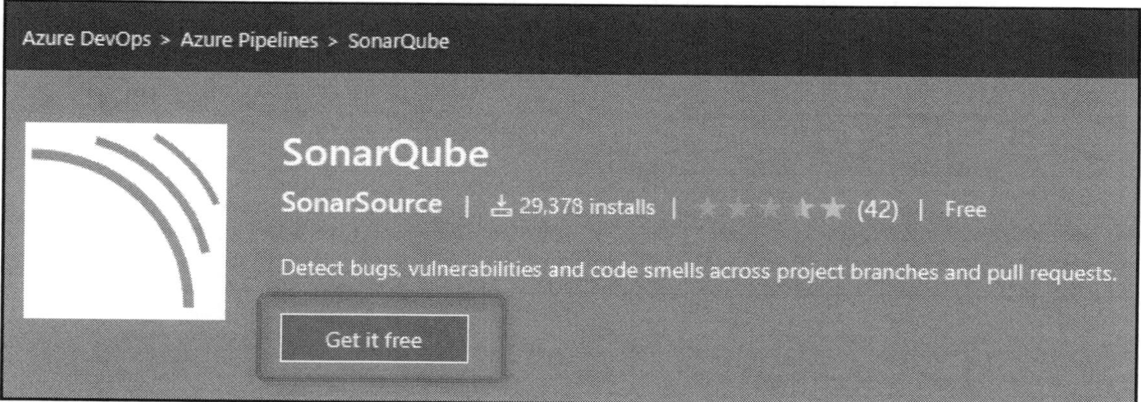

Once the extension has been installed, we can configure our continuous integration build.

In Azure Pipelines, we will create a new build definition with the following configuration:

1. In the **Get Sources** tab, select the repository and the branch that contains the source code of the application, as shown in the following screenshot:

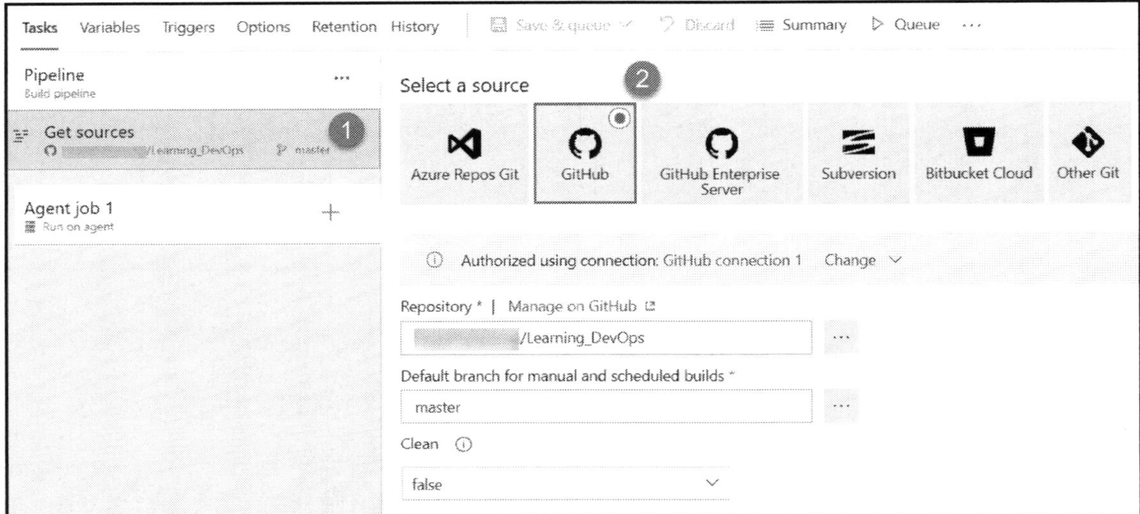

2. Then, in the **Tasks** tab, configure the schedule of the tasks, as follows:

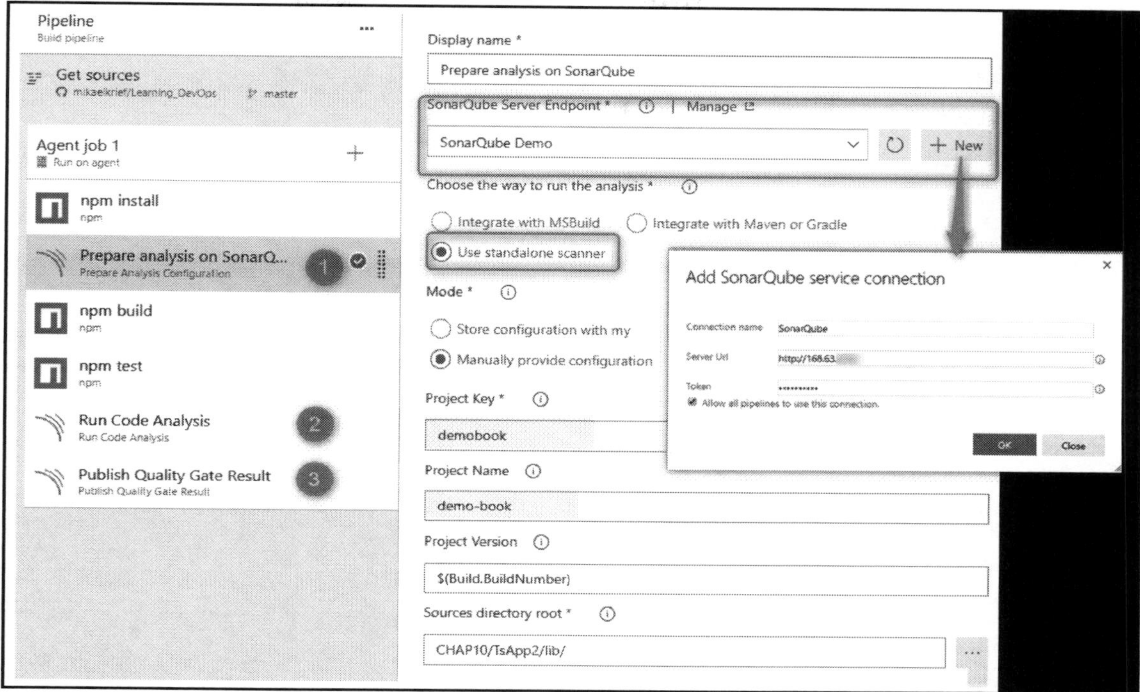

Here are the details of the configuration of these tasks:

1. The **Prepare analysis for SonarQube** task includes configuring SonarQube with the following:
    - An endpoint service, which is the connection to SonarQube with its URL and token that we generated previously in the SonarQube configuration
    - The key and name of the SonarQube project
    - The version number of the analysis

2. Then, we build and execute the unit tests of the application with `npm build` and `npm test`.
3. The **Run Code Analysis** task retrieves the test results, analyzes the TypeScript code of our application, and sends the data from the analysis to the SonarQube server.

Then, we save, start executing the CI build, and wait for it to finish.

The SonarQube dashboard has been updated with the code analysis, as shown in the following screenshot:

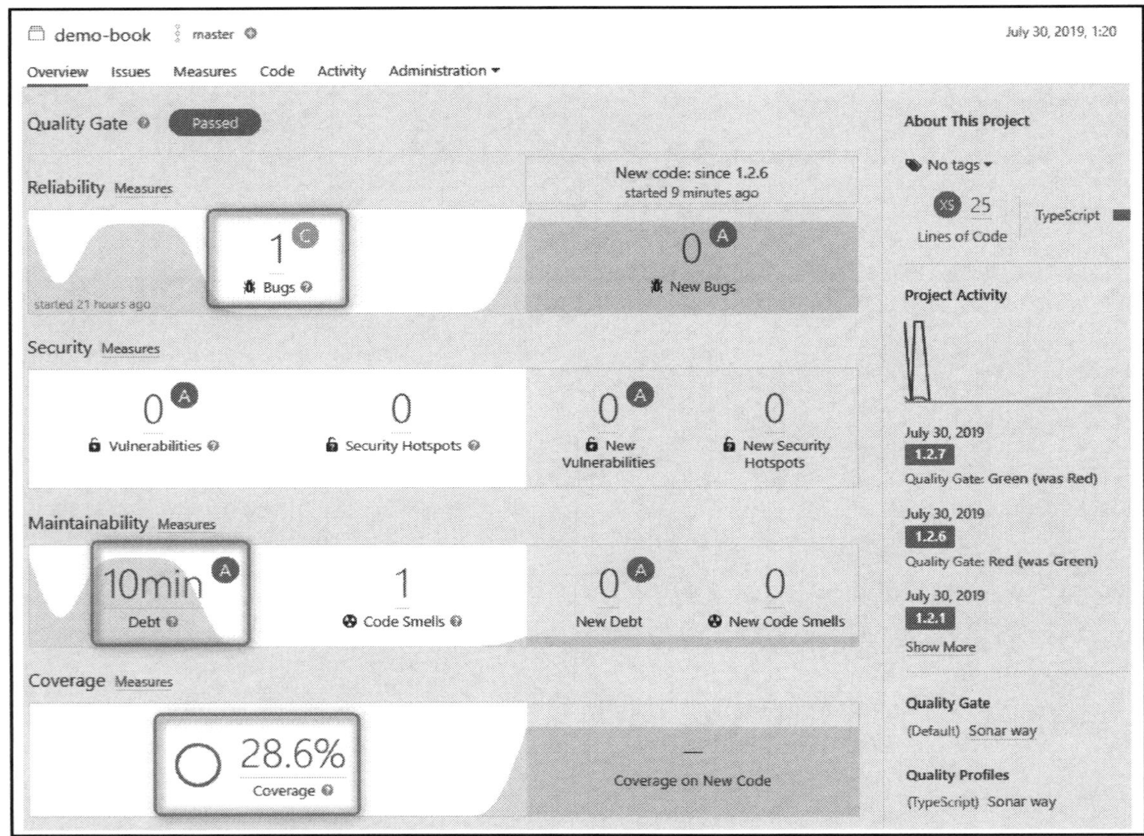

Here, we can see the measurements of bug numbers, code maintainability, and also code coverage. By clicking on each of these pieces of data, we can access the details of the element.

In this section, we have looked at how to integrate SonarQube analysis into a continuous integration pipeline that will provide a dashboard, along with the results and reports of the code analysis of each code commit.

Summary

In this chapter, we have looked at how to analyze the static code of an application using SonarQube. This analysis can detect and prevent code syntax problems, vulnerabilities in the code, and also indicate the code coverage provided by unit tests.

Then, we detailed the use of SonarLint, which allows developers to check their code in real time as they write their code.

Finally, we looked at the configuration of SonarQube and its integration into a continuous integration process to ensure continuous analysis that will be triggered at each code commit of a team member.

In the next chapter, we will look at some security practices by performing security tests with the **Zed Attack Proxy** (**ZAP**) tool, executing performance tests with Postman, and launching load tests with Azure DevOps.

Questions

1. What language is SonarQube developed in?
2. What are the requirements for installing SonarQube?
3. What is the role of SonarQube?
4. What is the name of the tool that allows real-time analysis by developers?

Further reading

If you want to find out more about SonarQube, here is the resource:

- SonarQube documentation: `https://docs.sonarqube.org/latest/`

11
Security and Performance Tests

In `Chapter 9`, *Testing APIs with Postman,* and `Chapter 10`, *Static Code Analysis with SonarQube,* we talked about test automation with API tests with Postman on the one hand, and with static code analysis with SonarQube on the other hand.

In this chapter, we will discuss how to perform security and penetration tests on a web application using the ZAP tool based on the OWASP recommendations. Then, we will add to our Postman skills, with which we will perform performance tests on APIs.

This chapter covers the following topics:

- Applying web security and penetration testing with ZAP
- Running performance tests with Postman

Technical requirements

To use ZAP, we need to install the **Java Runtime Environment** (**JRE**), which is available at `https://www.oracle.com/technetwork/java/javase/downloads/jre8-downloads-2133155.html` (an Oracle account is required).

In this chapter, we'll talk about Postman, which we already discussed in `Chapter 9`, *Testing APIs with Postman*.

Check out the following video to see the Code in Action:
`http://bit.ly/32M3BHk`

Applying web security and penetration testing with ZAP

Today, application security must be at the heart of companies' concerns. Indeed, as soon as a web application (or website) is publicly exposed on the internet, it is a candidate for an attack by malicious persons. In addition, it is important to note that application security is even more important if it is used to store sensitive data such as bank accounts or your personal information.

To address this problem, **OWASP** (short for **Open Web Application Security Project**, `https://www.owasp.org/index.php/Main_Page`) is a worldwide organization that studies application security issues. The goal of this organization is to publicly highlight the security problems and vulnerabilities that can be encountered in an application system. In addition to this valuable security information, OWASP provides recommendations, solutions, and tools for testing and protecting applications.

One of the important and useful projects and documents provided by OWASP is the top 10 application security issues. This document is available at https://www.owasp.org/images/7/72/OWASP_Top_10-2017_%28en%29.pdf.pdf. It is very detailed, with an explanation, examples, and a solution for each security issue. In this document, we find that the top security vulnerability that applications are most vulnerable to is injection vulnerability, such as SQL injection, which consists of injecting code or requests into an application in order to collect, delete, or corrupt data from the application.

We also have in this document another known security flaw, which is **Cross-Site Scripting (XSS)**, which consists of executing HTML or malicious JavaScript code on a user's web browser.

The challenge for companies is to be able to automate the security tests of their applications in order to protect them and take steps as quickly as possible when a flaw is discovered.

There are many security and penetration testing tools available. A very complete list is available at https://www.owasp.org/index.php/Appendix_A:_Testing_Tools. Among them, we learned about SonarQube in the previous chapter, which allows the analysis of code in order to detect security vulnerabilities.

Another tool in this list that is very interesting is **ZAP** (short for **Zed Attack Proxy**, https://www.owasp.org/index.php/OWASP_Zed_Attack_Proxy_Project), which was developed by the OWASP community.

Let's see how to use ZAP to perform security tests on our applications.

Using ZAP for security testing

ZAP is a graphical, free, and open source tool that allows you to scan websites and perform a multitude of security and penetration tests.

Unlike SonarQube, which also performs security analysis, ZAP runs the application and performs security tests. While SonarQube performs a security analysis in the application's source code, it does not execute it.

When running, ZAP will act as a proxy between the user and the application by scanning all the URLs of the application, then performing a series of penetration tests on these different URLs. It is currently one of the most widely used tools in application testing because, in addition to being free, it provides many very interesting features, such as the ability to configure Ajax penetration tests, and also advanced test configurations. In addition, it integrates very well with many CI/CD pipeline platforms, and finally, it is possible to control it using REST APIs, whose documentation is at `https://github.com/zaproxy/zaproxy/wiki/ApiDetails`.

What I propose to do is to make a small lab using ZAP on a public demonstration website that has security holes. As mentioned in the *Technical requirements* section of this chapter, one prerequisite of using ZAP is to have Java installed on the machine that will perform the tests. It can be a local machine or a build agent.

We can download ZAP at `https://github.com/zaproxy/zaproxy/wiki/Downloads`; download the package corresponding to your OS.

Then, install ZAP following the software installation procedures of your OS and, once the installation is complete, we can open ZAP and access its interface.

This screenshot shows the default ZAP interface:

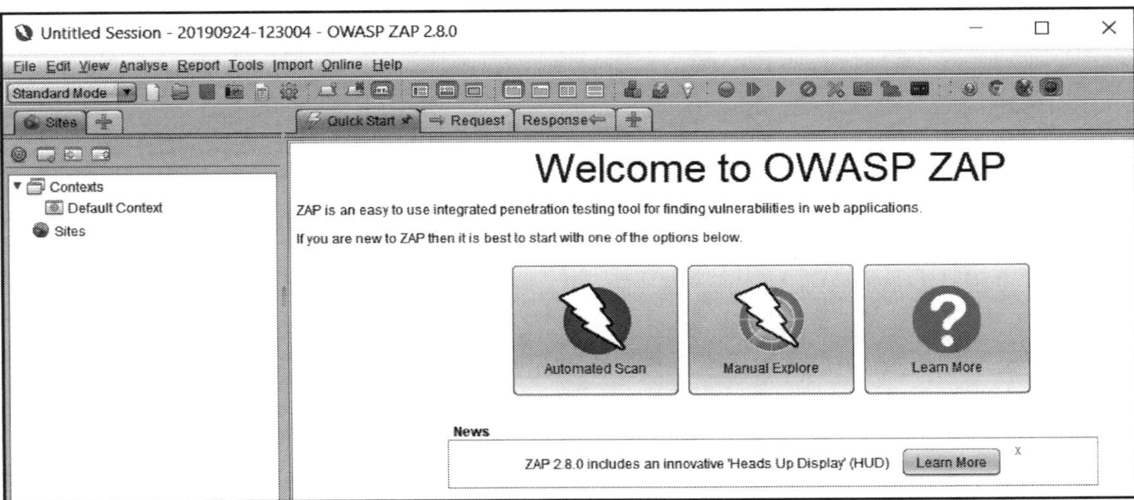

We will do our first security analysis with ZAP by following these steps:

1. In the right-hand panel, click on the **Automated Scan** button, which opens a form in which we enter the URL to be scanned.
2. In the **URL to attack** field, enter the URL of the website to analyze. In our example, we enter the URL of a demo site: `http://demo.guru99.com/Security/SEC_V1/`
3. Then, to start the analysis, we click on the **Attack** button.

   The following screenshot shows the preceding steps visually:

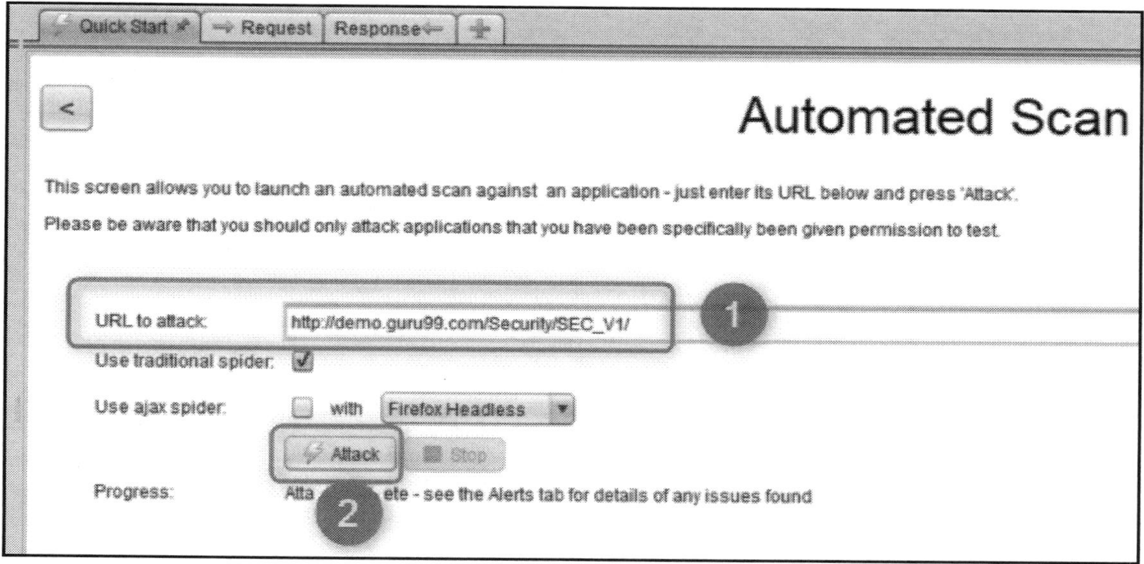

   We wait for the analysis of the security test of this website to be completed.

4. As soon as the analysis is completed, we can see the result of the security problems encountered in the panel in the bottom left.
5. Finally, clicking on one of the alerts displays the details of the problem and helps us to solve it.

The following screen shows the display of the analysis results that we have just detailed:

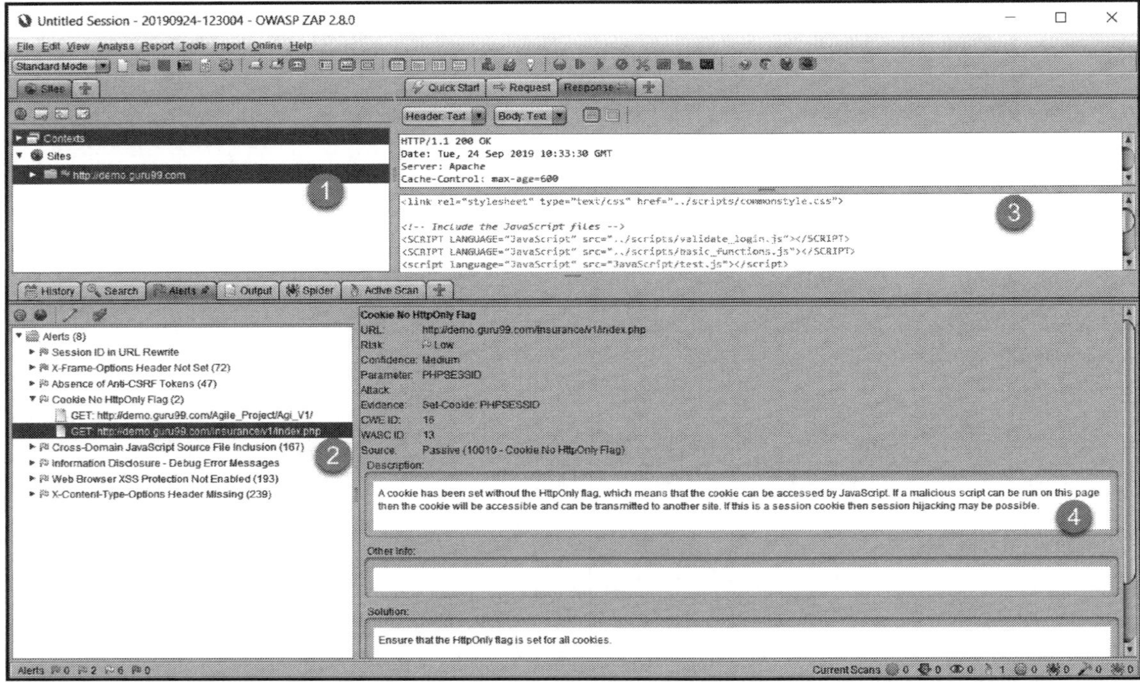

We have just seen how to use ZAP, which is a graphical tool that's used to analyze the security vulnerabilities of a website very quickly.

Let's now look at the different ways to automate the execution of ZAP.

Ways to automate the execution of ZAP

We can also automate this ZAP analysis by installing it on an agent server of our CI/CD pipeline and using the `zap-cli` tool, available at `https://github.com/Grunny/zap-cli`, which is used in the command line and calls the ZAP APIs.

The following screenshot shows using `zap-cli` on the command line to analyze our demo website:

```
C:\Users\MikaelKRIEF>zap-cli --api-key "8rqgl7                    " -p "8080" -v active-scan http://demo.guru99.com/Security/SEC_V1/
[1m[36m[INFO]            [0mRunning an active scan...
[1m[35m[DEBUG]           [0mScanning target http://demo.guru99.com/Security/SEC_V1/...
[1m[35m[DEBUG]           [0mStarted scan with ID 5...
[1m[35m[DEBUG]           [0mScan progress %: 13
[1m[35m[DEBUG]           [0mScan #5 completed

C:\Users\MikaelKRIEF>zap-cli --api-key "8rqgl71j                  " -p "8080" report -o owasp-quick-scan-report.html --output-format html
[1m[36m[INFO]            [0mReport saved to "owasp-quick-scan-report.html"
```

In the preceding execution, two commands are used:

- The first, `zap-cli active-scan`, analyzes the website passed as a command parameter.
- The second, `zap-cli report`, generates a report of the scan result in HTML format.

In these preceding commands we used `--api-key` parameter, and for retrieve our API Key value go to **Tools** | **Options** | **API** menu in our ZAP tool instance.

If we use Azure DevOps as a CI/CD pipeline platform, we can use the OWASP Zed Attack Proxy Scan task of Visual Studio Marketplace, available at `https://marketplace.visualstudio.com/items?itemName=kasunkodagoda.owasp-zap-scan`. And if we also have an Azure subscription, Azure Pipelines can also run ZAP in a Docker container, hosted in an Azure Container instance, as explained and detailed at `https://devblogs.microsoft.com/premier-developer/azure-devops-pipelines-leveraging-owasp-zap-in-the-release-pipeline/`.

If we use Jenkins as a build factory, this article explains how to integrate and use the ZAP plugin when running a job: `https://www.breachlock.com/integrating-owasp-zap-in-devsecops-pipeline/`.

We have just seen how to perform security tests on our web applications with ZAP, which is developed by the OWASP community. We looked at its basic use via its graphical interface and performed security tests on a demonstration application. Then, we saw that it is also possible to automate its execution with `zap-cli` and thus be able to integrate it into a DevOps CI/CD pipeline.

We will now see how to do performance tests with Postman.

Running performance tests with Postman

Among the tests that need to be done to guarantee the quality of our applications are functional, code analysis, and security tests, but there are also performance tests. The purpose of performance testing isn't to detect bugs in applications; it's to ensure that the application (or API) responds within an acceptable time frame to provide a good user experience.

The performance of an application is determined by metrics such as the following:

- Its response time
- The use of resources (CPU, RAM, and network)
- The error rate
- The number of requests per second

Performance tests are divided into several types of tests, such as load tests, stress tests, and scalability tests.

There are many tools available to perform performance tests. This article, `https://www.softwaretestinghelp.com/performance-testing-tools-load-testing-tools/`, lists the 15 best ones. Among the tools already seen in this book, Postman is not really a dedicated tool for performance testing, especially since it focuses mainly on APIs and not on monolithic web applications. However, Postman can provide a good indication of the performance of our API.

We have already discussed its usage in detail for API testing in `Chapter 9`, *Testing APIs with Postman*. When executing a request that tests an API in a unitary way, Postman provides the execution time of that API, as we can see in the following screenshot:

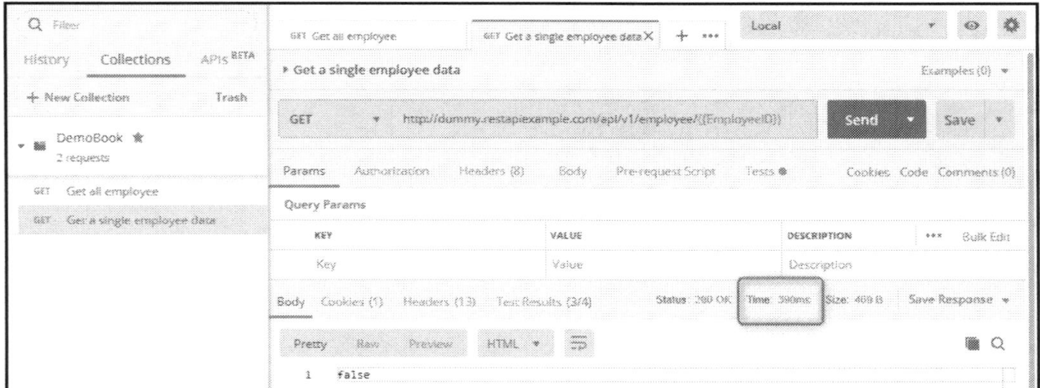

In addition, in Postman's **Collection Runner**, it is possible to execute all the requests of a collection, indicating the number of iterations; that is, the number of times the runner will execute the loop tests. This simulates several connections that call the API, and it is also where the execution time rendered by Postman becomes very interesting.

The following screenshot shows the configuration of Postman's **Collection Runner** with several iterations in the input parameters:

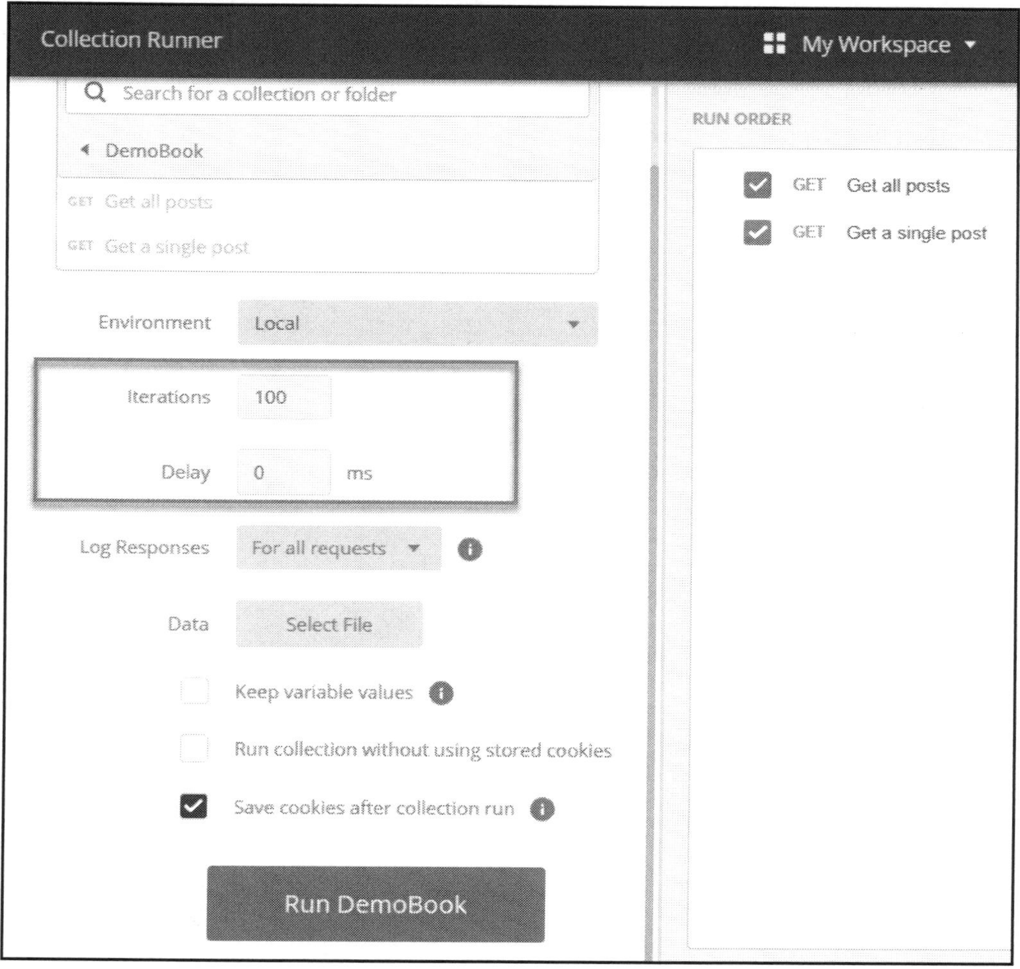

And the following screenshot shows the results of the **Collection Runner**:

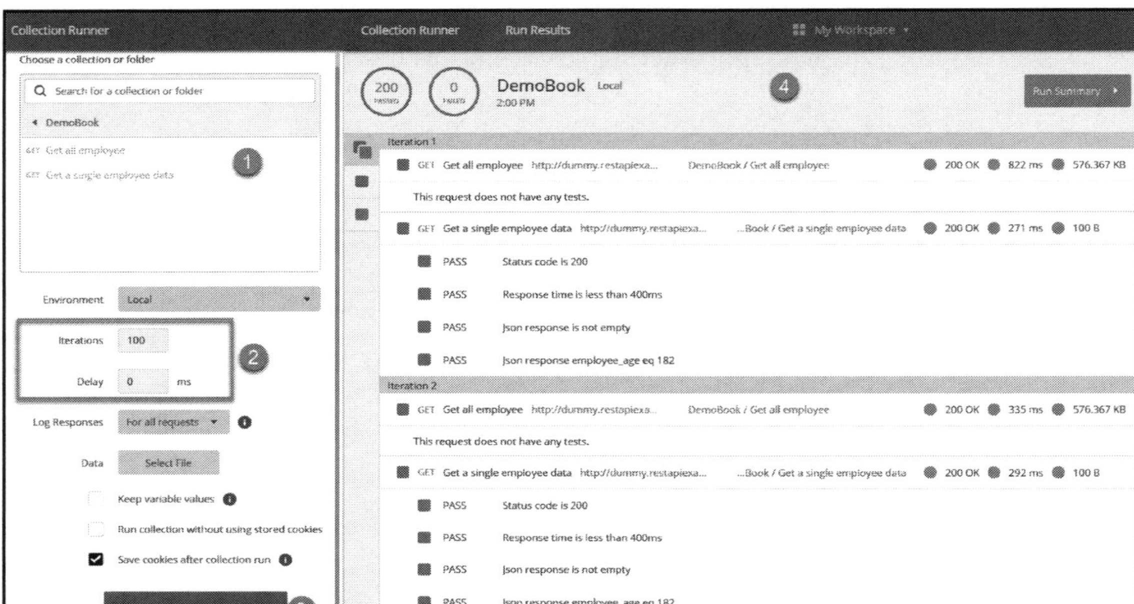

We see that the runner displays the execution time of each request, and thus it will be possible to identify overload problems on an API.

We have just seen how to perform performance tests with Postman.

Summary

In this chapter, we looked at how to use ZAP, which is a tool developed by the OWASP community to automate the execution of web application security tests. And we also saw how Postman can provide information on API performance.

In the next chapter, we will continue to talk about security and DevSecOps with the automation of infrastructure testing with Inspec, secret protection with Hashicorp's Vault, and the Secure DevOps Kit for Azure for checking the security compliance of Azure infrastructures.

Questions

1. Is ZAP a tool that analyzes the source code of an application?
2. In Postman what is the metric that allows us to have performance information?

Further reading

- Learn Penetration Testing: `https://www.packtpub.com/networking-and-servers/learn-penetration-testing`
- Pluralsight videos about OWASP and ZAP: `https://www.pluralsight.com/search?q=owasp`

5

Section 5: Taking DevOps Further

This section explains the advanced topic of DevOps processes with security integration in DevOps (DevSecOps), some techniques for blue-green deployment, and how to apply DevOps in an open source project.

We will have following chapters in this section:

- `Chapter 12`, *Security in the DevOps Process with DevSecOps*
- `Chapter 13`, *Reducing Deployment Downtime*
- `Chapter 14`, *DevOps for Open Source Projects*
- `Chapter 15`, *DevOps Best Practices*

12
Security in the DevOps Process with DevSecOps

In this book, we have discussed, in detail, DevOps culture as well as the DevOps tools that will facilitate communication and collaboration between developers and operational people (IT Ops).

However, in this union, we have noticed that a very important aspect is often missing, which is security. Indeed, CI/CD pipelines and **Infrastructure as Code** (**IaC**) allow faster deployment of infrastructure and applications, but the problem is that to deploy faster, we do not include security teams, which causes the following:

- Security teams block or slow down deployments and therefore lead to longer deployment cycles.
- Security problems are detected very late in the infrastructure and in applications.

This is why, for some time now, security has been included in the DevOps culture by becoming a **DevSecOps** culture more broadly. There is nothing outside the ambit of security. Since we are developing at rapid speeds, it makes ample sense to make it part of the process rather than outside it.

The DevSecOps culture or approach is, therefore, the union of developers and operations with the integration of security as early as possible in the implementation and design of projects. The DevSecOps approach is also the automation of compliance and security verification processes in CI/CD pipelines, to guarantee constant security and not slow down application deployment cycles.

Today, DevOps culture must absolutely integrate security teams but also all security processes, whether on tools, infrastructure, or applications. This is to provide better quality but also more secure applications.

In this chapter, we'll focus on the DevSecOps approach; we'll see how to test the compliance of an Azure infrastructure using InSpec. Then, we'll use the Secure DevOps Kit for Azure tool to test the security of Azure infrastructure. Finally, we'll learn how to protect all infrastructure and application secrets with Vault from HashiCorp.

This chapter will cover the following:

- Testing Azure infrastructure compliance with InSpec
- Using the Secure DevOps Kit for Azure
- Keeping your sensitive data safe with HashiCorp Vault

Technical requirements

In this chapter, we'll see the use of InSpec, which requires **Ruby** version 2.4 or later to be installed on the local machine. To install Ruby according to our OS, read this documentation: `https://www.ruby-lang.org/en/documentation/installation/`.

In the section on Vault, we'll discuss the integration between Vault and Terraform, without looking into the details of Terraform, so I suggest you first read `Chapter 2`, *Provisioning Cloud Infrastructure with Terraform*.

We'll also see the use of the Secure DevOps Kit for Azure, which requires to be on a Windows OS and to have **PowerShell** version 5 installed.

To use Security Azure DevOps Kit in a CI/CD pipeline, we'll use the **Azure DevOps** platform, which is free to use, and you can register here: `https://azure.microsoft.com/en-us/services/devops/`.

The complete source code for this chapter is available here: `https://github.com/PacktPublishing/Learning_DevOps/tree/master/CHAP12`.

Check out the following video to see the Code in Action:
`http://bit.ly/35WwFhf`

Testing Azure infrastructure compliance with Chef InSpec

One of the important practices of DevOps culture is IaC, detailed in `Chapter 1`, *The DevOps Culture and Practices*, which consists of coding the configuration of an infrastructure and then being automatically deployed via CI/CD pipelines. IaC allows cloud infrastructure to be deployed and provisioned very quickly, but the question that often arises is: *Does the automatically-provisioned infrastructure meet functional compliance and security requirements?*

To answer this question, we'll have to write and automate infrastructure tests that will verify the following:

- The infrastructure deployed corresponds well to the application and enterprise architecture specifications.
- The company's security policies are properly applied to the infrastructure.

These tests can be written in any scripting language that can interact with our cloud provider and if we have an Azure subscription, we can use, for example, the Azure CLI or PowerShell Azure command to code the tests of our Azure resources. Also, if we use PowerShell, we can use **Pester** (`https://github.com/pester/Pester/wiki/Pester`), which is a library that allows us to perform PowerShell tests and, combined with Azure PowerShell, allows us to perform infrastructure compliance tests.

To get an example of how to use Pester to test an Azure infrastructure, I suggest you read this article: `https://dzone.com/articles/azure-security-audits-with-pester`, and this blog post: `https://dev.to/omiossec/unit-testing-in-powershell-introduction-to-pester-1de7`.

The problem with these scripting tools is that they require a lot of code to be written. Also, these tools are dedicated to a specified cloud provider and they require learning a new scripting language.

One of the IaC tools is **InSpec** (`https://www.inspec.io/`), which performs infrastructure compliance tests.

In this section, we'll see in detail the use of InSpec to test the compliance of Azure infrastructure and, to start with its implementation, I will provide you with an overview of InSpec.

Overview of InSpec

InSpec is an open source tool written in Ruby that runs on the command line, and is produced by one of the leading DevOps tools, **Chef**, whose website is `https://www.chef.io/`. It allows users writing declarative-style code, to test the compliance of a system or infrastructure.

To use InSpec, it's not necessary to learn a new scripting language; we should already have enough knowledge to write the desired state of the infrastructure resources or the system we want to test.

With InSpec, we can test the compliance of remote machines and data and, since the latest version, it is also possible to test a cloud infrastructure such as Azure, AWS, and GCP.

After this little overview of InSpec, let's look at how to download and install it.

Installing InSpec

We have seen in the technical prerequisites section that InSpec needs to have **Ruby** (>2.4) installed on our machine.

InSpec can be installed either manually or via a script:

- **Manually**: This can be done by downloading the package corresponding to our OS from `https://downloads.chef.io/inspec`.
- **With a script**: By using gem (`https://rubygems.org/?locale=en`), which is the Ruby package manager, we can install InSpec by executing the following command in a Terminal:

```
gem install inspec-bin
```

The following screenshot shows the installation of InSpec via the `gem` command:

```
root@mkrief:/home/mikaelkrief# gem install inspec-bin
Fetching inspec-4.10.4.gem
Fetching inspec-bin-4.10.4.gem
Successfully installed inspec-4.10.4
Successfully installed inspec-bin-4.10.4
Parsing documentation for inspec-4.10.4
Installing ri documentation for inspec-4.10.4
Parsing documentation for inspec-bin-4.10.4
Installing ri documentation for inspec-bin-4.10.4
Done installing documentation for inspec, inspec-bin after 12 seconds
2 gems installed
```

For more information on the installation of InSpec by script, refer to this documentation: `https://github.com/inspec/inspec`.

To verify that InSpec has been correctly installed and is working, we run the `inspec --version` command to display its version, and the `inspec` command to display the list of available commands.

The following screenshot shows the execution of these commands:

```
root@mkrief:/home/mikaelkrief# inspec --version
4.10.4
root@mkrief:/home/mikaelkrief# inspec
Commands:
  inspec archive PATH                # archive a profile to tar.gz (default) or zip
  inspec artifact SUBCOMMAND         # Manage Chef InSpec Artifacts
  inspec check PATH                  # verify all tests at the specified PATH
  inspec compliance SUBCOMMAND       # Chef Compliance commands
  inspec detect                      # detect the target OS
  inspec env                         # Output shell-appropriate completion configuration
  inspec exec LOCATIONS              # run all test files at the specified LOCATIONS.
  inspec habitat SUBCOMMAND          # Manage Habitat with Chef InSpec
  inspec help [COMMAND]              # Describe available commands or one specific command
  inspec init SUBCOMMAND             # Generate InSpec code
  inspec json PATH                   # read all tests in PATH and generate a JSON summary
  inspec nothing                     # does nothing
  inspec plugin SUBCOMMAND           # Manage Chef InSpec and Train plugins
  inspec shell                       # open an interactive debugging shell
  inspec supermarket SUBCOMMAND ...  # Supermarket commands
  inspec vendor PATH                 # Download all dependencies and generate a lockfile in a `vendor` directory
  inspec version                     # prints the version of this tool

Options:
  l, [--log-level=LOG_LEVEL]                         # Set the log level: info (default), debug, warn, error
     [--log-location=LOG_LOCATION]                   # Location to send diagnostic log messages to. (default: $stdout or Inspec::Log.error)
     [--diagnose], [--no-diagnose]                   # Show diagnostics (versions, configurations)
     [--color], [--no-color]                         # Use colors in output.
     [--interactive], [--no-interactive]             # Allow or disable user interaction
     [--disable-core-plugins]                        # Disable loading all plugins that are shipped in the lib/plugins directory of InSpec. Useful in development.
     [--disable-user-plugins]                        # Disable loading all plugins that the user installed.
     [--enable-telemetry], [--no-enable-telemetry]   # Allow or disable telemetry
     [--chef-license=CHEF_LICENSE]                   # Accept the license for this product and any contained products: accept, accept-no-persist, accept-silent
```

Also, like many of the tools already detailed in this book, InSpec has been integrated into the **Azure Cloud Shell** tool suite, as shown in the following screenshot:

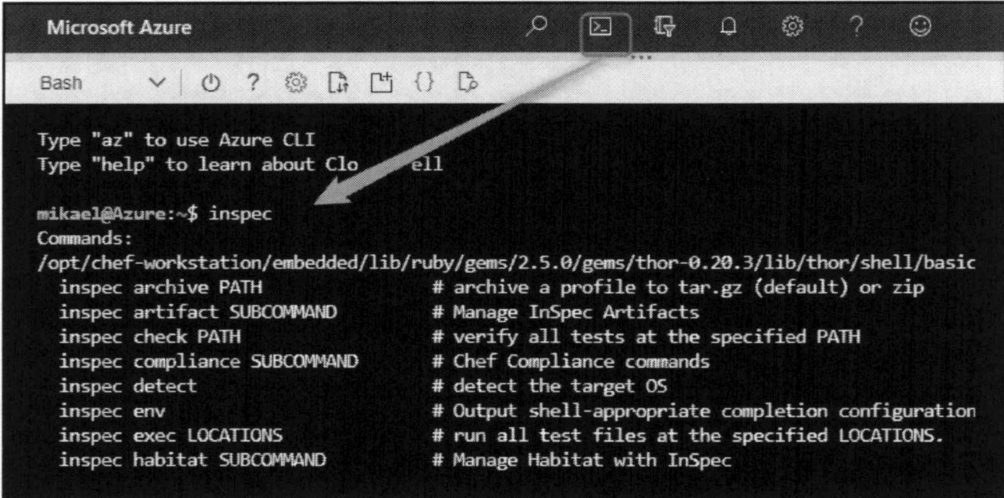

We have just seen the different ways to install InSpec, and we'll now see the configuration of Azure for InSpec.

Configuring Azure for InSpec

Before writing test cases to test the compliance of our Azure infrastructure, we need to create an Azure service principal that has read permission on the Azure resources that will be tested.

To create this Azure service principal, we'll use the same procedure that we already detailed in the *Configuring Terraform for Azure* section of `Chapter 2`, *Provisioning Cloud Infrastructure with Terraform*.

Using the **Azure CLI** tool, we execute the following `az cli` command:

```
az ad sp create-for-rbac -name="<SP name> -role="Reader" -
scopes="/subscriptions/<subscription Id>"
```

This preceding command requires the following parameters:

- `--name` is the name of the Azure service principal to be created.
- `--scopes` is the ID of the subscription (or other scopes) in which the Azure resources will be present.
- `--role` is the role name that the **service principal** (**SP**) will have on the specified resource scope.

The execution of this preceding command returns the three pieces of authentication information of the created service principal:

- The client ID
- The client secret
- The tenant ID

For more details on the service principal, see the documentation: `https://docs.microsoft.com/en-us/cli/azure/create-an-azure-service-principal-azure-cli?view=azure-cli-latest`.

We'll see how to use this authentication information when running InSpec. But, before it is executed, we should have enough knowledge of writing InSpec tests. Let's see how to write InSpec tests.

Writing InSpec tests

After installing InSpec and configuring authentication for Azure, we can start using InSpec. To show an example of InSpec tests, we'll write tests that will check that the Azure infrastructure we provisioned with Terraform in `Chapter 2`, *Provisioning Cloud Infrastructure with Terraform*, is compliant with the specifications of our Azure infrastructure, which must be composed of the following:

- One resource group named `bookRg`
- One VNet with one subnet inside it named `book-subnet`
- One virtual machine named `bookvm`

As a first step in writing our InSpec tests, we'll create an InSpec profile file.

Creating an InSpec profile file

To create an InSpec profile file, we'll generate the test directory structure, then modify the InSpec profile file that was generated.

To perform this manipulation, we perform the following steps:

1. Create a test folder structure and an InSpec profile that defines some metadata and the InSpec configuration. To create a structure and an InSpec profile file, on our machine, go to the directory of our choice and execute the following command:

   ```
   inspec init profile azuretests
   ```

   This command initializes a new profile by creating a new folder, `azuretests`, that contains all of the artifacts needed for InSpec tests, with the following:

   - Controls (tests)
   - Libraries
   - A profile file, `inspec.yml`, with some default metadata

2. Then, we modify this `inspec.yml` profile file with some personal metadata and add the URL link from the InSpec Azure Resource pack to the sample code, as follows:

   ```
   name: azuretests
   title: InSpec Profile
   maintainer: Your name
   copyright: Your name
   copyright_email: you@example.com
   license: All Rights Reserved
   summary: An InSpec Compliance Profile
   version: 0.1.0
   inspec_version: '>= 4.6.9'

   depends:
    - name: inspec-azure
      url: https://github.com/inspec/inspec-azure.git
   ```

In this code, we have entered some personal information, such as your name, and information about the license type of this code. Then, finally, in the last part of this file, we indicated a dependency with the URL of the InSpec Azure Resources library pack.

In fact, since version 2.2.7 of InSpec, we can use a set of InSpec libraries that use the Azure API and hence allow us to access all Azure resources. From there, the InSpec team creates an Azure Resource Pack that contains a lot of libraries to test a wide range of Azure resources such as Azure Users, Azure Monitor, Azure networking (VNet and Subnet), Azure SQL Server, Azure VM, and many other Azure resources.

For a complete list of available Azure resources, refer to the InSpec documentation for the InSpec Azure Resource Pack: `https://www.inspec.io/docs/reference/resources/#azure-resources`.

After generating our directories that contain the tests and profile file updates, we'll write our infrastructure compliance tests.

Writing compliance InSpec tests

To answer our example specification, we'll write tests that will verify our provisioned infrastructure contains a resource group, virtual machine, and subnet.

All of the tests we'll be drafting are located in the `controls` folder and are written in a Ruby file (`.rb`) with very simple code that's human-readable.

To begin, we'll write the test that checks the existence of the resource group, and for this, we'll delete the `example.rb` example file in the `controls` folder, which is an example of tests provided in the test templates, and create a new file named `resourcegroup.rb` that contains the following code content:

```
control 'rg' do
 describe azurerm_resource_groups do  #call the azurerm_resource_groups of
Azure Resource Pack
     its('names') { should include 'bookRg' } #test assert
 end
end
```

In this declarative code, the desired state of the resources is described and the following actions have been taken:

1. Create a control (or test) called `test_rg`.
2. In this control, we'll create a method of the `describe` type, in which we use the `azurerm_resource_groups` library of the Azure resource packs, which allows testing the existence of a resource group.
3. In this `describe` method, we write a test assert that checks whether there is a resource group, `bookRg`, in the Azure subscription.

Then, we'll continue to write our tests for the VM and subnet. To do this, we manually create a `subnet.rb` file in the `controls` directory that contains the following code:

```
control "subnet" do
 describe azurerm_subnet(resource_group: 'bookRg', vnet: 'book-vnet', name:
'book-subnet') do
     it { should exist }
     its('address_prefix') { should eq '10.0.10.0/24' }
 end
end
```

In this code, we use the `azurerm_subnet` library, which allows testing the existence of a subnet in a VNet. In this test, we check that the `book-subnet` subnet exists in the `book-vnet` VNet, and it has the IP range `10.0.10.0/24`.

Finally, here we finish with writing the tests that allow us to check our VM with the following code in the `vm.rb` file:

```
control 'vm' do
 describe azurerm_virtual_machine(resource_group: 'bookRg', name: 'bookvm')
 do
 it { should exist }
 its('properties.location') { should eq 'westeurope'}
 its('properties.hardwareProfile.vmSize') { should eq 'Standard_DS1_v2'}
 its('properties.storageProfile.osDisk.osType') { should eq 'Linux' }
}
 end
end
```

In this code, we use the `azurerm_virtual_machine` library and test that the VM named `demovm` exists in the resource group called `bookRg`. We also check some of its properties, such as region, OS type, and the size of the VM.

We have finished writing the InSpec tests that will be used to check the compliance of our Azure infrastructure, and we'll now see the execution of InSpec with these tests that we have just written.

Executing InSpec

To execute InSpec, we'll perform the following steps:

1. We'll configure the InSpec authentication to Azure; for this, we'll create environment variables with the values of the Azure Service Principal information that we created previously in the *Configuring Azure for InSpec* section.

The four environments variables and their values are as follows:

- `AZURE_CLIENT_ID` with the client ID of the service principal
- `AZURE_CLIENT_SECRET` with the secret client of the service principal
- `AZURE_TENANT_ID` with the tenant ID
- `AZURE_SUBSCRIPTION_ID` with the ID of the subscription that contains the resources and whose service principal has reader permissions

Here is an example of how to create these variables in Linux OS:

```
export AZURE_SUBSCRIPTION_ID="<Subscription ID"
export AZURE_CLIENT_ID="<Client Id>"
export AZURE_CLIENT_SECRET="<Secret Client>"
export AZURE_TENANT_ID="<Tenant Id>"
```

2. Then, in a Terminal, we'll place ourselves in the directory containing the profile file, `inspec.yml`, then execute the following `inspec` command to check that the syntax of the tests is correct:

```
inspec check .
```

The argument to be provided to this command is the path to the directory that contains the `inspec.yml` file. Here, in this command, we use `.` (dot) in the argument to indicate that the `inspec.yml` file is in the current directory, and the following screenshot shows the result of the execution of this command:

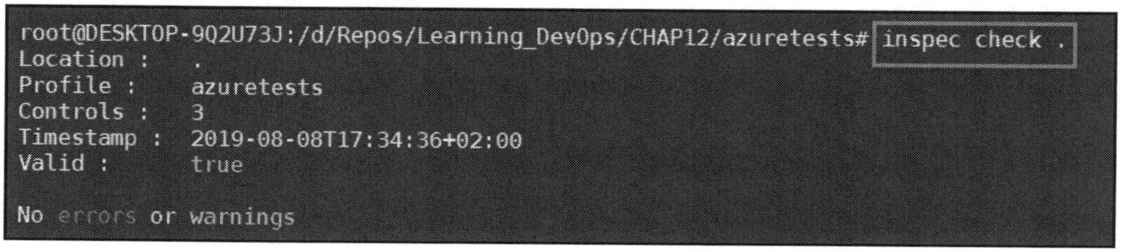

```
root@DESKTOP-9Q2U73J:/d/Repos/Learning_DevOps/CHAP12/azuretests# inspec check .
Location :    .
Profile :     azuretests
Controls :    3
Timestamp :   2019-08-08T17:34:36+02:00
Valid :       true

No errors or warnings
```

3. Finally, we execute InSpec to execute the tests with the `inspec exec` command, as follows:

```
inspec exec . -t azure://
```

This command takes as argument the path of the directory that contains the `inspec.yml` file (here, it is the dot). We also add the `-t` option, which takes the value of the tests to the target, that is, `azure`.

The following screenshot shows the result of the execution of this command:

```
root@DESKTOP-9Q2U73J:/d/Repos/Learning_DevOps/CHAP12/azuretests# inspec exec . -t azure://

Profile: InSpec Profile (azuretests)
Version: 0.1.0
Target:  azure://8a7aace5-74aa-416f-b8e4-2c292b6304e5

  ✔  rg: InspecTest RG
     ✔  Resource Groups names should include "bookRg"
  ✔  subnet: InspecTest Subnet
     ✔  'book-subnet' subnet should exist
     ✔  'book-subnet' subnet address_prefix should eq "10.0.10.0/24"
  ✔  vm: InspecTest VM
     ✔  'bookvm' Virtual Machine should exist
     ✔  'bookvm' Virtual Machine location should eq "westeurope"
     ✔  'bookvm' Virtual Machine properties.hardwareProfile.vmSize should eq "Standard_DS1_v2"
     ✔  'bookvm' Virtual Machine properties.storageProfile.osDisk.osType should eq "Linux"


Profile: Azure Resource Pack (inspec-azure)
Version: 1.3.0
Target:  azure://8a7aace5-74aa-416f-b8e4-2c292b6304e5

     No tests executed.

Profile Summary: 3 successful controls, 0 control failures, 0 controls skipped
Test Summary: 7 successful, 0 failures, 0 skipped
```

We can see from this result that all of the tests are green and therefore successful, so the compliance of the infrastructure is very successful.

We have just explained in this section how to write InSpec tests and run them to verify the compliance of our Azure infrastructure.

InSpec is a very powerful tool; it also allows you to test the configuration of a virtual machine. That's why I invite you to see the documentation, `https://www.inspec.io/docs/`.

We have just seen here the installation of InSpec, then we saw the writing of InSpec tests and finally how to use it with its command lines to test the compliance of Azure infrastructure. In the next section, we'll discuss how to analyze and verify the security of an Azure infrastructure using the Secure DevOps Kit for Azure tool.

Using the Secure DevOps Kit for Azure

We'll now talk about another very interesting tool that allows you to check the security of Azure infrastructure resources. It is a tool provided by Microsoft, called **Secure DevOps Kit for Azure** (**AzSK**) and its complete documentation is here: `https://azsk.azurewebsites.net/README.html`.

Unlike InSpec, AzSK does not verify the compliance of your Azure infrastructure with architectural requirements but rather will verify that the recommendations and good security practices are applied to your Azure subscription and resources.

AzSK also integrates seamlessly into a CI/CD pipeline and thus allows developers and operational staff to continuously ensure that their Azure resources are secure and do not open security vulnerabilities to unwanted people.

We'll see how to install AzSK; then we'll look at how it is used to verify the security of Azure provisioned resources; and finally, we'll integrate into a CI/CD pipeline in Azure Pipeline.

To begin, let's proceed with installing the Azure DevOps Security Kit.

Installing the Azure DevOps Security Kit

To install the Azure DevOps Security Kit, the following actions must be performed:

1. In a PowerShell Terminal, execute the following command:

    ```
    Install-Module AzSK -Scope CurrentUser
    ```

    Executing this command installs the AzSK module from the PowerShell public package manager called PowerShell Gallery available here: `https://www.powershellgallery.com/`.

If you have already installed the Azure PowerShell module, add the `AllowClobber` option, which allows you to reinstall the commands already present in this module.

2. Then, for importing the module, we execute the following command:

```
Import-Module -Name AzSK
```

This command imports the module into the user context of the PowerShell execution.

3. Finally, to check that the AzSK module is installed correctly, we run the following PowerShell command:

```
Get-Module -Name AzSK
```

Executing this command displays the details of the AzSK module, as shown in the following screenshot:

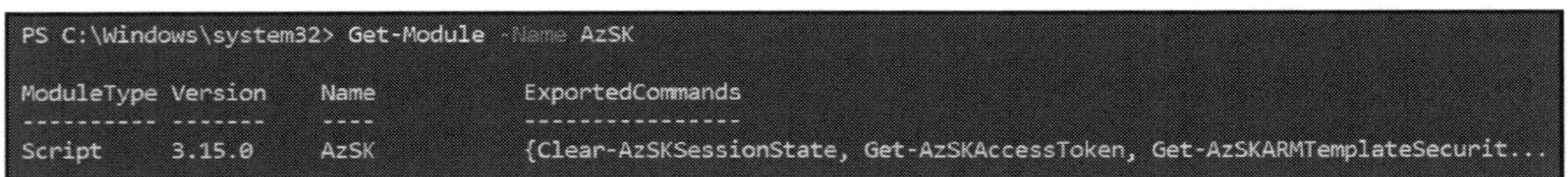

For more information on installing the Azure DevOps Security Kit, see the documentation at https://azsk.azurewebsites.net/00a-Setup/Readme.html.

We have now installed the PowerShell AzSK module, and we'll use it to check the security of the Azure resources.

Checking the Azure security using AzSK

AzSK is a PowerShell module that uses the Azure PowerShell module to call the Azure API, which allows a good interaction between these two modules.

To start using the AzSK to check Azure security, you must connect to Azure using the Azure PowerShell commands.

For this, we go to a PowerShell Terminal (or PowerShell ISE) to execute the following command:

```
Connect-AzAccount
```

This command opens a window that asks you for your Azure IDs as follows:

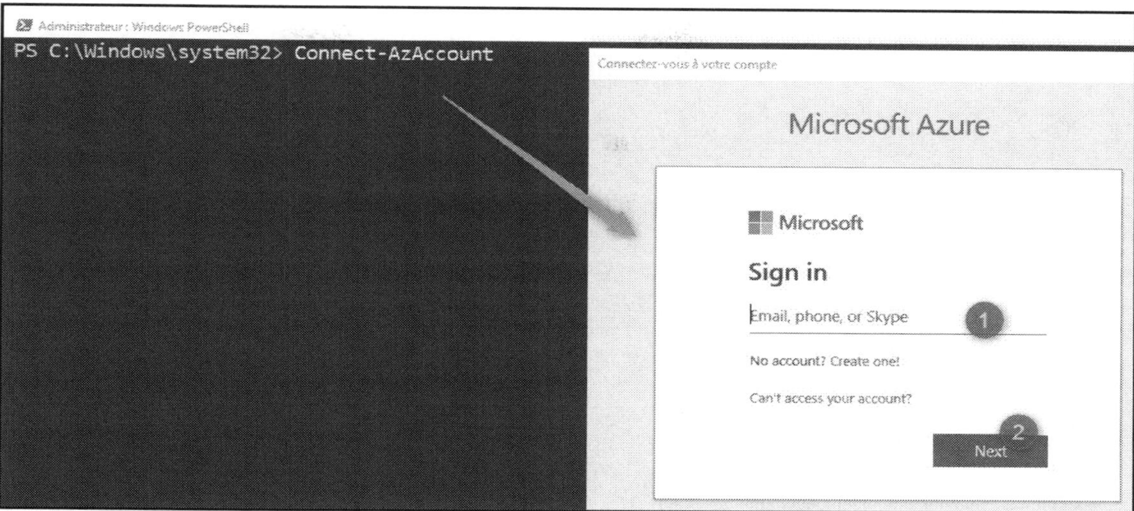

Once our Azure authentication account is validated, we can then use AzSK.

AzSK can be used in the following two security checking use cases.

The first scenario is to test the security of the subscription as a *Subscription Health Scan* mode by checking, for example, the roles, permissions, and administrator and owner accounts of this subscription. This type of check is intended more for Azure administrators and architects who are responsible for the overall security of Azure subscription.

To perform this check, we go to our PowerShell Terminal and execute the following command:

```
Get-AzSKSubscriptionSecurityStatus -SubscriptionId <Subscription ID>
```

This command takes the subscription ID as a parameter, and its execution does the following actions:

1. It checks the security rules on the subscription.
2. It displays the execution of the errors that are detected in a table.
3. Finally, it generates a file in CSV format that consists of the complete execution report.

For more information on the *Subscription Health Scan* scenario, see the documentation, `https://azsk.azurewebsites.net/01-Subscription-Security/Readme.html`.

The second scenario of using AsZK is to test the security of the resources present in an Azure resource group, which are as follows:

- The network type resources, such as a VNet
- The resources that contain data from company applications such as web apps, storages accounts, or databases

This scenario is intended more for developers who need to create and modify this type of Azure resource on a regular basis.

To check Azure resources with AzSK, we go to a PowerShell Terminal and execute the following command:

```
Get-AzSKAzureServicesSecurityStatus -SubscriptionId <Subscription ID> -
ResourceGroupNames <Resourcegroup name>
```

This command takes the ID of the subscription as a parameter as well as the name of the resource group that contains the resources to be verified.

Executing this command performs the following actions:

1. It checks each of the sensitive resources of the group's resources.
2. It displays a summary table of the errors found.
3. It generates a CSV file that constitutes the complete execution report.

For more information on resource security, which is called **Security Verification Tests**, see the complete documentation, `https://azsk.azurewebsites.net/02-Secure-Development/Readme.html`.

We have just seen how to use AzSK to verify the security of Azure resources, and the next step is to see its integration into a CI/CD pipeline in Azure Pipeline.

Integrating AzSK in Azure Pipelines

Using AzSK locally to check the security of our Azure infrastructure is already a good improvement in integrating security into DevOps practices.

However, if you want to ensure continuous checking and avoid doing it manually, you must integrate the execution of AzSK into the CI/CD pipelines that provision your infrastructure. One of the advantages of AzSK is that it integrates very well with Microsoft's DevOps tool, that is, **Azure Pipelines**.

We have already discussed Azure Pipelines in this book, in particular in `Chapter 6`, *Continuous Integration and Continuous Delivery*; `Chapter 7`, *Containerizing Your Application with Docker*; and `Chapter 10`, *Static Code Analysis with SonarQube*, so the purpose of this section is just to see the main steps of integrating AsZK into Azure Pipelines.

Let's look at how to use AzSK in a CI/CD pipeline:

1. To start, we must install the *Secure DevOps Kit (AzSK) CICD Extensions for Azure* extension in our Azure DevOps organization from the Visual Studio Marketplace, `https://marketplace.visualstudio.com/items?itemName=azsdktm.AzSDK-task`.
2. Then, in a release definition, we add the **AzSK Security Verification Tests** task that is present in the catalog, as shown in the following screenshot:

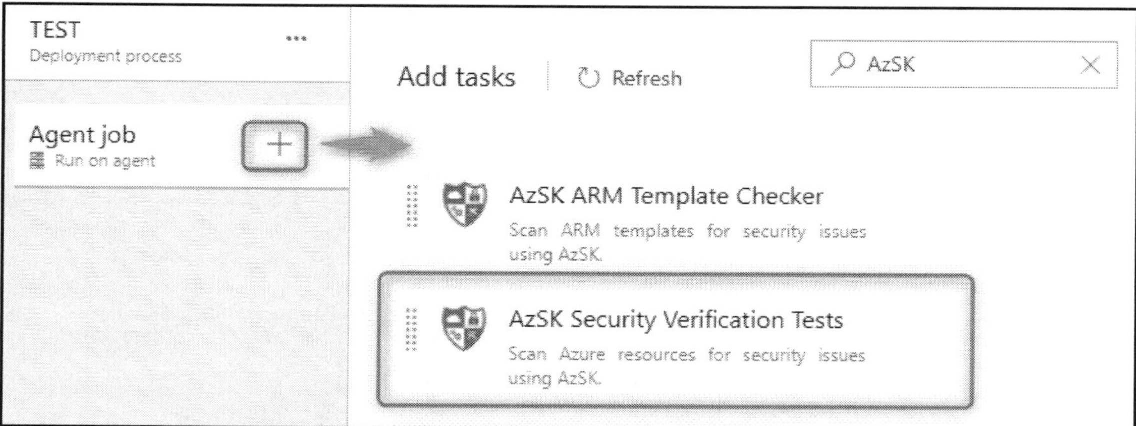

3. Then, we configure the task parameters, as follows:

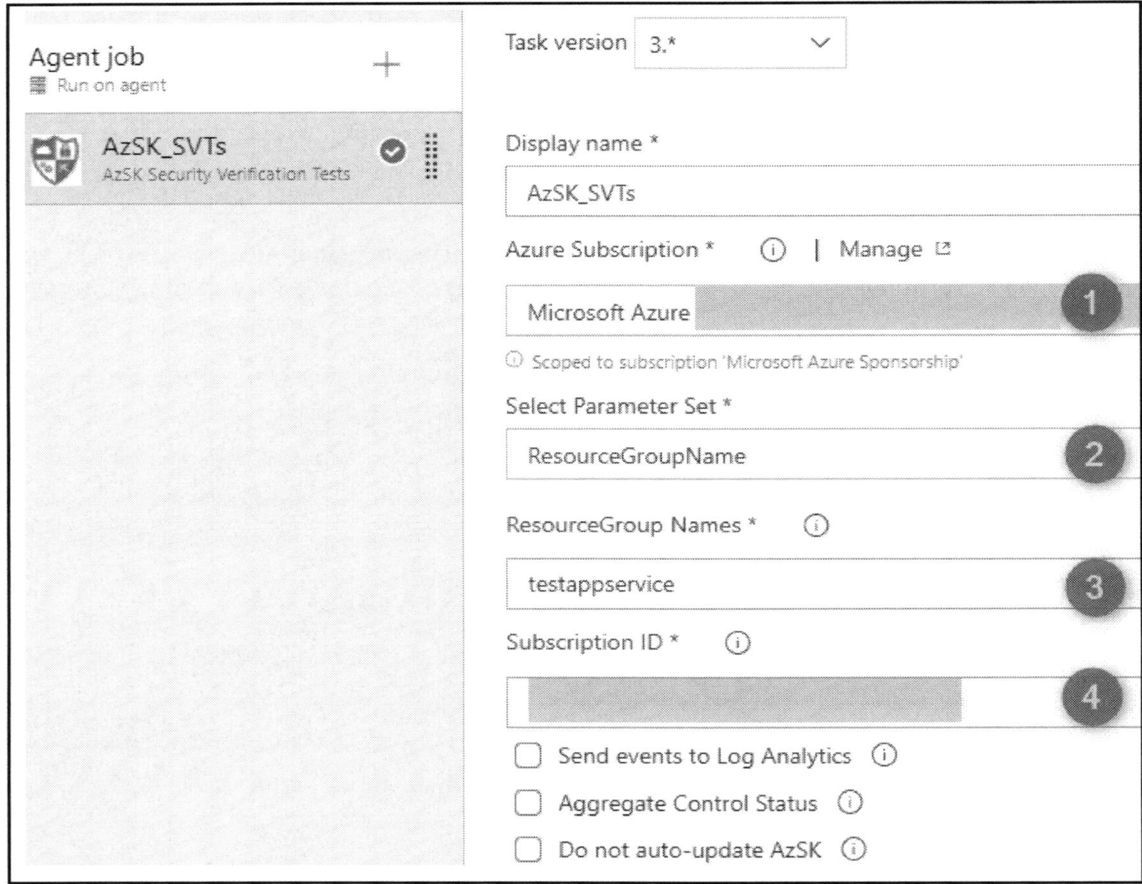

The task parameters are as follows:

- The choice of the Azure subscription that is necessary for the execution of the PowerShell
- In the **Select Parameter Set** property, the `ResourceGroupName` value from the drop-down list
- The name of the resource group that contains the resources
- The subscription ID in which the resources to be checked is located

4. We save the definition of the release, then we trigger it by creating a new release.
5. At the end of its execution, if AsZK detects security problems in the scanned Azure resources, its execution fails the release and displays the different problems in the logs, as shown in the following screenshot:

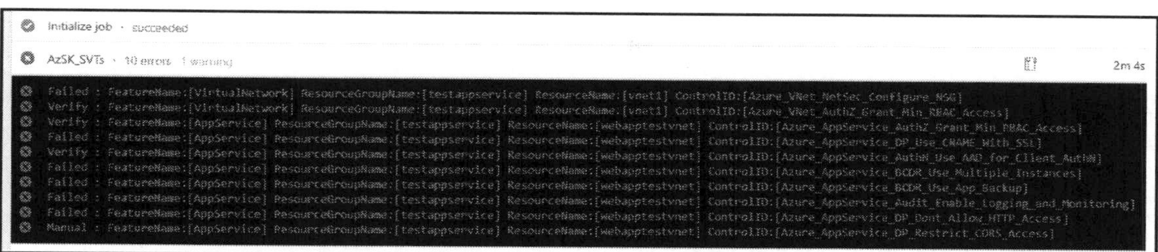

6. Finally, we download the logs of the release execution by clicking on the **Download all logs** button:

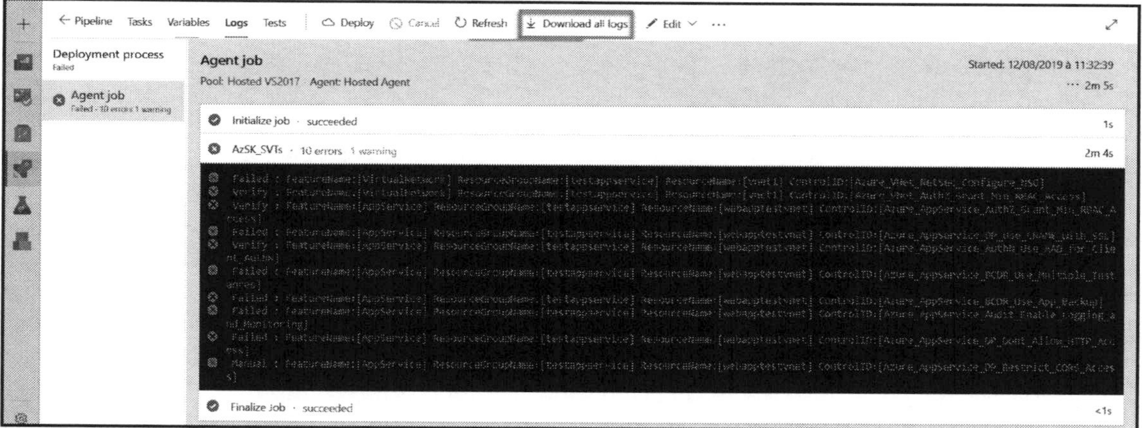

We'll retrieve the CSV report generated by AzSK inside the downloaded ZIP file:

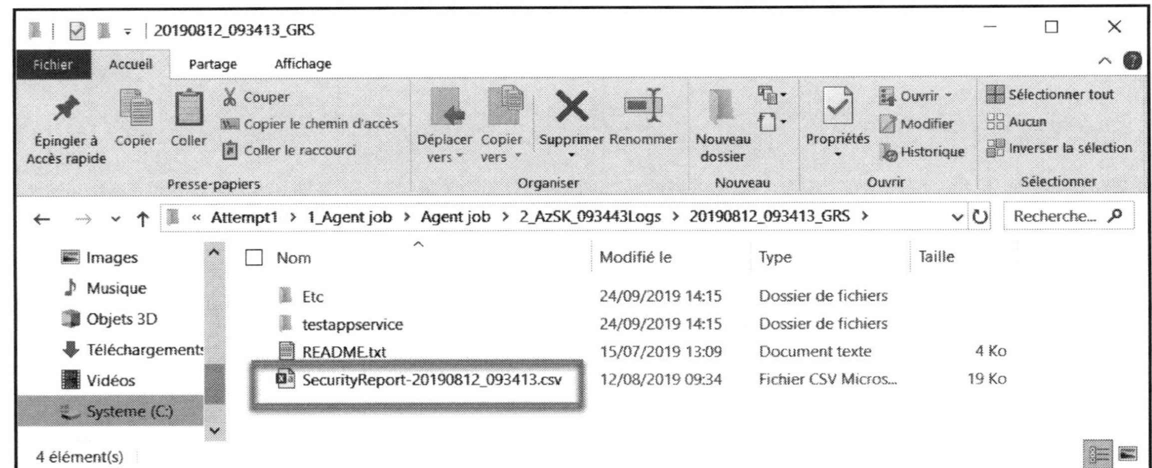

To learn more about AzSK's integration into the pipeline, you can read the complete documentation at `https://azsk.azurewebsites.net/03-Security-In-CICD/Readme.html`.

In this section, we gave an overview of the Azure Secure DevOps Kit tool, we installed it via the PowerShell commands, then we analyzed the security of the resources we had provisioned in an Azure infrastructure using its PowerShell commands.

We have seen that, in addition to the AzSk security analysis, the process also generates a CSV report of its analysis at the end of its execution.

We completed this part with the implementation of AzSK automation in a DevOps CI/CD pipeline using the Azure Pipelines platform as well as the Azure Security extension from the Visual Studio Marketplace.

In the next section, we'll look at another aspect of security with the protection of sensitive data using the secret manager, Vault, from HashiCorp.

Preserving data with HashiCorp's Vault

Today, when we talk about security in information systems, the most expected topic is the protection of sensitive data between different components of the system. This sensitive data that needs to be protected includes server access passwords, database connections, API authentication tokens, and application user accounts. Indeed, many security attacks occur because this type of data is decrypted in the source code of applications or in poorly protected files that are exposed to local workstations. There are many known tools that can be used to secure this sensitive data, such as these:

- KeyPass (`https://keepass.info/`)
- LastPass (`https://www.lastpass.com/`)
- Ansible Vault, the use of which we discussed `Chapter 3`, *Using Ansible for Configuring IaaS Infrastructure*
- Vault from HashiCorp

Also, cloud providers offer secret protection services such as the following:

- Azure Key Vault: `https://azure.microsoft.com/en-us/services/key-vault/`
- KMS for Google Cloud: `https://cloud.google.com/kms/`
- AWS Secrets Manager for AWS: `https://aws.amazon.com/secrets-manager/?nc1=h_ls`

Out of all of the tools we have mentioned, we'll see the use of **Vault** from HashiCorp, which is free and open source and can be installed on any type of OS as well as on Kubernetes.

The main features and benefits of Vault are as follows:

- It allows the storage of static secrets as well as dynamic secrets.
- It also has a system for rotating and revoking secrets.
- It allows data to be encrypted and decrypted without having to store it.
- It also has a web interface that allows the management of secrets.
- It integrates with a multitude of authentication systems.
- All secrets are stored in a single centralized tool.
- It allows you to be independent of your architecture by being accessible with all major cloud providers, Kubernetes, or even on internal data centers (on-premises).

For more information on Vault features, see the product page, `https://www.vaultproject.io/docs/what-is-vault/index.html`.

With all of these very interesting features, Vault is, therefore, a tool that I recommend for any company that wants to protect its sensitive information and integrate its secure accessibility into a CI/CD pipeline. Indeed, in addition to being very efficient for data protection, Vault integrates very well in CI/CD pipelines. These pipelines will be able to use this protected data when provisioning an infrastructure and deploying applications.

After an overview of Vault, we'll proceed with the installation of Vault on a local machine and use it for encrypting and decrypting data. We'll also give an overview of the Vault UI, and finally, we'll expose the process of how to retrieve data from Vault in Terraform.

Installing Vault locally

If you decide to use Vault, it is important to know that Vault is a tool that is responsible for the security of your sensitive infrastructure and application data. In practice, Vault is not just a tool, and before installing it in production, you need to understand its concepts and its different architectural topologies.

To learn more about Vault architecture topologies, please refer to the documentation, `https://learn.hashicorp.com/vault/operations/ops-reference-architecture`.

The purpose of this chapter is therefore not to go into the details of the concepts and architecture of Vault but to explain the installation and use of Vault in development mode. In other words, we'll install Vault on a local workstation to have a small instance that's used for tests and development.

We have already detailed the use of HashiCorp tools in this book with Terraform and Packer, and similarly, Vault can be installed either manually or via a script:

- To **install Vault manually**, the procedure is exactly the same as that of installing Terraform and Packer, so you need to do the following:
    1. Navigate to the download page: `https://www.vaultproject.io/downloads.html`.
    2. Download the package related to your OS in the folder of your choice.
    3. Unzip the package and update the `PATH` environment variable with its path to this folder.
- To **install Vault automatically**, we'll use a script, the code of which depends on our OS.

For **Linux**, in a Terminal, run the following script:

```
VAULT_VERSION="1.2.1"
curl --silent --remote-name
https://releases.hashicorp.com/vault/${VAULT_VERSION}/vault_${VAULT_VERSION
}_linux_amd64.zip
unzip vault_${VAULT_VERSION}_linux_amd64.zip
sudo mv vault /usr/local/bin/
```

This script performs the following actions:

1. It initializes a variable that contains the version of the vault to download.
2. It downloads Vault with curl. Curl is a tool for downloading any URL content, and can be downloaded from `https://curl.haxx.se/`.
3. It unzips the package.
4. It copies the vault binary into the `/usr/local/bin` folder (which is already filled in the `PATH` environment variable).

This script is also available here: `https://github.com/PacktPublishing/Learning_DevOps/blob/master/CHAP12/vault/install_vault.sh`.

For **Windows**, to install vault via a script, we'll use **Chocolatey**, which is the Windows software package manager (`https://chocolatey.org/`), by executing the following command in a Terminal:

```
choco install vault -y
```

This command downloads and installs Vault from Chocolatey.

Apart from these scripts that allow Vault to be installed on a local workstation, HashiCorp also provides the Terraform code that allows you to create a complete Vault infrastructure on different cloud providers.

This Terraform code is available for Azure at `https://github.com/hashicorp/terraform-azurerm-vault`; for AWS, it is available at `https://github.com/hashicorp/terraform-aws-vault`; and for **Google Cloud Platform (GCP)**, it is here: `https://github.com/hashicorp/terraform-google-vault`.

After installing Vault, we'll test its installation by running the following command in the Terminal:

```
vault --version
```

This command displays the installed version of Vault. We can also execute the `vault --help` command to display a list of the available commands.

We have just seen the different ways to install Vault on a local machine; the next step is to start the Vault server.

Starting the Vault server

Vault is a client-server tool that consists of a client component that's used by developers for applications and a server component that is responsible for protecting data in remote backends.

Vault supports a very large number of backends, a list of which is available here: `https://www.vaultproject.io/docs/configuration/storage/index.html`.

After installing Vault locally, we only have access to the client part, and to be able to use Vault, we'll start the server component.

To start the Vault server component in development mode, we'll execute this command in a second Terminal:

```
vault server -dev
```

This command starts and configures the Vault server with a minimal configuration that contains an authentication token and a default backend called **in-memory**, which store all secrets data in the memory of the server, as shown in the following screenshot:

```
root@mkrief:/home/mikaelkrief# vault server -dev
==> Vault server configuration:

             Api Address: http://127.0.0.1:8200
                     Cgo: disabled
         Cluster Address: https://127.0.0.1:8201
              Listener 1: tcp (addr: "127.0.0.1:8200", cluster address: "127.0.0.1:8201", max_request_duration: "1m30s", max_request_size: "33554432", tls: "disabled")
               Log Level: info
                   Mlock: supported: true, enabled: false
                 Storage: inmem
                 Version: Vault v1.2.1

WARNING! dev mode is enabled! In this mode, Vault runs entirely in-memory
and starts unsealed with a single unseal key. The root token is already
authenticated to the CLI, so you can immediately begin using Vault.

You may need to set the following environment variable:

    $ export VAULT_ADDR='http://127.0.0.1:8200'

The unseal key and root token are displayed below in case you want to
seal/unseal the Vault or re-authenticate.

Unseal Key: UvEy9upmBT+h
Root Token: s.rwgO0i2Jsr

Development mode should NOT be used in production installations!

==> Vault server started! Log data will stream in below:
```

Important considerations: This Terminal must remain open to keep the Vault server running.

Also, since we are in development mode and the backend storage is just in the memory, as soon as the Vault server stops, all of the secret data is deleted from the memory.

Then, as indicated during this execution, we'll export the `VAULT_ADDR` environment variable with this command in another Terminal:

```
export VAULT_ADDR='http://127.0.0.1:8200'
```

Finally, to check the execution status of the Vault server, we execute the following command:

```
vault status
```

Here is the command output, which displays the properties of the Vault server:

```
root@mkrief:/home/mikaelkrief# vault status
Key             Value
---             -----
Seal Type       shamir
Initialized     true
Sealed          false
Total Shares    1
Threshold       1
Version         1.2.1
Cluster Name    vault-cluster-9e7d6ef4
Cluster ID      82bfd424-73ce-9c3d-1dd1-dae6d88bb604
HA Enabled      false
```

To learn more about the Vault server started in development mode, read the documentation at `https://www.vaultproject.io/docs/concepts/dev-server.html`.

Now that Vault is installed and the server is started, we'll see how to write data to Vault to protect it, and read that data so it can be used from a third-party application.

Writing secrets in Vault

When you want to protect sensitive data that will be used by an application or infrastructure resources, the first step is to store this data in the secret data manager that has been chosen by the company. We'll see in practice the steps for writing data in Vault.

To protect data in Vault, we go to a Terminal and execute the following command:

```
vault kv put secret/vmadmin vmpassword=admin123*
```

The following screenshot shows its execution:

```
root@mkrief:/home/mikaelkrief# vault kv put secret/vmadmin vmpassword=admin123*
Key              Value
---              -----
created_time     2019-08-13T13:56:14.5200652Z
deletion_time    n/a
destroyed        false
version          1
```

The command, with the `put` operation, creates a new secret data in memory with the title `vmadmin` of the `key-value` type, which in this example is the admin account of a VM, in the `secret/` path.

In Vault, all protected data is stored in a path that corresponds to an organizational location in Vault. The default path for Vault is `secret/`, and it is possible to create custom paths that will allow better management of secret rights and better organization by domain, topic, or application.

About the secrets stored in Vault, one of its advantages is that it is possible to store multiple data in the same secret; for example, we'll update the secret data that we have created with another secret, which is the login admin of the VM.

For this, we'll execute the following command that adds another key-value secret in the same Vault data:

```
vault kv put secret/vmadmin vmpassword=admin123* vmadmin=bookadmin
```

As we can see in this execution, we used exactly the same command with the same secret, and we added new key-value data, that is, `vmadmin`.

For more information on this `kv put` command, read the documentation at `https://www.vaultproject.io/docs/commands/kv/put.html`.

We learned how to use commands to create a secret in Vault and the various uses of secrets, and we'll now have a look at the command to read this secret in order to use it inside an application or in infrastructure resources.

Reading secrets in Vault

Once we have created secrets in Vault, we'll have to read them to use them in our applications or infrastructure scripts.

To read a key that is stored in a vault, we go to a Terminal to execute this command:

```
vault kv get secret/vmadmin
```

In this command, we use the `kv` operation with the `get` operator, and we indicate in the parameter the complete path of the key to get the protected value within our example, `secret/vmadmin`.

The following screenshot shows the command execution, as well as its output:

```
root@mkrief:/home/mikaelkrief# vault kv get secret/vmadmin
====== Metadata ======
Key              Value
---              -----
created_time     2019-08-13T14:09:20.4661459Z
deletion_time    n/a
destroyed        false
version          2

======= Data =======
Key           Value
---           -----
vmadmin       bookadmin
vmpassword    admin123*
```

What we notice in the output of this command is the following:

- The version number of the secret here is `2` because we executed the `kv put` command twice, so the version number was incremented at each execution.
- There are two key-value data items that we protected in secret in the previous section, *Writing secrets in Vault*.

If you want to access the data stored in this secret but from an earlier version, we can execute the same command by optionally specifying the desired version number, as in this example:

```
vault kv get -version=1 secret/vmadmin
```

The following screenshot shows its execution:

```
root@mkrief:/home/mikaelkrief# vault kv get -version=1 secret/vmadmin
====== Metadata ======
Key              Value
---              -----
created_time     2019-08-13T13:56:14.5200652Z
deletion_time    n/a
destroyed        false
version          1

======= Data =======
Key           Value
---           -----
vmpassword    admin123*
root@mkrief:/home/mikaelkrief#
```

We can see in this output, in the **Data** section, version 1 of the key-value data we had during the first execution of the `kv put` command.

For more information on the `kv get` command, read the documentation at `https://www.vaultproject.io/docs/commands/kv/get.html`.

We have just seen the use of the `kv get` Vault command to retrieve all or specific versions of the values of a secret; we'll now briefly see how to use the Vault UI web interface for better management of secrets.

Using the Vault UI web interface

One of the interesting features of Vault is that, apart from the client-side tool that allows you to perform all operations on the Vault server, Vault has a UI web interface that allows you to manage secrets but more visually and graphically.

To open and use the Vault web interface to visualize the secrets that we have created with the client tool, we must follow these steps:

1. In a browser, enter the URL provided when starting the server, that is, `http://127.0.0.1:8200/ui`, which is the default local Vault URL.
2. In the authentication form, enter the token that was provided in the Terminal in the root token information, as shown in the following screenshot:

```
root@mkrief:/home/mikaelkrief# vault server -dev
==> Vault server configuration:

             Api Address: http://127.0.0.1:8200
                     Cgo: disabled
         Cluster Address: https://127.0.0.1:8201
              Listener 1: tcp (addr: "127.0.0.1:8200", cluster address: "127.0.0.1:8
               Log Level: info
                   Mlock: supported: true, enabled: false

    $ export VAULT_ADDR='http://127.0.0.1:8200'

The unseal key and root token are displayed below in case you want to
seal/unseal the Vault or re-authenticate.

Unseal Key: fh7vj+J8lHHXr+AtTva4DU2Ru4kcPtqQuM9jKBoO1BA=
Root Token: s.6MGUVmH1bnhD36aWf0Fb9oR4

Development mode should NOT be used in production installations!
```

3. Click on the **Sign In** button to authenticate:

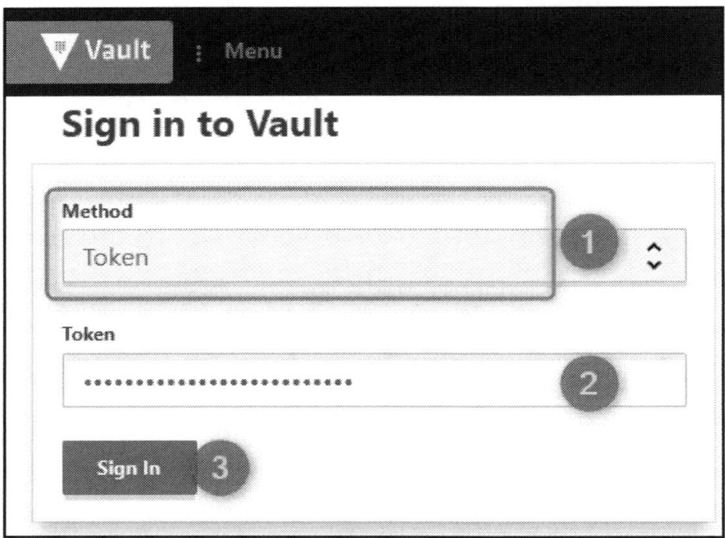

4. The home of the interface displays the list of secret paths, called **Secrets Engines**, containing the secrets that have been stored:

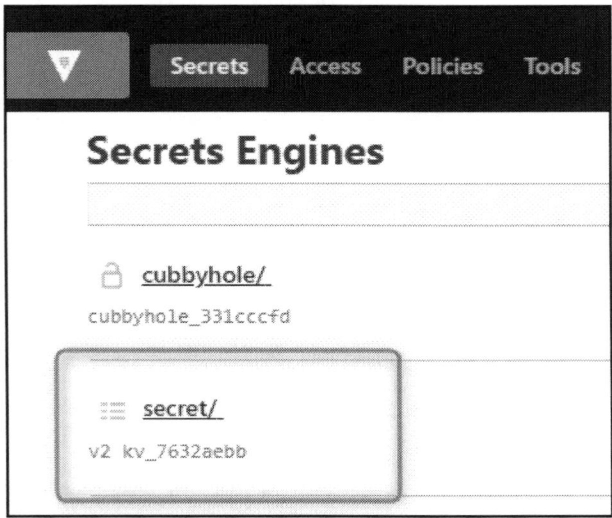

5. By clicking on each secret engine, you can see the list of secrets that have been saved. The following screenshot shows the secret engine page, which displays the secret we created on the command line in the previous section:

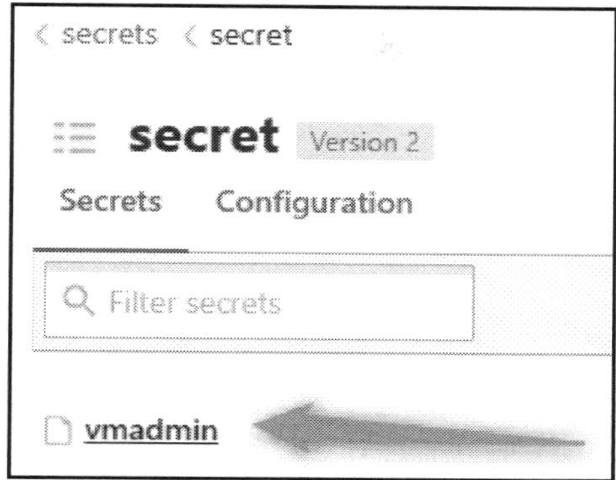

6. By clicking on a specific secret, you can access the list of data that we have protected with the possibility of viewing the values of each key in cleartext. We also can display the history of the content of a secret by selecting the desired version in the **History** drop-down menu:

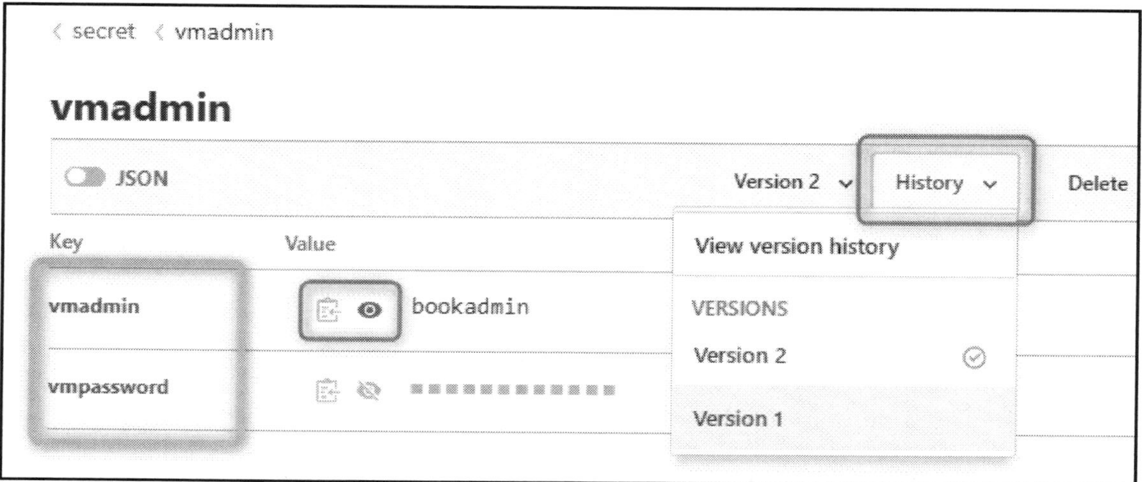

The Vault web interface also allows you to perform all management operations on secrets and other Vault components.

If you want to know more about this web interface, read this article: `https://www.hashicorp.com/resources/vault-oss-ui-introduction`.

We have seen that the Vault web interface is a very good alternative to the client tool that allows you to view and manage Vault elements. After this overview of Vault and its operations, I propose a small Vault use case that shows you how to get secrets in Terraform code.

Getting Vault secrets in Terraform

As we have already seen in `Chapter 2`, *Provisioning Cloud Infrastructure with Terraform*, it is very important to protect the infrastructure configuration information that we write in Terraform code. One way to protect this sensitive data is to store it in a secret manager such as Vault and recover it directly with Terraform dynamically.

Here is an example of Terraform code that allows you to retrieve the password of a VM that you want to provision from Vault. This example of Terraform code is composed of three blocks, which are as follows:

1. First, we use the Vault provider for configuring the Vault URL:

```
provider "vault" {
    address = "http://127.0.0.1:8200" #Local Vault Url
}
```

   The Vault provider is configured with the Vault server configuration and its authentication.

   In our case, we configure the Vault server URL in the Terraform code, and for the authentication of a token, we'll use an environment variable when running Terraform after the explanation of the code.

For more details on the Terraform Vault provider and its configuration, see the documentation: `https://www.terraform.io/docs/providers/vault/index.html`.

2. Then, we add the Terraform data block, `vault_generic_secret`, which is used for retrieving a secret from a Vault server:

```
data "vault_generic_secret" "vmadmin_account" {
    path = "secret/vmadmin"
}
```

This data block allows us to retrieve (in read only mode) the content of a secret stored in Vault. Here, we ask Terraform to retrieve the secret that is in the Vault `secret/vmadmin` path that we created earlier in this section.

For more details on the `vault_generic_secret` data and its configuration, see the documentation: `https://www.terraform.io/docs/providers/vault/d/generic_secret.html`.

3. Finally, we add an `output` block to use the decrypted value of the secret:

```
output "vmpassword" {
 value =
"${data.vault_generic_secret.vmadmin_account.data["vmpassword"]}"
 sensitive = true
}
```

This block provides an example of the exploitation of the secret.

The `data.vault_generic_secret.vmadmin_account.data["vmpassword"]` expression is used to get the secret returned by the previously used data block. In the `data` array, we add the name of only those keys for which we need the encrypted values to be recovered. Also, this output is considered *sensitive* so that Terraform does not display its value in plain text when it is executed.

The complete source code of this Terraform code is also available here: `https://github.com/PacktPublishing/Learning_DevOps/blob/master/CHAP12/vault/terraform_usevault/main.tf`.

We have finished writing the Terraform code; we'll now quickly execute it to see the recovery of the secret.

For the execution of the code, which we have already detailed in `Chapter 2`, *Provisioning Cloud Infrastructure with Terraform*, in this section, we'll only quote the commands without further detailing them.

To execute Terraform, we go to a Terminal in the folder that contains the Terraform code, and then we proceed in this order:

1. Export the `VAULT_TOKEN` environment variable with the value of the Vault token. In our development mode case, this token is provided at the start of the Vault server.

   The following command shows the export of this environment variable on a Linux OS:

   ```
   export VAULT_TOKEN=xxxxxxxxxxxx
   ```

2. Then, we'll execute Terraform with these commands:

   ```
   terraform init
   terraform plan
   terraform apply
   ```

   Here is a quick summary of the details of these commands:

   - The `terraform init` command initializes the context and downloads all necessaries providers.
   - The `terraform plan` command displays a preview of all changes that will be applied by Terraform.
   - The `terraform apply` command applies all changes on the infrastructure and displays the outputs values.

To learn all the details about the main Terraform commands and the Terraform life cycle, read `Chapter 2`, *Provisioning Cloud Infrastructure with Terraform*, of this book.

The following screenshot shows the execution of the `plan` and `apply` commands:

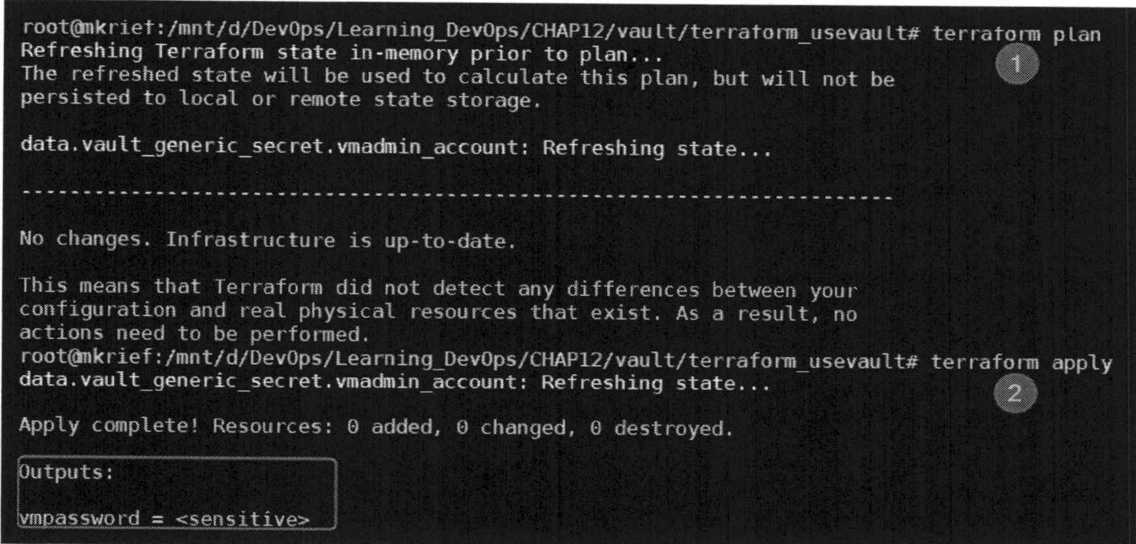

We can see that the value of Terraform output named `vmpassword` is not displayed in clear text in the Terminal.

3. Finally, we display the Terraform output value in JSON format with the `terraform output` command with the `-json` option:

```
terraform output -json
```

The result of execution is shown in this screenshot:

```
root@mkrief:/mnt/d/DevOps/Learning_DevOps/CHAP12/vault/terraform_usevault# terraform output -json
{
  "vmpassword": {
    "sensitive": true,
    "type": "string",
    "value": "admin123*"
  }
}
```

We see that Terraform has displayed the value of the key that was in the secret, which we had inserted in Vault. This value can now be used for any sensitive data that should not be stored in Terraform code, such as VM passwords.

However, be careful: we have protected our Terraform code by outsourcing all sensitive data to a secret manager, but it should not be forgotten that Terraform stores all information including data and output information, in the `tfstate` file. It is therefore very important to protect it by storing the `tfstate` file in a protected remote backend, as we saw in the *Protecting tfstate in a remote backend* section of `Chapter 2`, *Provisioning Cloud Infrastructure with Terraform.*

In this section, we have studied the use of HashiCorp's Vault, which is a secret data manager. We have seen how the installation on different OSes can be done manually and automatically. We used its command lines to protect and read data that we have protected inside. Then, we discussed how to manage secrets in Vault using its web interface. Finally, we wrote and executed a Terraform code that uses the Vault provider and allows us to retrieve secrets that we stored in a Vault server.

Summary

This chapter is dedicated to integrating security into DevOps practices. We presented three tools to verify and secure your data and cloud infrastructure. We discussed how to check the compliance of an Azure infrastructure using **InSpec** from Chef.

To do this and check infrastructure compliance, we installed InSpec and then detailed the writing of InSpec tests. We used its command lines to verify the compliance of Azure infrastructure. Then, we explored the use of the **Secure DevOps Kit for Azure** tool provided by Microsoft to analyze the security applied on a subscription and to Azure resources. Then we looked at its automation and integration into a CI/CD pipeline with Azure Pipelines.

In the last section, we saw how to protect sensitive data with **Vault** from HashiCorp. In this section, we looked at data encryption and decryption in Vault and wrote Terraform code that will dynamically retrieve the secrets stored in Vault.

In the next chapter, we'll present the concept of **blue-green deployment** with its patterns for reducing the deployment downtime. Then, we'll learn how to implement it in an application as well as in the deployment of an Azure infrastructure.

Questions

1. What is the role of InSpec?
2. What is the name of the package manager that allows you to download InSpec via a command line?
3. Which InSpec command allows you to execute InSpec tests?
4. Who is the publisher of Vault?
5. Which command starts Vault in development mode?
6. When Vault is installed locally, can it be used for production?
7. In local mode, where is Vault's encrypted data stored?
8. What are the requirements to install the Secure DevOps Kit for Azure tool?
9. What is the format of the report generated by Secure DevOps Kit for Azure?

Further reading

If you want to know more about DevSecOps with InSpec, Vault, and Secure DevOps Kit for Azure, here are some resources:

- InSpec documentation: `https://www.inspec.io/`
- HashiCorp Vault documentation: `https://www.vaultproject.io/docs/`
- Learn about HashiCorp Vault: `https://learn.hashicorp.com/vault`
- Secure DevOps Kit for Azure (AzSK): `https://azsk.azurewebsites.net/index.html`

13
Reducing Deployment Downtime

So far, we have discussed DevOps practices such as Infrastructure as Code, CI/CD pipelines, and the automation of different types of tests.

In `Chapter 1`, *DevOps Culture and Practices*, we saw that these DevOps practices will improve the quality of applications and thus improve the financial gain of a company. We will now go deeper into DevOps practices by looking at how to ensure the continuous availability of your applications even during your deployments, and how to deliver new versions of these applications more frequently in production.

Often, what we see is that deployments require your applications to be interrupted by, for example, infrastructure changes or service shutdowns. Moreover, what we also see is that companies are still reluctant to deliver more frequently in production. They are not equipped to test the application in the production environment or they are waiting for other dependencies.

In this chapter, we will look at several practices that will help you improve application delivery processes. We'll start with a way to reduce the downtime of your infrastructure and applications during Terraform deployments. Then, we will discuss the concept and patterns of blue-green deployment and how to configure it with some Azure resources. Finally, we will present the details of implementing a feature flag in your application, which will allow you to modify the operation of an application without having to redeploy it in production.

You will also learn how to configure Terraform code to reduce application downtime. You'll be able to configure Azure resources with blue-green deployment and implement feature flags in your applications with either an open source component or the LaunchDarkly platform.

In this chapter, we will cover the following topics:

- Reducing deployment downtime with Terraform
- Understanding blue-green deployment concepts and patterns
- Applying blue-green deployments on Azure
- Introducing feature flags
- Using an open source framework for feature flags
- Using the LaunchDarkly solution

Technical requirements

In order to understand the Terraform concepts that will be presented in this chapter, you need to have read `Chapter 2`, *Provisioning Cloud Infrastructure with Terraform*.

We will look at an example of how to implement blue-green deployment in Azure. If you don't have an Azure subscription, you can create a free Azure account here: `https://azure.microsoft.com/en-gb/free/`.

Then, we will look at an example of how to use feature flags in an ASP.NET Core application. To use our example, you will need to install the .NET Core SDK, which can be downloaded from `https://dotnet.microsoft.com/download`.

For code editing, we used the free Visual Studio Code editor, which is available for download here: `https://code.visualstudio.com/`.

The complete source code for this chapter can be found at `https://github.com/PacktPublishing/Learning_DevOps/tree/master/CHAP13`.

Check out the following video to see the code in action:

`http://bit.ly/2p9Jh3X`

Reducing deployment downtime with Terraform

In `Chapter 2`, *Provisioning Cloud Infrastructure with Terraform*, we detailed the use of Terraform by looking at its commands and life cycle and put it into practice with an implementation in Azure.

One of the problems with Terraform is that, depending on the infrastructure changes that need to be implemented, Terraform may automatically destroy and rebuild certain resources.

To fully understand this behavior, let's look at the output of this following Terraform execution, which provisioned a web app in Azure and has been modified with a name change:

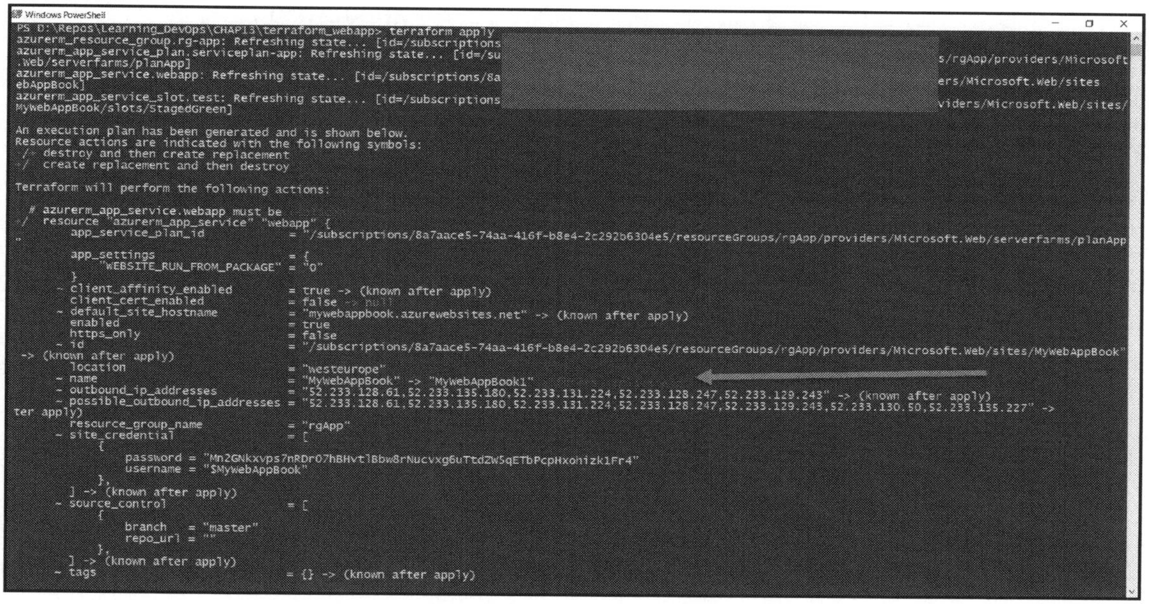

Here, we can see that Terraform will destroy the web app and then rebuild it with the new name. Although destruction and reconstruction are done automatically, during this period of time in which Terraform will destroy and rebuild the web app, the application will be inaccessible to users.

To solve this problem of downtime, we can add the Terraform `create_before_destroy` option:

```
resource "azurerm_app_service" "webapp" {
    name = "MyWebAppBook1" #new name
    location = "West Europe"
    resource_group_name = "${azurerm_resource_group.rg-app.name}"
    app_service_plan_id = "${azurerm_app_service_plan.serviceplan-app.id}"
    app_settings = {
       WEBSITE_RUN_FROM_PACKAGE = var.package_zip_url"
    }
```

```
        lifecycle { create_before_destroy = true}
    }
```

By adding this option, Terraform will do the following:

1. First, Terraform creates the new web app with a new name.
2. During the provisioning of the new web app, it uses the URL for the application package in ZIP format that's provided in the `app_settings` property. Use `WEBSITE_RUN_FROM_PACKAGE` to launch the application.
3. Then, Terraform will destroy the old web app.

Using the `Terraform create_before_destroy` option will ensure the viability of our applications during deployments.

Be careful, though: this option will only be useful if the new resource that's being created allows us to have the application running very quickly at the same time as it's provisioning so that a service interruption doesn't occur.

In our example of a web app, this worked when we used the `WEBSITE_RUN_FROM_PACKAGE` property of the web app. For a virtual machine, we can use a VM image created by **Packer**. As we saw in `Chapter 4`, *Optimizing Infrastructure Deployment with Packer*, Packer contains information regarding the VM applications that have already been updated inside the VM image.

For more information on the `create_before_destroy` option, please view the following Terraform documentation: `https://www.terraform.io/docs/configuration/resources.html#lifecycle-lifecycle-customizations`.

We have just seen that, with Terraform and Infrastructure as Code, it is possible to reduce downtime during deployments in the case of resource changes.

We will now look at the concepts and patterns of a practice called **blue-green deployment**, which allows us to deploy and test an application in production with great confidence.

Understanding blue-green deployment concepts and patterns

Blue-green deployment is a practice that allows us to deploy a new version of an application in production without impacting the current version of the application. In this approach, the production architecture must be composed of **two identical environments**; one environment is known as the **blue** environment while the other is known as the **green** environment.

The element that allows routing from one environment to another is a router, that is, a load balancer.

The following diagram shows a simplified schematic of a blue-green architecture:

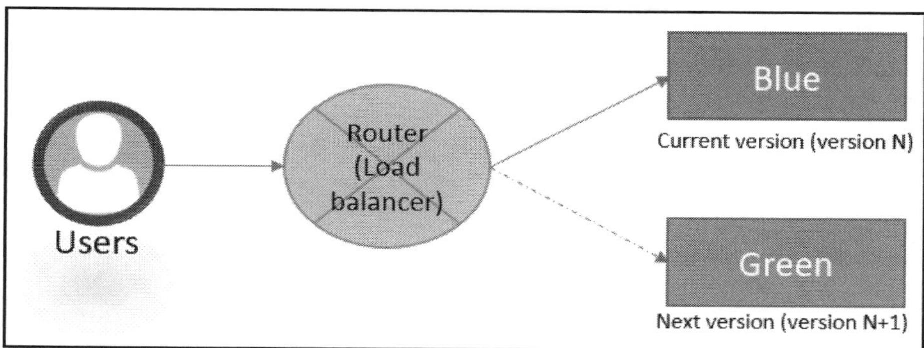

As we can see, there are two identical environments: the environment called **blue**, which is the current version of the application, and the environment called **green**, which is the new version or the next version of the application. We can also see a router, which redirects users' requests either to the blue environment or the green environment.

Now that we've introduced the principle of blue-green deployment, we will look at how to implement it in practice during deployment.

Using blue-green deployment to improve the production environment

The basic usage pattern of the blue-green deployment is that when we're deploying new versions of the application, the application is deployed in the blue environment (version *N*) and the router is configured in this environment.

When deploying the next version (version *N+1*), the application will be deployed in the green environment, and the router is configured in this environment.

The blue environment becomes unused and idle until the deployment of version *N*+2. It also will be used in the case of rapid rollback to version *N*.

This practice of blue-green deployment can also be declined on several patterns, that is, the canary release and dark launch patterns. Let's discuss the implementation of each of these patterns in detail. We will start with the canary release pattern.

Understanding the canary release pattern

The canary release technique is very similar to blue-green deployment. The new version of the application is deployed in the green environment, but only for a small restricted group of users who will test the application in real production conditions.

This practice is done by configuring the router (or load balancer) to be redirected to both environments. On this router, we apply redirection restrictions of a user group so that it can only be redirected to the green environment, which contains the new version.

The following is a sample diagram of the canary release pattern:

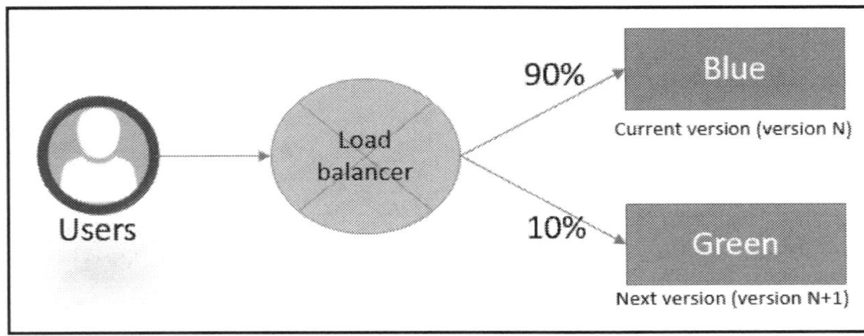

In the preceding diagram, the router redirects 90% of users to the blue environment and 10% of users to the green environment, which contains the new version of the application.

Then, once the tests have been performed by this user group, the router can be fully configured in the green environment, thus leaving the blue environment free for testing the next version (*N*+2).

As shown in the following diagram, the router is configured to redirect all users to the green environment:

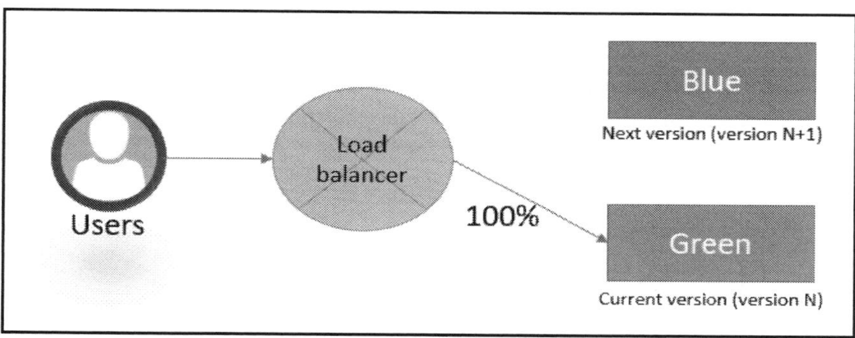

This deployment technique, therefore, makes it possible to deploy and test the application in the real production environment without having to impact all users.

We will look at a practical implementation of this blue-green deployment pattern in Azure later in this chapter, in the *Applying blue-green deployments on Azure* section.

But before that, let's look at another blue-green deployment pattern, that is, the dark launch pattern.

Exploring the dark launch pattern

The dark launch pattern is another practice related to blue-green deployment that consists of deploying new features in hidden or disabled mode (so that they're inaccessible) into the production environment. Then, when we want to have access to these features in the deployed application, we can activate them as we go along without having to redeploy the application.

Unlike the canary release pattern, the dark launch pattern is not a blue-green deployment that depends on the infrastructure but is implemented in the application code. To set up the dark launch pattern, it is necessary to encapsulate the code of each feature of the application in elements called **feature flags** (or feature toggle), which will be used to enable or disable these features remotely.

We will look at the implementation and use of feature flags with an open source framework and a cloud platform in the last few sections of this chapter.

In this section, we have presented the practice of blue-green deployment, along with its concepts and patterns, such as the canary release and dark launch patterns.

We have discussed that this practice requires changes to be made in the production infrastructure since it's composed of two instances of the infrastructure – one blue and one green – as well as a router that redirects users' requests.

Now that we've talked about blue-green deployment patterns, we will look at how to implement one in practice in an Azure cloud infrastructure.

Applying blue-green deployments on Azure

Now that we've looked at blue-green deployment, we'll look at how to apply it to an Azure infrastructure using two types of components: App Service slots and Azure Traffic Manager.

Let's start by looking at the most basic component: App Service slots.

Using App Service with slots

If we have an Azure subscription and want to use blue-green deployment without investing a lot of effort, we can use App Service slots (web apps or Azure Functions).

In App Services such as a web app, we can create a second instance of our web app by creating a slot for it (up to 20 slots, depending on the application service plan). This slot is a secondary web app but is attached to our main web app.

In other words, the main web app represents the blue environment and the slot represents the green environment.

To use this web app and its slot as a blue-green architecture, we will perform the following configuration steps:

1. Once the web app slot has been created, the new version of the application will be deployed in this slot and we can assign a percentage of traffic, as shown in the following screenshot:

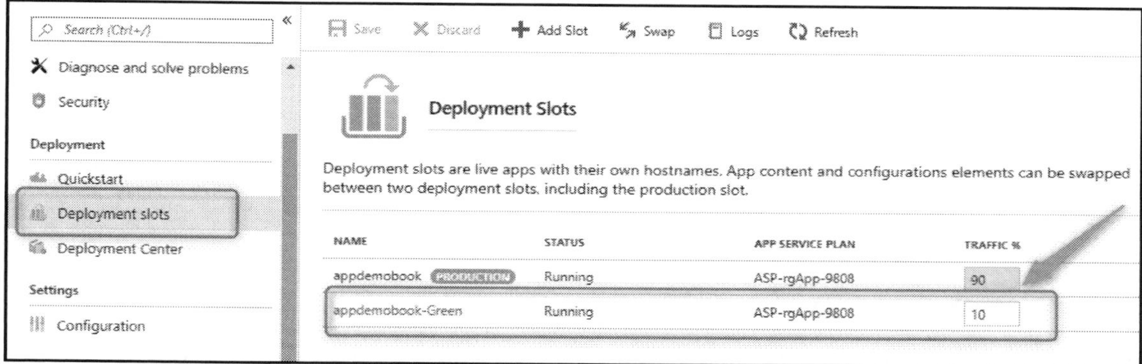

   Here, we've assigned 10% of traffic to the web app slot, which includes the changes of the new version of the application.

2. As soon as the new version of the application has been tested on the slot, we can swap the slot to the main web app (the blue environment), as shown in the following screenshot:

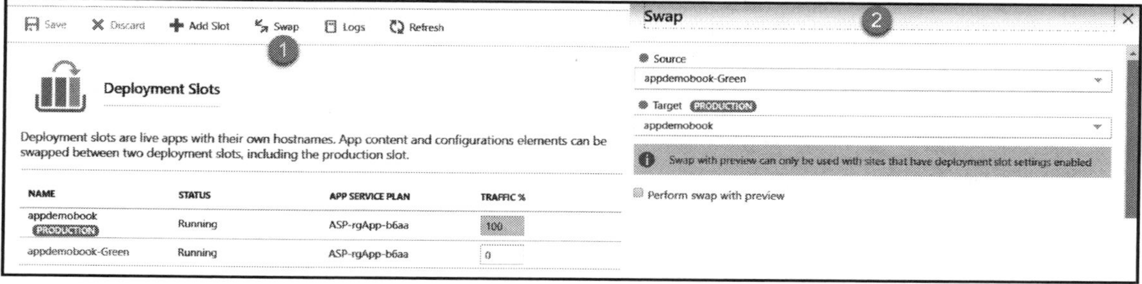

With this swap, the web app takes the content of its slot, and vice versa.

The web app now contains the new version (*N+1*) of the application and the slot contains the older version (*N*). In the case of an urgent problem, we can recover the previous version of the application by redoing a swap.

To learn more about configuring and using web app slots, please read the following documentation: `https://docs.microsoft.com/en-us/azure/app-service/deploy-staging-slots`.

This is exactly what we saw in the canary release pattern, which allows us to distribute production traffic for a group of users as well as route the application to the environment that has the *N+1* version of the application.

Now that we've discussed the use of slots, we'll take a look at the Azure Traffic Manager component, which also allows us to implement blue-green deployment.

Using Azure Traffic Manager

In Azure, there is a component called **Azure Traffic Manager** that allows us to manage traffic between several resource endpoints, such as two web apps.

To do this, we need to have two web apps: one for the blue environment and another for the green environment.

Then, in our Azure subscription, we have to create a Traffic Manager that we'll configure with the following steps:

1. In the Traffic Manager, first, we will configure a profile that determines the traffic routing method. In our case, we will configure a weight profile, that is, configure it according to a weight that we will assign in our web app. The following screenshot shows the configuration of the profile by weight:

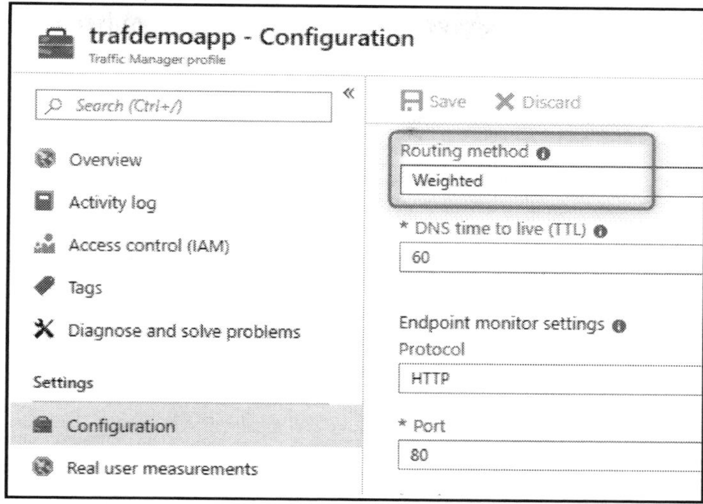

To find out about the other profile configuration options and how they work, please read the following documentation: `https://docs.microsoft.com/en-us/azure/traffic-manager/traffic-manager-routing-methods`.

2. Then, we will record the endpoints that make up our two web apps. For each of them, we'll configure a preponderance weight, as shown in the following screenshot:

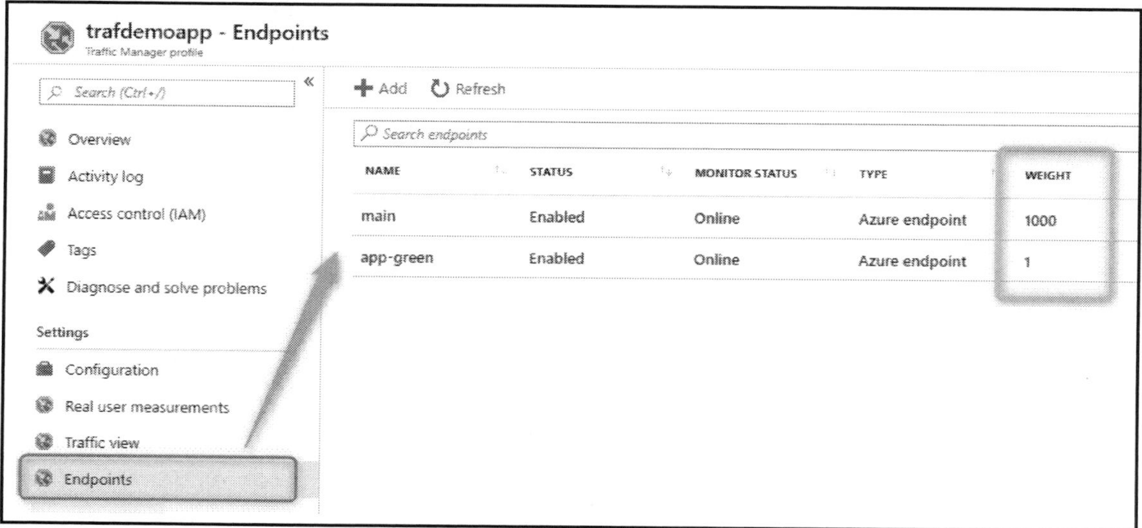

Thus, the **main** endpoint, that is, the blue environment, has the maximum weight, which is equivalent to 100% traffic.

If you would like to know more about configuring Traffic Manager, please follow this tutorial: `https://docs.microsoft.com/en-us/azure/traffic-manager/tutorial-traffic-manager-weighted-endpoint-routing`.

As for the App Service slots, with Traffic Manager, we can adjust this weight according to the traffic we want on each endpoint and then apply blue-green deployment.

We have just discussed the implementation of blue-green deployment and, more specifically, with the canary release pattern in an Azure infrastructure using a couple of solutions:

- For the first solution, we used a slot that was under a web app and we configured their percentage of user traffic.
- For the second solution, we used and configured an Azure Traffic Manager resource that acts as a router between two web apps.

Now, let's look at the dark launch pattern in detail, starting with an introduction to feature flags and their implementation.

Introducing feature flags

Feature flags (also called **feature toggle**) allow us to dynamically enable or disable a feature of an application without having to redeploy it.

Unlike the blue-green deployment with the canary release pattern, which is an architectural concept, feature flags are implemented in the application's code.

Their implementation is done with a simple encapsulation using conditional `if` rules, as shown in the following code example:

```
if(activateFeature("addTaxToOrder")==True) {
    ordervalue = ordervalue + tax
}else{
    ordervalue = ordervalue
}
```

In this example code, the `activateFeature` function allows us to find out whether the application should add the tax to order according to the `addTaxToOrder` parameter, which is specified outside the application (such as in a database or configuration file).

Features encapsulated in feature flags may be necessary either for the running of the application or for an internal purpose such as log activation or monitoring.

The activation and deactivation of features can be controlled either by an administrator or directly by users via a graphical interface.

The lifetime of a feature flag can be either of the following:

- **Temporary**: To test a feature. Once validated by users, the feature flag will be deleted.
- **Definitive**: To leave a feature flagged for a long time.

Thus, using feature flags, a new version of an application can be deployed to the production stage faster. This is done by disabling the new features of the release. Then, we will reactivate these new features for specific group of users like testers, who will test these features directly in production.

Moreover, if we notice that one of the application's functionalities is not working properly, it is possible for the features flags to disable it very quickly, without us having to redeploy the rest of the application.

Feature flags also allow A/B testing, that is, testing the behavior of new features by certain users and collecting their feedback.

There are several technical solutions when it comes to implementing feature flags in an application:

- You develop and maintain your custom feature flags system, which has been adapted to your business needs. This solution will be suitable for your needs but requires a lot of development time, as well as the necessary considerations of architecture specifications such as the use of a database, data security, and data caching.
- You use an open source tool that you must install in your project. This solution allows us to save on development time but requires a choice of tools, especially in the case of open source tools. Moreover, among these tools, few offer portal or dashboard administration that allows for the management of features flags remotely. There is a multitude of open source frameworks and tools for feature flags. Please go to `http://featureflags.io/resources/` to find them. Please refer to the following as well:
    - RimDev.FeatureFlags (`https://github.com/ritterim/RimDev.FeatureFlags`)
    - FlagR (`https://github.com/checkr/flagr`)
    - Unleash (`https://github.com/Unleash/unleash`)

    - Togglz 9(`https://github.com/togglz/togglz`),
    - Flip (`https://github.com/pda/flip`).
- You can use a cloud solution (PaaS) that requires no installation and has a back office for managing features flags, but most of them require a financial investment for large-scale use in an enterprise. Among these solutions, we can mention:
    - LaunchDarkly (`https://launchdarkly.com/`)
    - Rollout (`https://app.rollout.io/signup`)
    - Featureflag.tech (`https://featureflag.tech/`)
    - Featureflow (`https://www.featureflow.io/`).

In this section, we have talked about how the use of features flags is a development practice that allows you to test an application directly in the production stage.

We also mentioned the different feature flags usage solutions and illustrated their implementation. Let's discuss one of its implementations with an open source tool known as RimDev.FeatureFlags.

Using an open source framework for feature flags

As we've seen, there are a large number of open source tools or frameworks that allow us to use feature flags in our applications.

In this section, we will look at an example of implementing feature flags within a .NET (Core) application using a simple framework called **RimDev.FeatureFlags**. RimDev.FeatureFlags is a framework written in .NET that's free and open source (`https://github.com/ritterim/RimDev.FeatureFlags`), and is packaged and distributed via a NuGet package. It can be found here: `https://www.nuget.org/packages/RimDev.AspNetCore.FeatureFlags`.

To store the feature flag's data, **RimDev.FeatureFlags** uses a database that must be created beforehand. The advantage of RimDev.FeatureFlags is that, once implemented in our application, it provides a web user interface that allows us to enable or disable feature flags.

As a prerequisite for this example, we need to we have an ASP.NET MVC Core application already initialized. We will use a SQL Server database that has been created to store feature flags data.

To initialize RimDev.FeatureFlags in this application, we will perform the following steps:

1. The first step consists of referencing the NuGet `RimDev.FeatureFlags` package in our application and modifying (with any text editor) the `.csproj` file of the application, which is located at the root of the application's files and contains some application parameters, by adding a `PackageReference` element, as follows:

```
<ItemGroup>
    ...
    <PackageReference Include="RimDev.AspNetCore.FeatureFlags"
Version="1.2.0" />
</ItemGroup>
```

Alternatively, we can execute the following command in a Terminal to reference a NuGet package in the existing project:

```
dotnet add package RimDev.AspNetCore.FeatureFlags
```

2. Then, we'll go to the `appsettings.json` configuration file to configure the connection string to the database we created beforehand with the following code:

```
"connectionStrings": {
"localDb": "Data Source=<your database
server>;Database=FeatureFlags.AspNetCore;User ID=<your
user>;Password=<password data>
}
```

3. In the `startup.cs` file, which is located at the root of the application's files, we'll add the configuration to `RimDev.FeatureFlags` with these two blocks of code:

```
private readonly FeatureFlagOptions options;
public Startup(IConfiguration configuration)
{
    Configuration = configuration;
    options = new FeatureFlagOptions()
.UseCachedSqlFeatureProvider(Configuration.GetConnectionString("loc
alDb"));
}
```

In the preceding code, we initialized the options of `RimDev.FeatureFlags` by using the database connection. We can configure service loading with the following code:

```
public void ConfigureServices(IServiceCollection services)
{
...
    services.AddFeatureFlags(options);
}
public void Configure(IApplicationBuilder app,
IHostingEnvironment env)
{
...
    app.UseFeatureFlags(options);
    app.UseFeatureFlagsUI(options);
}
```

As soon as the application starts, `RimDev` will load the feature flags data into the application context. With this, we've configured `RimDev.FeatureFlags` in our project.

Now, we will create features flags and use them in the application. For this example, we will create a feature flag called `ShowBoxHome` that may or may not display the image in the middle of our application's home page. Let's look at how to create and manipulate these feature flags in our project:

1. First, we will create the feature flags by creating a new class that contains the following code:

```
using RimDev.AspNetCore.FeatureFlags;
    namespace appFeatureFlags.Models{
        public class ShowBoxHome : Feature {
            public override string Description { get; } = "Show the
home center box.";
        }
    }
```

This class contain the `ShowBowHome` feature flag. An override description is given to this feature flag.

2. Then, in our controller, we call the `ShowBoxHome` class with the following code:

```
public class HomeController : Controller {
    private readonly ShowBoxHome showboxHome;
    public HomeController (ShowBoxHome showboxHome){
        this.showboxHome = showboxHome;
    }
    public IActionResult Index() {
        return View(new HomeModel{ShowboxHome =
this.showboxHome.Value});
    }
...
}
```

The controller receives the values of the feature flags stored in the database, which were loaded when the application was started.

3. We'll also create a `HomeModel` class that will list all the features flags needed for the home page:

```
public class HomeModel
{
   public bool ShowBoxHome { get; set; }
}
```

4. Finally, in `Views/Home/index.chtml`, we'll use this model to display the image in the center of the home page, depending on the value of the feature flag.

Once the development process has come to an end, deploy and run our application. By default, there is no image in the middle of the home page, as shown in the following screenshot:

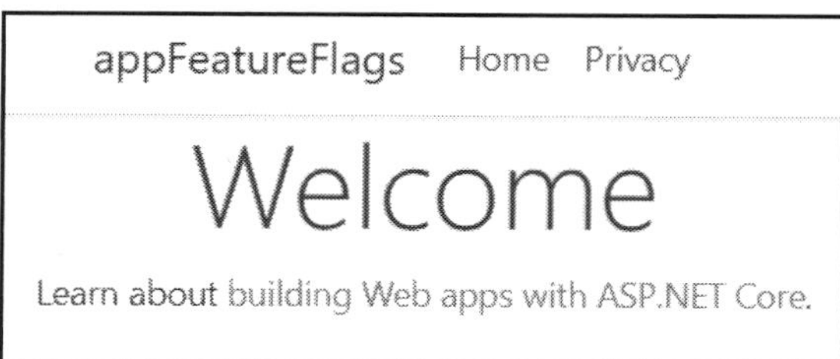

To display the image, we need to activate the feature dynamically, like so:

1. Go to the back office at `http://<yoursite>/_features`. We'll see a switch called `ShowHomeBow`.
2. We activate the flags by switching on the toggle, as shown in the following screenshot:

3. Reload the home page of our application. Here, we can see that the image is displayed in the center of the page:

By using the `RimDev.FeatureFlags` framework and feature flags, we were able to enable or disable a feature of our application without having to redeploy it.

The complete source code for this application can be found at `https://github.com/PacktPublishing/Learning_DevOps/tree/master/CHAP13/appFeatureFlags`.

We have just seen how to use an open source tool to implement basic feature flags in a .NET Core application. We noticed that, with the use of an open source tool, we can create a database in our infrastructure. This use of features flags is quite basic, and moreover, the access to the UI for managing features flags is not secure.

Finally, as with any open source tool, it is important to check whether it is maintained and updated regularly by its editor or community. However, the use of open source tools for feature flags remains appealing and inexpensive for small business projects.

Now, let's look at another tool solution for feature flags, which is to use a PaaS solution in the cloud. One example of such a solution is LaunchDarkly.

Using the LaunchDarkly solution

In the previous section, we discussed using open source tools for feature flags, which can be a good solution but requires some infrastructure components and is dependent on a development language (in our example, it was .NET Core).

For the better use and management of feature flags, we can use a cloud solution that does not require the implementation of an architecture and provides a lot of features around the feature flags.

Among these cloud solutions (SaaS), there is **LaunchDarkly** (`https://launchdarkly.com/`), which is a SaaS platform that is composed of a feature flags management back office and SDKs that allow us to manipulate the features flags in our applications.

The LaunchDarkly SDKs are available for many development languages, such as .NET, JavaScript, Go, and Java. The complete list of SDKs is available here: `https://launchdarkly.com/features/sdk/`.

In addition to the classic version of feature flags management with RimDev.FeatureFlags, LaunchDarkly allows feature flags to be managed by a user and also provides A/B testing features that are linked to feature flags. A/B testing has the ability to measure the use of the application's features through feature flags.

However, LaunchDarkly is a paid solution (`https://launchdarkly.com/pricing/`). Fortunately, it provides a 30-day trial so that we can test it out. Please take a look at how to use LaunchDarkly so that you can implement feature flags in a .NET application.

For that, we will start by creating some feature flags:

1. First, log in to your LaunchDarkly account by clicking on the **Sign In** button that is located on the top menu on LaunchDarkly site. Alternatively, you can go to `https://app.launchdarkly.com/`.

2. Once we're connected to our account, in the **Account settings** section, we can create a new project called `DemoBook`:

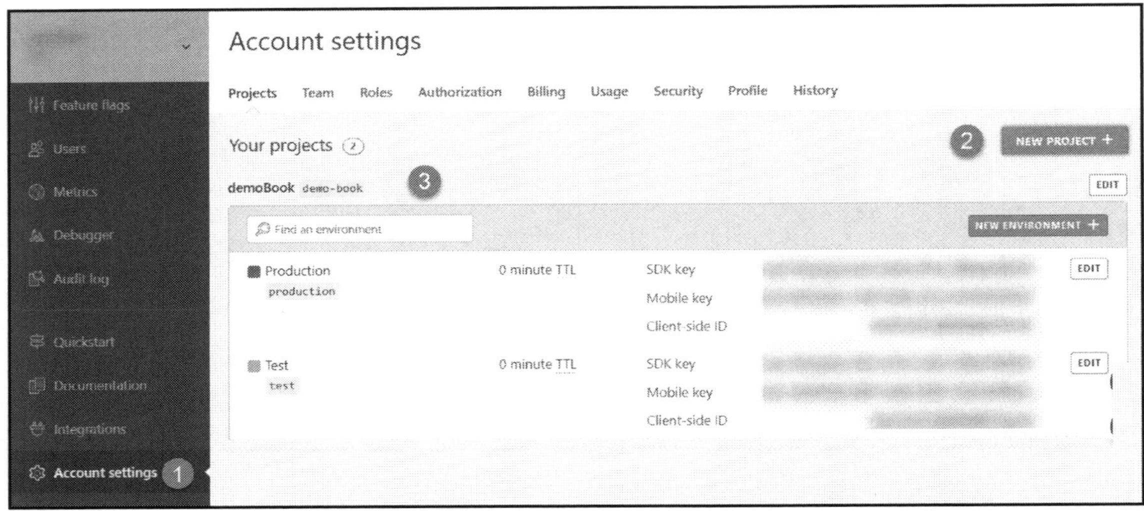

By default, there are two environments that are created in the project. We'll be able to create our own environments and then test the feature flags in different environments. In addition, each of these environments has a unique SDK key, which will serve as authentication for the SDK.

3. Then, in the environment called **Test**, we'll navigate on the **Feature flags** menu and click on the **NEW +** button, then we'll create a feature flag called `ShowBoxHome`, as shown in the following screenshot:

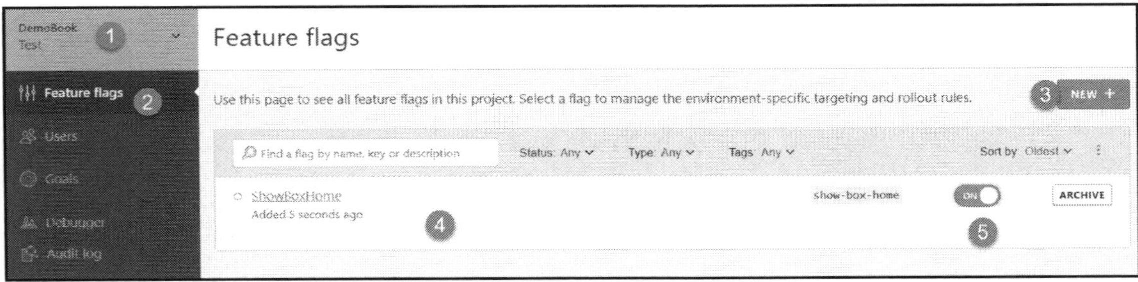

4. Once created, we can activate it by clicking on the toggle on/off switch.

Now that we have configured and created the feature flag in the LaunchDarkly portal, we will see how we can use the SDK in the application code.

In LaunchDarkly, the variations that are made to the features flags are made by users connected to the application. This means that the application must provide an authentication system.

To use the SDK and launch the application, follow these steps:

1. The first step is to choose the SDK that corresponds to the application development language. We can do this by going to `https://docs.launchdarkly.com/docs/getting-started-with-launchdarkly-sdks#section-supported-sdks`. In our case, we have a .NET application, so we will follow the following procedure: `https://docs.launchdarkly.com/docs/dotnet-sdk-reference`. Now, let's integrate the reference to the NuGet `LaunchDarkly.ServerSdk` (`https://www.nuget.org/packages/LaunchDarkly.ServerSdk/`) package in the `.csproj` file of our application by adding it to the reference packages, like so:

```
<ItemGroup>
 <PackageReference Include="LaunchDarkly.ServerSdk" Version="5.6.5"
/>
 ...
 </ItemGroup>
```

2. In the .NET code, we do this in the controller. To do this, we need to import the SDK with the `using` command:

```
using LaunchDarkly.Client;
```

3. Still in the controller code, we add the connection to LaunchDarkly, as invoked by `FeatureFlag`:

```
public IActionResult Index() {
 LdClient ldClient = new LdClient("sdk-eb0443dc-xxxx-xxx-xx-xxx");
 User user = LaunchDarkly.Client.User.WithKey(User.Identity.Name);
 bool showBoxHome = ldClient.BoolVariation("show-box-home", user,
false);
 return View(new HomeModel{ShowBoxHome = showBoxHome});
}
```

For the connection to LaunchDarkly, we need to use the SDK key that was provided when the project was created. Then, in this code, we connect the user who is connected to the application to the feature flag that we created previously in the portal.

4. Finally, in the `Home/Index.chtml` view, we add the following code to add the condition that will display the image, depending on the value of the feature flag:

```
<div class="text-center">
    @if(Model.ShowBoxHome){
        <div><img src="img/test.png"></div>
    }
<div>
```

5. Finally, we deploy and execute the application. The home page shows the central image because the feature flags are set to **true** by default:

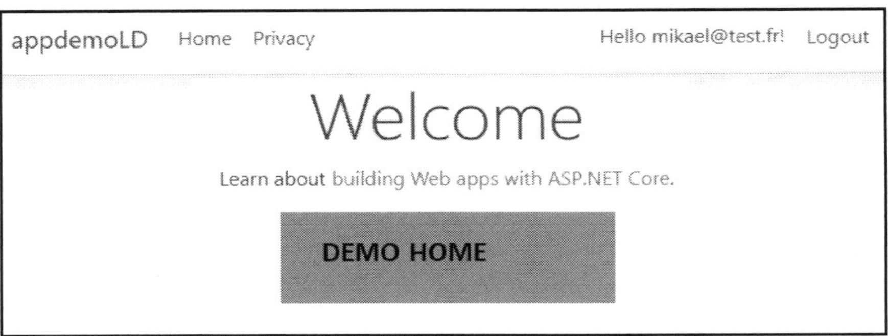

6. Then, we go to the LaunchDarkly portal and modify the configuration of this feature flag for the current user with the value **false**. In **Users** management, we select the user:

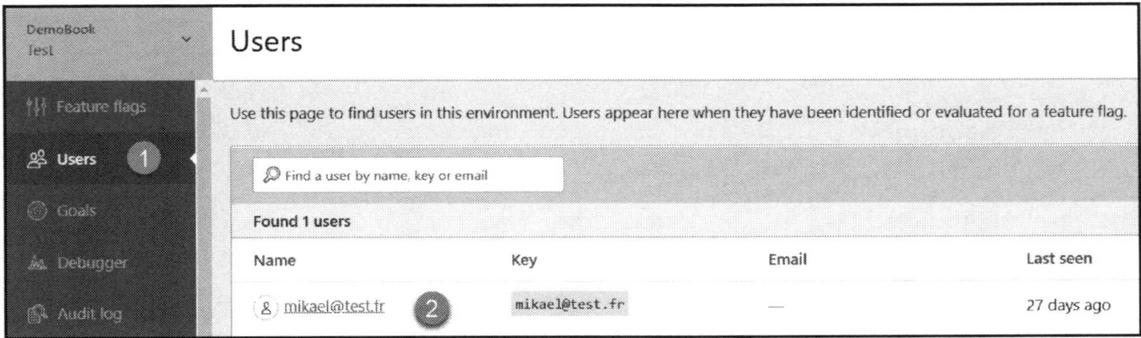

Then, we update the value of the feature flag to **false**:

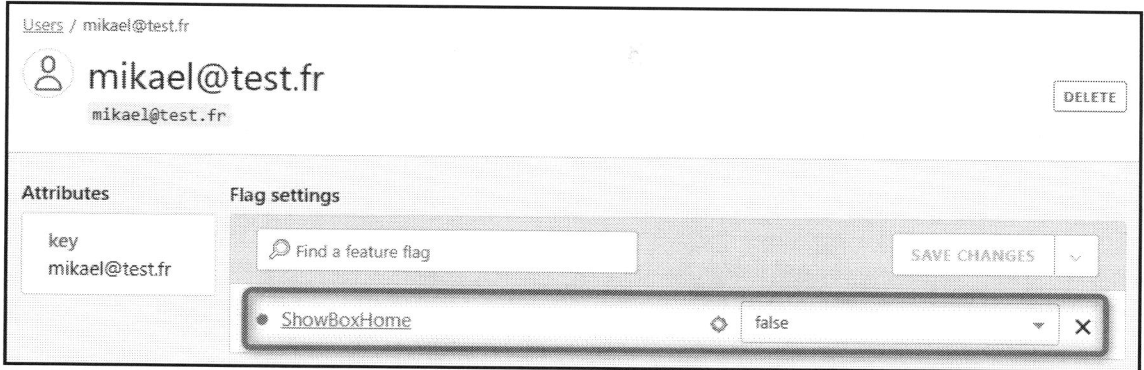

7. By reloading the page, we can see that the central image is no longer displayed:

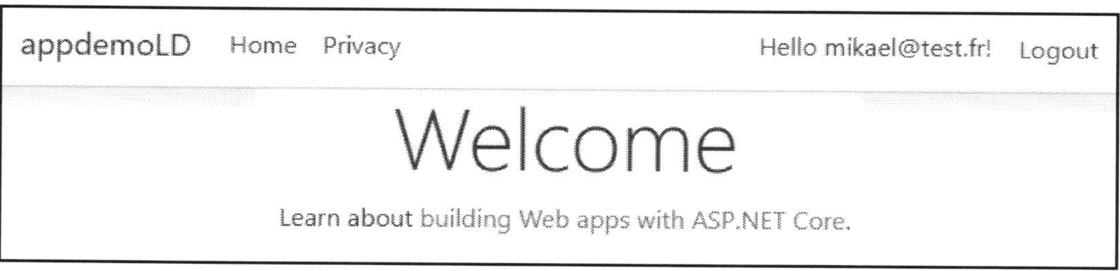

Here is an example of how to use LaunchDarkly, which has many other interesting features, such as a user management system, feature usage with A/B testing, integration with CI/CD platforms, and reporting.

The complete code source for this application can be found at `https://github.com/PacktPublishing/Learning_DevOps/tree/master/CHAP13/appFeatureFlags`.

What we have discussed in this section is an overview of LaunchDarkly, which is a feature flags cloud platform. We studied its implementation in a web application with the creation of feature flags in the LaunchDarkly portal.

Then, we manipulated this feature flag in the application code via the SDK provided by LaunchDarkly. Finally, in the LaunchDarkly portal, we enabled/disabled a feature for a user who wants to test a new feature of the application without having to redeploy it.

Summary

In this chapter, we focused on improving production deployments. We started with using Terraform to reduce downtime during provisioning and resource destruction.

Then, we focused on the practice of blue-green deployment and its patterns, such as canary release and dark launch. We looked at the implementation of a blue-green deployment architecture in Azure using App Service and the Azure Traffic Manager component.

Finally, we detailed the implementation of feature flags in a .NET application using two types of tools: RimDev.FeatureFlags, which is an open source tool that offers a basic feature flags system, and LaunchDarkly, which is a cloud-based solution. It's not free of charge but provides complete and advanced feature flags management.

The next chapter is dedicated to GitHub. Here, we will look at the best practices for contributing to open source projects.

Questions

1. In Terraform, what option can we use to reduce downtime?
2. What is a blue-green deployment infrastructure composed of?
3. What are the two blue-green deployment patterns that we looked at in this chapter?
4. In Azure, what are the components that allow us to apply a blue-green deployment practice?
5. What is the role of feature flags?
6. What is the RimDev.FeatureFlags tool?
7. Which feature flags tool that we discussed in this chapter is an SaaS solution?

Further reading

If you want to find out more about zero-downtime and blue-green deployment practices, take a look at the following resources:

- Zero Downtime Updates with HashiCorp Terraform: `https://www.hashicorp.com/blog/zero-downtime-updates-with-terraform`
- Blue-Green Deployment by Martin Flower: `https://martinfowler.com/bliki/BlueGreenDeployment.html`

- Feature Toggle by Martin Flower: https://martinfowler.com/articles/feature-toggles.html
- Blue-Green Deployment on Azure with Zero Downtime (article): http://work.haufegroup.io/Blue-Green-Deployment-on-Azure/
- Feature Flags Guide: http://featureflags.io/
- LaunchDarkly Uses Cases: https://launchdarkly.com/use-cases/

14
DevOps for Open Source Projects

Until a few years ago, the open source practice, which consists of delivering the source code of a product to the public, was essentially only used by the Linux community. Since then, many changes have taken place in relation to open source with the arrival of GitHub. Microsoft has since made a lot of its products open source and is also one of the largest contributors to GitHub.

Today, open source is a must in the development and enterprise world, regardless of whether we wish to use a project or even contribute to it.

Throughout this book, we can see many instances where open source tools such as Terraform, Ansible, Packer, Jenkins, SonarQube, and the Security DevOps Kit have been used. However, one of the great advantages of open source is not only the use of products, but also the fact that we can contribute to them.

To contribute to an open source project, we need to participate in its evolution by discussing issues with its use or by making suggestions regarding its improvement. In addition, if you are a developer, we can also modify its source code to make it evolve.

Finally, as a developer or a member of an operational team, we can share our own project in open source and make it available to the community.

In this chapter, we'll discuss open source contributions and why it's important to apply DevOps practices to all open source projects.

All of these practices, such as the use of Git, a CI/CD pipeline, and security analysis, have already been discussed throughout this book. However, in this chapter, we'll focus more on how to apply them in the context of our open source project.

We will start with how to share the code of a project in GitHub and how to initialize a contribution to another project. We will also discuss the management of pull requests, which is one of the most important features of the contribution. In addition, we will look at how to indicate version changes using release notes, and the topic of binary sharing in GitHub Releases. We'll detail two tools: Travis CI and GitHub Actions, both of which allow us to make CI/CD pipelines on open source projects hosted on GitHub. Finally, we will end this chapter by looking at the source code analysis of open source projects. We'll do this using SonarCloud, which is used for static code analysis, and WhiteSource Bolt, which is used for analyzing package security vulnerabilities contained in an open source project.

In this chapter, we will cover the following topics:

- Storing the source code in GitHub
- Contributing to open source projects using pull requests
- Managing the changelog and release notes
- Sharing binaries in GitHub releases
- Using Travis CI for continuous integration
- Getting started with GitHub Actions
- Analyzing code with SonarCloud
- Detecting security vulnerabilities with WhiteSource Bolt

Technical requirements

In this chapter, we will use GitHub as a Git repository platform to store our open source project, so you will need a GitHub account, which you can create for free here: `https://github.com/`. In order to fully understand the DevOps practices that will be used in this chapter, you should be well-versed with the following chapters of this book:

- `Chapter 5`, *Managing Your Source Code with Git*
- `Chapter 6`, *Continuous Integration and Continuous Delivery*
- `Chapter 10`, *Static Code Analysis with SonarQube*

The complete code for this chapter can be found at `https://github.com/PacktPublishing/Learning_DevOps/tree/master/CHAP14`.

Check out the following video to see the code in action:

`http://bit.ly/2Jlm1Xz`

Storing the source code in GitHub

If we want to share one of our projects in an open source fashion, we must version its code in a Git platform that allows the following elements:

- Public repositories; that is, we need to have access to the source code contained in this repository, but without necessarily being authenticated on this Git platform.
- Features and tools for code collaboration between the different members of this platform.

There are two main platforms that allow us to host open source tools: **GitLab**, which we looked at in the *Using GitLab CI* section in `Chapter 6`, *Continuous Integration and Continuous Delivery*, and **GitHub**, which is now the most used platform for open source projects.

Let's look at how to use GitHub so that we can host our project or contribute to another project.

Creating a new repository on GitHub

If we want to host our project on GitHub, we need to create a repository. Follow these steps to do so:

1. The first step is to log in to our GitHub account or create an account if we are a new user by going to `https://github.com/`.
2. Once connected, we will go to the **Repositories** tab inside our account.
3. Click on the **New** button, as shown in the following screenshot:

4. Create a new repository form that can be filled in, as shown in the following screenshot:

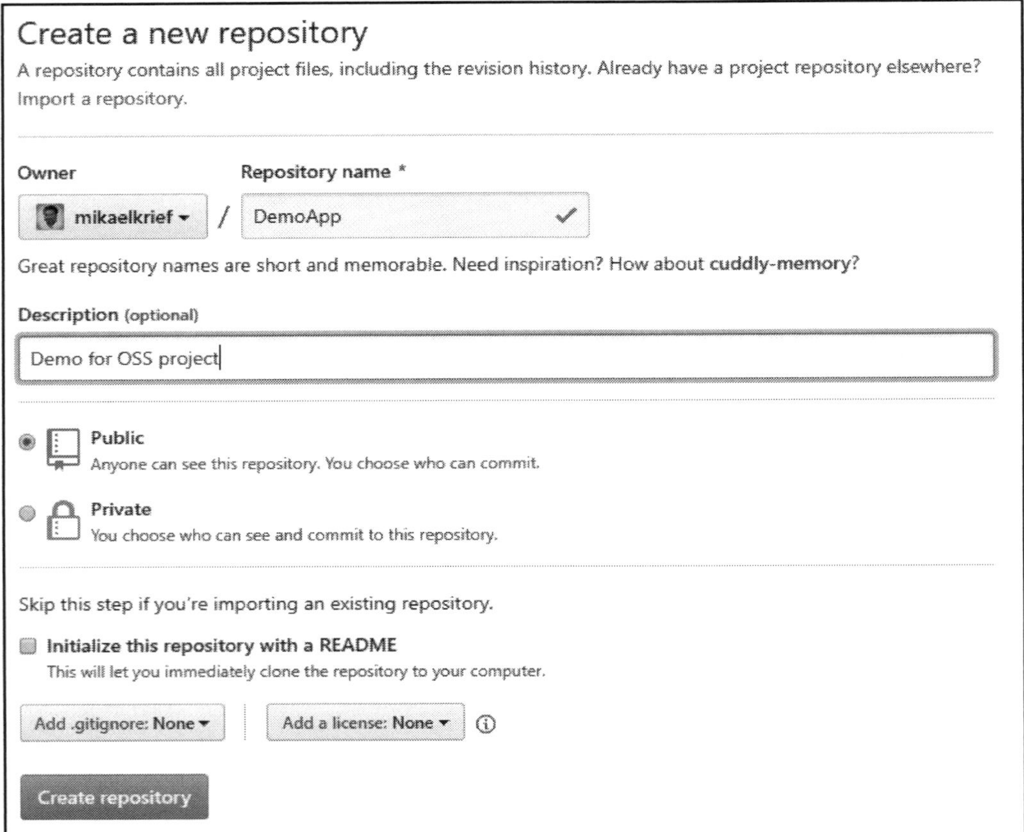

The information that needs to be filled in for the creation of a repository is as follows:

- The name of the repository.
- A description (which is optional).
- We need to specify whether the repository is **Public** (accessible by everyone, even if they're not authenticated) or **Private** (available only to the members we give it access to).
- We can also choose to initialize the repository with an empty `README.md` file, as well as a `.gitignore` file.

5. Then, validate the form by clicking on the **Create repository** button.
6. As soon as the repository is created, the home page will display the first Git instructions so that we can start archiving its code. The following screenshot shows part of the instruction page of a new GitHub repository:

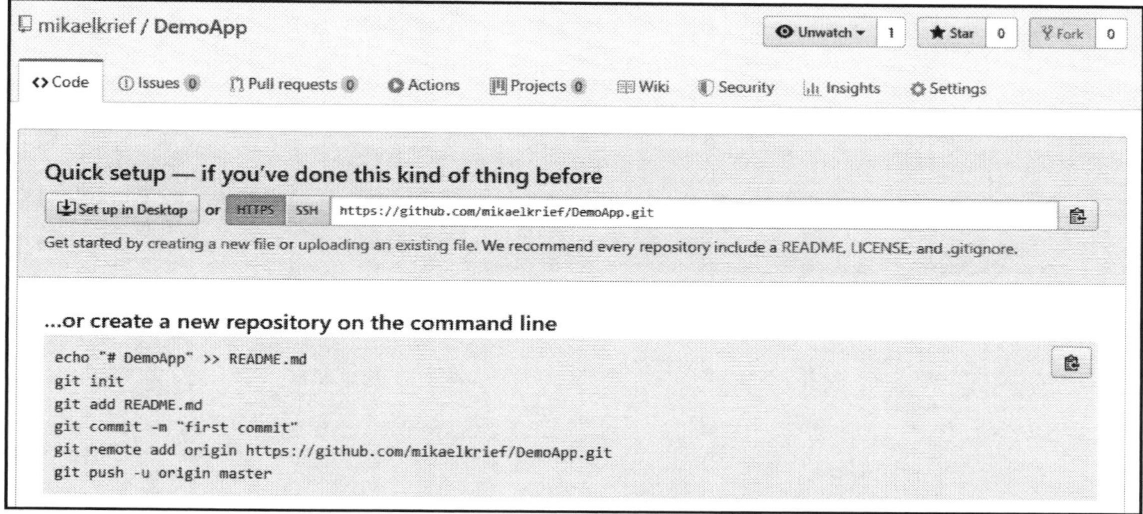

Everything is ready for archiving the code in GitHub. We do this by using the workflow and Git commands we looked at in `Chapter 5`, *Managing Your Source Code with Git*.

This procedure is valid for creating a repository in GitHub. Now, let's look at how to contribute to GitHub by using a project from another repository.

Contributing to the GitHub project

We have just looked at how to create a repository on GitHub. However, what we need to know is that, by default, only the owner of the repository is allowed to modify the code of this repository.

We can add people as collaborators to this repository so that they can make changes to the code. For more information about this procedure, please go to `https://help.github.com/en/articles/inviting-collaborators-to-a-personal-repository`.

With the help of this principle, we don't have the right to modify the code of another repository.

To contribute to the code of another repository, we will need to create a **fork** of the initial repository that we want to contribute to. A fork is a duplication of the initial repository that is performed in our GitHub account, thus creating a new repository in our account.

Follow these steps to learn how to create a fork of a repository:

1. First of all, let's navigate to the initial repository that we want to contribute to.
2. Then, click on the **Fork** button at the top of the page, as shown in the following screenshot:

3. After a few seconds, this repository will be forked and duplicated with all its content in our account. By doing this, we get a new repository in our account that is linked to the initial repository, as shown in the following screenshot:

4. Now, we have an exact copy of the repository that we want to contribute to in our GitHub account. We are free to modify the code and make commits of our changes, all of which will be archived in our repository.

However, even if there is a link between the initial repository and the fork, the code for each repository is completely uncorrelated and there is no synchronization of the automatic code.

In this section, we've discussed the steps we need to take to create a GitHub repository or to make a fork of another repository so that we can contribute to it. We will now look at how to propose code changes and merge our code into another repository using a pull request.

Contributing using pull requests

When we want to contribute to an open source project in GitHub, we need to make changes to the source code of the application that is in the repository of our GitHub account. To merge these code changes to the initial repository, we have to perform a **Merge** operation.

In GitHub, there is an element called **pull request** that allows us to perform a Merge operation between repositories. In addition to performing a simple and classic merge between code branches, a pull request also adds a whole new aspect of collaboration by providing features that allow different contributors to discuss code changes.

Let's look at how to carry out a pull request step by step:

1. After making changes to the code source in the repository in our account, we archive these changes by making a commit.
2. The changes that have been made are now ready to be merged with the remote repository. To do this, we go in our repository, either in the **Code** tab or in the **Pull requests** tab, and click on the **New pull request** button, as shown in the following screenshot:

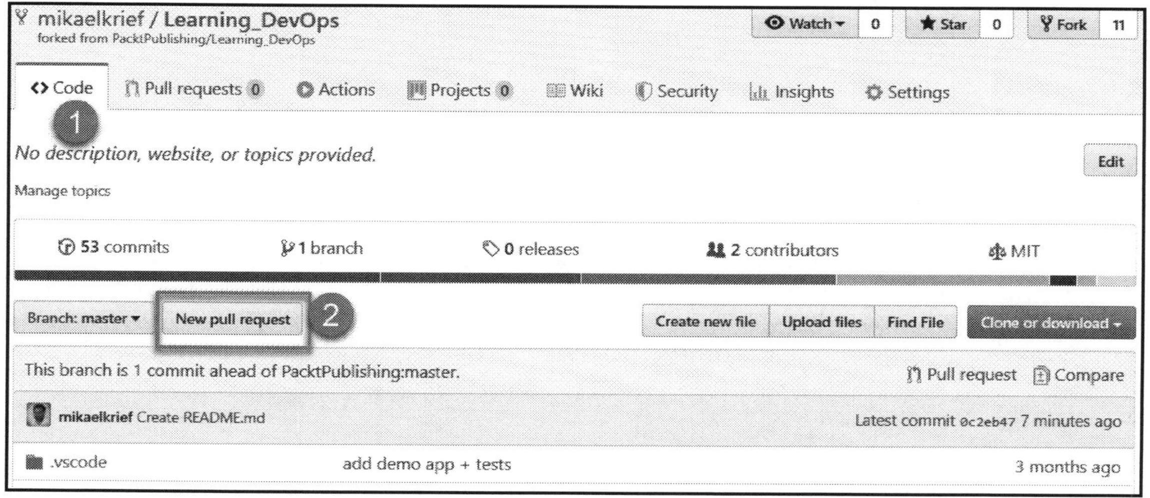

3. The page that appears specifies all of the information regarding the pull request that will be created, as shown in the following screenshot:

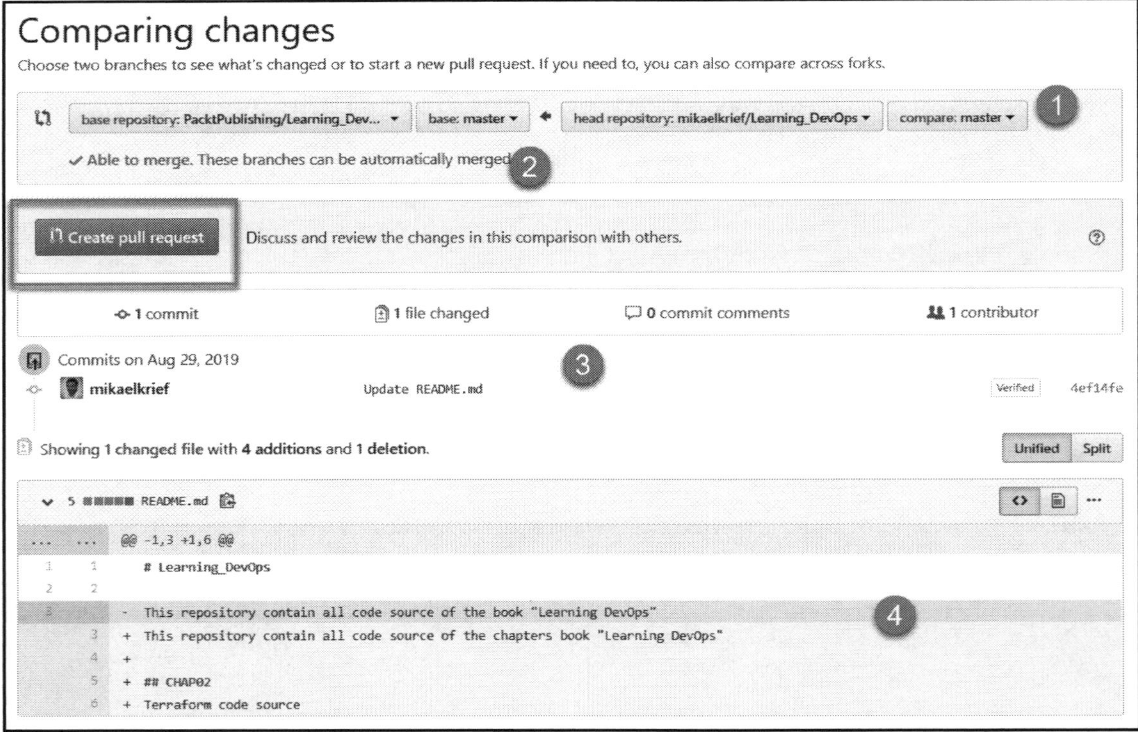

The information that is displayed on the screen is as follows:

- The source repository/branch and the target repository/branch
- An indicator that shows whether there are any code conflicts
- The list of commits that are included in this pull request
- The code differences of modified files

4. To validate the creation of the pull request, click on the **Create pull request** button.
5. Enter the name and description of the pull request in the form that appears, as shown in the following screenshot:

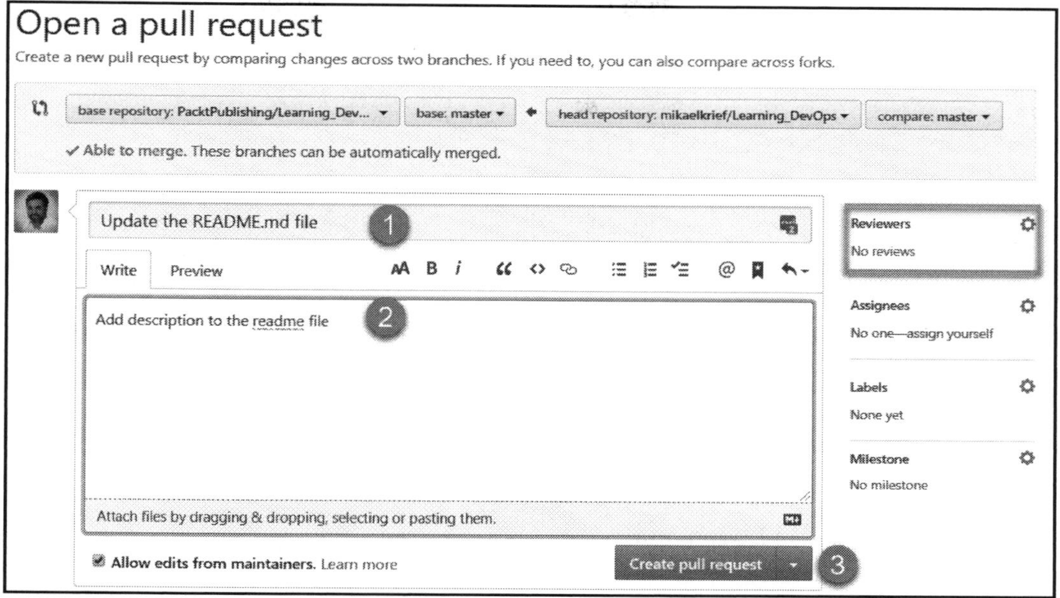

This information is important because it will help the repository target owner quickly understand the objectives of the code changes included in this **pull request**. In addition, from the right-hand panel, it is possible to select reviewers who will be notified of the pull request by email. This will be done by the person who is in charge of reviewing the code changes and validating or rejecting them.

6. Finally, we validate the creation of the pull request by clicking on the **Create pull request** button.

After the creation of the pull request, the owner of the original repository will see that a fresh pull request has been opened (with the title we provided) in the **Pull requests** tab of their repository. They can click on the **Pull requests** to access it and check all of its details:

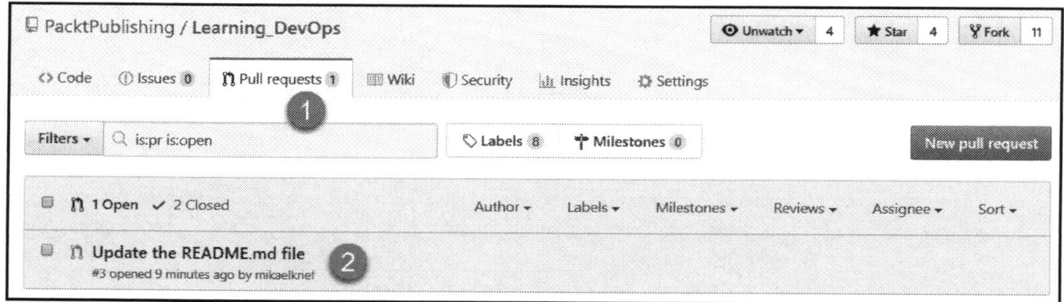

In the following screenshot, we can see the different options that have been proposed for this pull request:

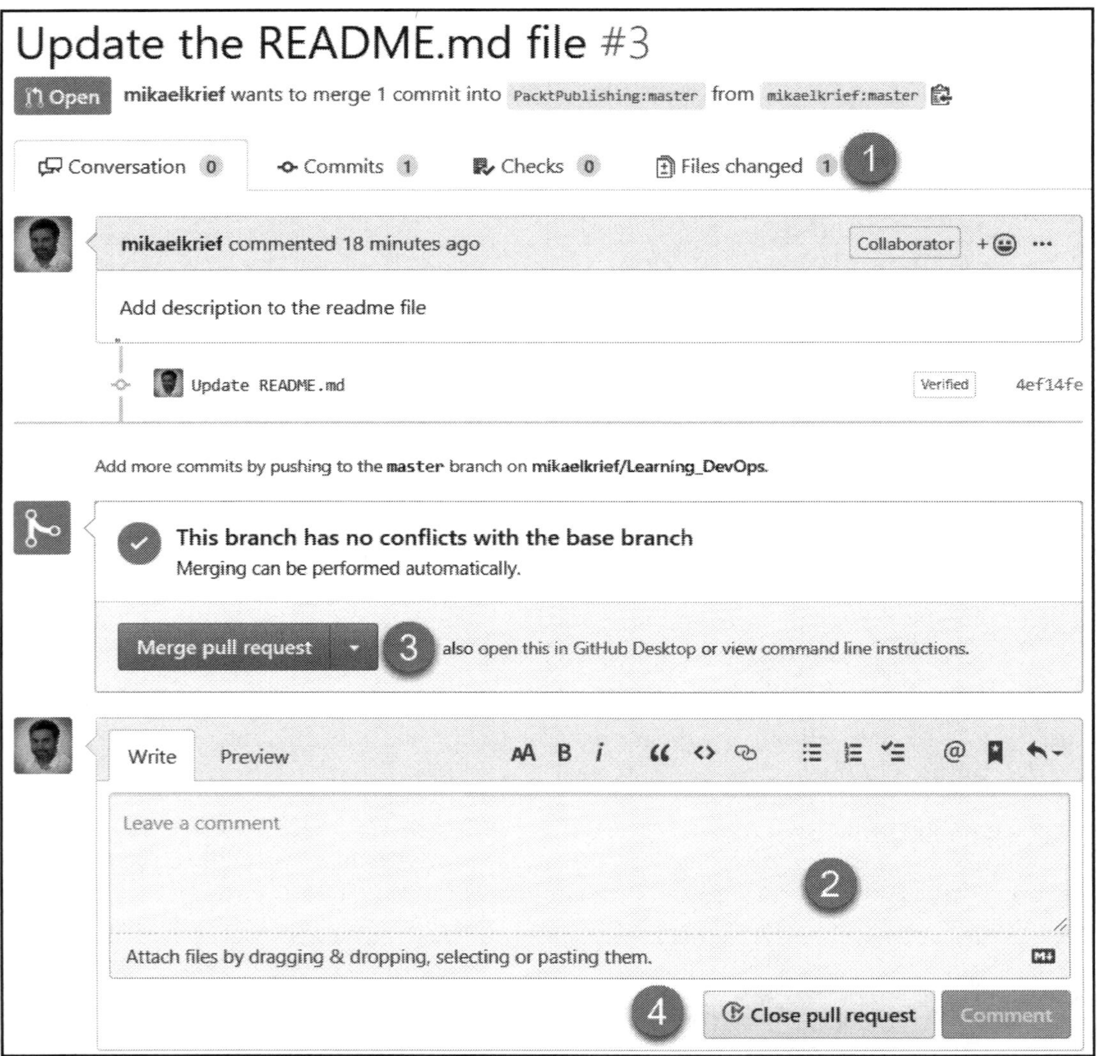

The following are the different operations that the repository owner can perform on this pull request:

- By clicking on the **Files changed** tab, the reviewer can see the changes that have been made in the code and leave notes on each of the modified lines.
- The reviewer or repository owner can initiate a discussion on code changes and click on the **Comment** button to validate their comments.
- If the owner is satisfied with the code changes, they can click on the **Merge pull request** button to perform the merge.
- On the other hand, if the owner is not satisfied and refuses the pull request, they can click on the **Close pull request** button, where the request is then closed.

Once merged, the pull request will have a **Merged** status and the code of the original repository will be updated with the code changes we have made.

In this section, we have seen that, with a pull request, we have a simple way of contributing to an open source project in GitHub by proposing code changes. Then, the project owner can either accept this pull request and merge the code or refuse the changes. In the next section, we will see how we can manage the changes that we've made to our project using the changelog file.

Managing the changelog and release notes

When we host a project as open source, it is good practice to provide information to users about the changes that are being applied to it as they occur. This change logging system (also called **release notes**) is all the more important if, in addition to the source code, our repository also publicly provides a binary of the application since the use of this binary is dependent on its different versions and code changes.

Logically, we could find the history of code changes by navigating through the history of Git commits. However, this would be too tedious and time-consuming for Git novices. For these reasons, we will indicate the change in history with the code versions in a text file that can be read by everyone. This file has no fixed nomenclature or formalism, but for simplicity, we have decided to call it `CHANGELOG.md`.

This changelog file is, therefore, a text file in markdown format, which is easy to edit with simple formatting and is placed at the root of the repository. In this file, the history of the changes is indicated in a list form without giving too much detail on each change.

For better visibility, this history of the most recent changes will be ordered from newest to oldest so that we can quickly access the latest changes. To give you an idea of the shape of the changelog file, here is a screenshot that shows an extract from the changelog file of the Terraform provider for Azure:

1.30.1 (June 07, 2019)

BUG FIXES:

- Ensuring the authorization header is set for calls to the User Assigned Identity API's (#3613)

1.30.0 (June 07, 2019)

FEATURES:

- **New Data Source:** `azurerm_redis_cache` (#3481)
- **New Data Source:** `azurerm_sql_server` (#3513)
- **New Data Source:** `azurerm_virtual_network_gateway_connection` (#3571)

IMPROVEMENTS:

- dependencies: upgrading to Go 1.12 (#3525)
- dependencies: upgrading the `storage` SDK to `2019-04-01` (#3578)
- Data Source `azurerm_app_service` - support windows containers (#3566)
- Data Source `azurerm_app_service_plan` - support windows containers (#3566)
- `azurerm_api_management` - rename `disable_triple_des_chipers` to `disable_triple_des_ciphers` (#3539)
- `azurerm_application_gateway` - support for the value `General` in the `rule_group_name` field within the `disabled_rule_group` block (#3533)
- `azurerm_app_service` - support for windows containers (#3566)
- `azurerm_app_service_plan` - support for the `maximum_elastic_worker_count` property (#3547)
- `azurerm_managed_disk` - support for the `create_option` of `Restore` (#3598)
- `azurerm_app_service_plan` - support for windows containers (#3566)

The complete content of this file is available here: `https://github.com/terraform-providers/terraform-provider-azurerm/blob/master/CHANGELOG.md`.

The important information to mention in this file is the version history of the changes that have been delivered in this application. For each version, we write the list of new features, improvements, and bug fixes that were delivered in this version.

For each change, there is a very short description, and the commit number is assigned as a link that allows us to view all the details of the changes by clicking on it.

For full details on the format of the changelog file, take a look at the following documentation: `https://keepachangelog.com/en/1.10/`.

Finally, for integration into a DevOps process, it is also possible to automate the generation of the changelog file using Git commits and tags.

There are many scripts and tools that allow for the generation of this changelog, for example, a GitHub account (`https://github.com/conventional-changelog`) that offers this type of tool. However, if you hesitate between writing or generating this file, then here is a very interesting article explaining the pros and cons of these two methods: `https://depfu.com/blog/changelogs-to-write-or-to-generate`.

In this section, we have looked at how to inform users and contributors about the history of code changes on an open source project using a changelog file. Then, we looked at the useful information we should mention in this changelog file so that users can find out exactly what changes the application is undergoing between each version.

In the next section, we will look at how to share binaries in an open source project in GitHub releases.

Sharing binaries in GitHub releases

The purpose of an open source project is not only to make the source code of a project visible but also to share it with public users. For each new version of the project (called a **release**), this share contains a release note as well as the binary resulting from the compilation of the project.

Thus, for a user who wishes to use this application, it's not necessary for them to retrieve the entire source code and compile it—they just have to retrieve the shared binary from the desired release and use it directly.

Note that a release is linked to a Git tag, which is used to position a label at a specific point in the source code's history. A tag is often used to provide a version number to the source code; for example, a tag can be v1.0.1.

To learn more about tag handling in Git, read the documentation here: `https://git-scm.com/book/en/v2/Git-Basics-Tagging`.

In GitHub, in each repository, it is possible to publish releases from Tags Git, which will have a version number (from the Git tag), a description with the list of changes, and the application binaries.

Following this introduction to releases in GitHub, we will see how we can create a release in GitHub using the web interface:

1. To create a GitHub release, go to the repository that contains the application code.
2. Click on the **releases** menu, which can be found in the **Code** tab, as shown in the following screenshot:

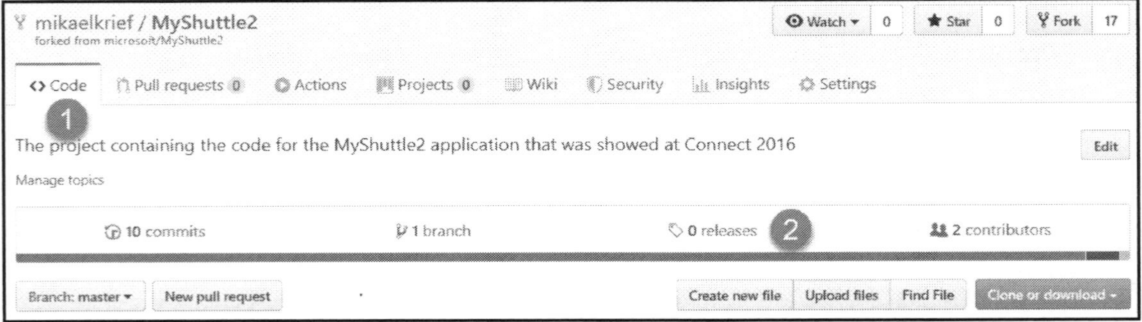

3. On the next page that appears, click on the **Create a new release** button to create a new release.
4. The release form is displayed and the release information is filled in, as shown in the following screenshot:

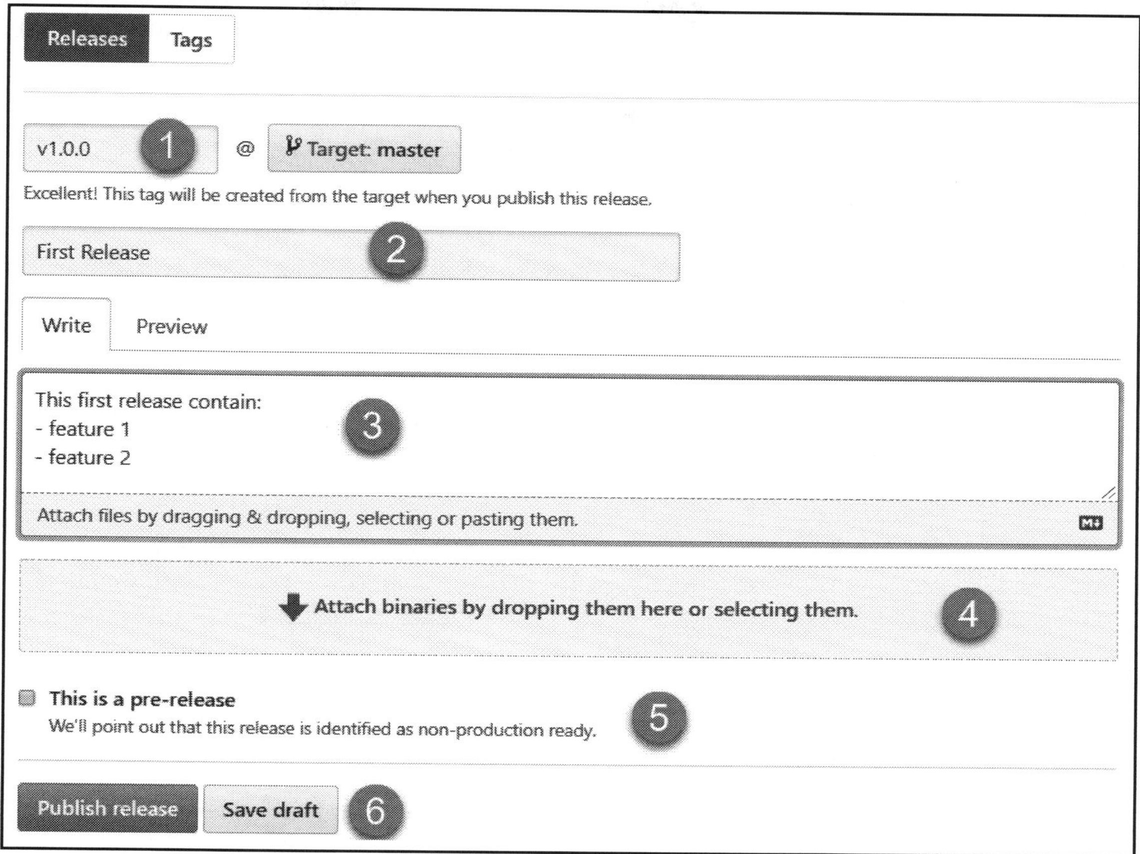

In this form, we have entered the following information:

- The tag associated with the release.
- The title of the release.
- The description of the release, which may contain the list of changes (release notes).
- We upload the application binary in a ZIP file format that corresponds to this release.
- We also mark the checkbox regarding whether it's a prerelease.

5. We then validate the new release by clicking on the **Publish release** button.

6. Finally, we are redirected to the list of releases of our project that we have just created, as shown in the following screenshot:

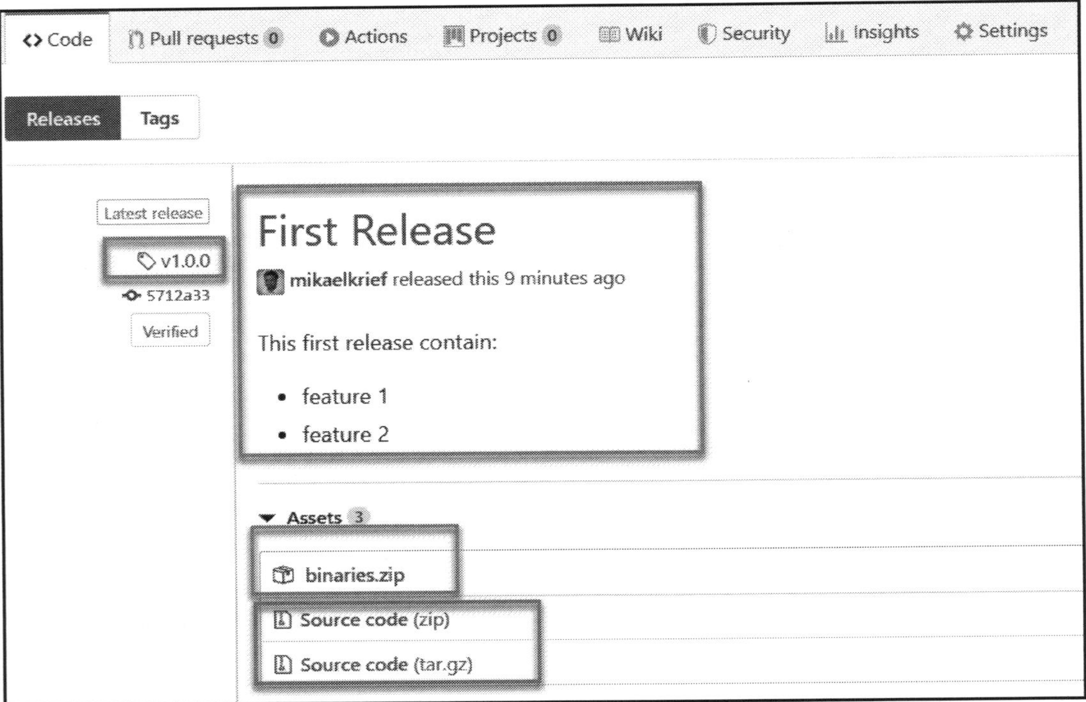

In the preceding screenshot, we can observe the tag (v1.0.0) associated with the release, the information we entered, and the `binaries.zip` file we uploaded. In addition, GitHub has been automatically added to the release of other assets, that is, the package (ZIP) that contains the source code of the application associated with this tag.

We have just seen that, via the GitHub web interface, we can create a GitHub release that allows us to share the release notes and binary files of a project to all users.

It is also possible to integrate all of these steps into the CI/CD pipeline with an automatic script using various GitHub APIs. The documentation for this can be found at `https://developer.github.com/v3/repos/releases/`.

In the next section, we will discuss the creation of a CI pipeline for an open source project using a cloud solution known as Travis CI.

Using Travis CI for continuous integration

When we have an open source project, it is important to set up a continuous integration pipeline that has two roles:

- To check the code changes that were proposed during a pull request that was opened by a contributor to the project. This verification will allow the code to be compiled and the unit tests from the pull request to be performed. The advantage of using this continuous integration process is that it blocks a pull request even before it has been opened and taken into account by other reviewers.
- Compile the code and execute the unit tests present in the main branch and publish the binaries of the compilation result. These binaries can be published publicly either in a GitHub release, as we saw in the previous section, *Sharing binaries in a GitHub release*, in a package manager such as NuGet or npm, or in any other public marketplace.

In this section, we will see how we can set up the simple continuous integration of a project that is in GitHub using a cloud platform known as **Travis CI** (`https://travis-ci.org/`).

Travis CI is a free cloud platform that allows us to run CI pipelines for open source projects that are hosted on GitHub. I recommend taking a look at the following, very simple, example of using Travis CI to create a CI pipeline in a Node.js application: `https://github.com/PacktPublishing/Learning_DevOps/tree/master/CHAP14/appdemo`.

To create this CI pipeline with Travis CI, follow these steps:

1. To use Travis CI, we need to connect to the Travis CI platform (`https://travis-ci.org/`) and click on **Sign in with GitHub**:

The Travis CI dashboard is displayed

2. Then, in the Travis CI portal, we will activate our GitHub repository by going to the **Settings** option of our account in the **Repositories** tab, as shown in the following screenshot:

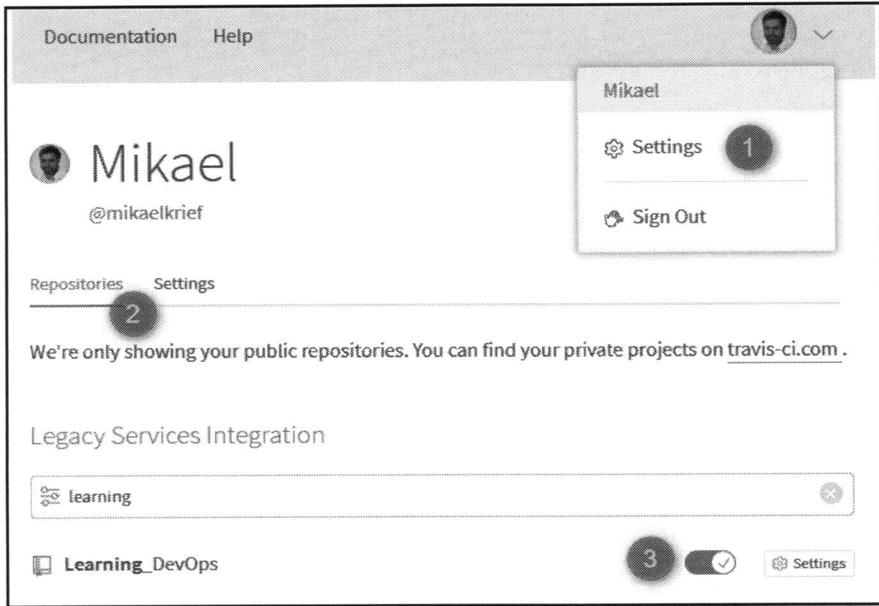

3. At the root of our GitHub repository, we create a `.travis.yml` file that contains the configuration and steps of our CI pipeline.

As a reminder, we have already seen this type of process, known as Pipeline as Code, which describes the configuration of a CI/CD pipeline in code, in the *Using the GitLab CI* section of `Chapter 6`, *Continuous Integration and Continuous Delivery*.

4. In the `.travis.yml` file, insert the following code:

```
language: node_js
node_js:
 - "7"
install:
 - cd ./CHAP14/appdemo/
 - npm install
script:
 - npm run build
 - npm run test
```

The complete source code for this file can be found at https://github.com/PacktPublishing/Learning_DevOps/blob/master/.travis.yml.

In the preceding code, in the `language` property, we indicate the development language of the application, that is, Node.js. Then, in the `install` section, we write the actions to be performed in the dependency's installation steps. Here, we used the `npm install` command.

Finally, in the `script` section, we describe the list of tasks to be performed in our pipeline, which are as follows:

- Execution of the `npm run build` command to compile the project
- Execution of the `npm run test` command to execute the tests contained in the project

5. We commit this file to our GitHub repository.
6. In the Travis CI portal, which can be found in the **Dashboard** section, we will see a build process that has been automatically triggered. This happens because it is a CI pipeline that is triggered at each commit. After executing for a few minutes, the job ends successfully. The following screenshot shows the successful completion of the Travis CI build:

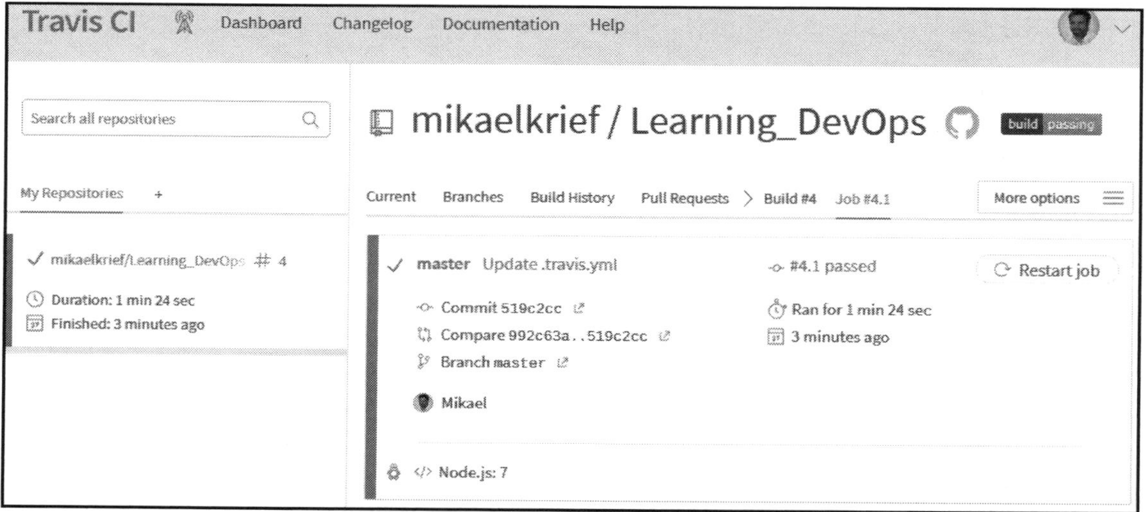

In this way, when a contributor requests a pull request, the code that's proposed by the contributor will be merged into the initial repository. This merge will, therefore, trigger a Travis CI build that will indicate whether the application code is valid with its compilation and test execution via the CI pipeline.

In this section, we have looked at the use of Travis CI, a SaaS solution, to ensure that the code of an open source project in GitHub is valid (that is to say that this code is written properly, that it compiles well and that the unit tests are successfully executed) for each commit of a contributor through a pull request.

In the next section, we will create this same pipeline but in GitHub using GitHub Actions.

Getting started with GitHub Actions

At the time of writing, for the past few months, GitHub has been integrating several other DevOps features into its repository source platform. This offers the advantage of being fully integrated with the repository's code.

Today, these new features are as follows:

- A GitHub Package registry, which is a package manager, whose presentation documentation can be found at `https://github.com/features/package-registry`
- GitHub Actions, which is a CI/CD pipeline manager, whose presentation documentation can be found at `https://github.com/features/actions`

At the time of writing this book, these two new features are still in preview on GitHub. They will be released live in the coming months during the event known as **GitHub universe** (November 13, 2019).

To be able to use GitHub Actions, we must have previously registered for the Beta program at `https://github.com/features/actions` and wait for this feature to be activated in our account by the GitHub team.

In this section, we will provide an overview of the use of GitHub Actions, which allows for the creation of CI/CD pipelines directly within GitHub. This will check and deploy the source code that is hosted in our GitHub repository.

For this demonstration, we will create a CI pipeline in GitHub that will compile and run the tests of our Node.js application. The resources for this can be found at `https://github.com/PacktPublishing/Learning_DevOps/tree/master/CHAP14/appdemo`.

Follow these steps to create a CI pipeline with GitHub Actions:

1. Once the **Actions** feature has been activated on our GitHub account, go to the repository that contains the source code to be deployed and click on the **Actions** tab, as shown in the following screenshot:

2. Then, the GitHub interface proposes pipeline templates, called **workflows**, according to the different development languages, or the creation of a workflow by starting with an empty template. The following screenshot shows the choices of workflow creation:

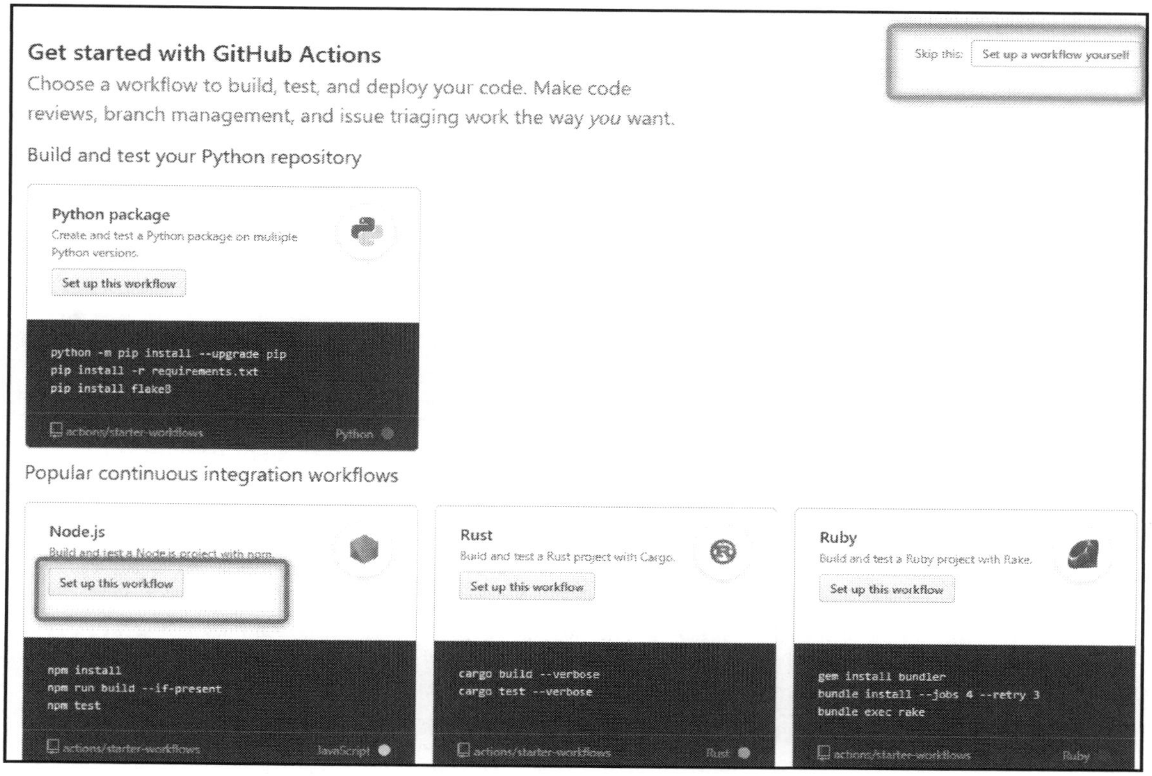

Get started with GitHub Actions

Choose a workflow to build, test, and deploy your code. Make code reviews, branch management, and issue triaging work the way *you* want.

Skip this: Set up a workflow yourself

Build and test your Python repository

Python package
Create and test a Python package on multiple Python versions.
Set up this workflow

```
python -m pip install --upgrade pip
pip install -r requirements.txt
pip install flake8
```

actions/starter-workflows Python

Popular continuous integration workflows

Node.js
Build and test a Node.js project with npm.
Set up this workflow

```
npm install
npm run build --if-present
npm test
```

actions/starter-workflows JavaScript

Rust
Build and test a Rust project with Cargo.
Set up this workflow

```
cargo build --verbose
cargo test --verbose
```

actions/starter-workflows Rust

Ruby
Build and test a Ruby project with Rake.
Set up this workflow

```
gem install bundler
bundle install --jobs 4 --retry 3
bundle exec rake
```

actions/starter-workflows Ruby

3. For this demonstration, we'll create a workflow from the workflow template designed for Node.js applications by clicking on the **Set up this workflow** button in the Node.js template box.
4. Then, GitHub displays the YAML code of the workflow that will be commuted in a `nodejs.yml` file, which will automatically be created in the `.github/workflows` folder tree. In this YAML code, which is also available at `https://github.com/PacktPublishing/Learning_DevOps/blob/master/.github/workflows/nodejs.yml`, we see the following:
    - The `runs-on` property, which specifies the Ubuntu agent of the pipeline provided by GitHub
    - A list of steps regarding the use of an action block (`actions/checkout`) that allows for the retrieval of the GitHub code, followed by a script block (`npm`) that will be executed on the Ubuntu agent
5. Before archiving this file, we'll add a small piece of code to it indicating the execution path of the scripts, as shown in the following screenshot:

```
steps:
- uses: actions/checkout@v1
- name: Use Node.js ${{ matrix.node-version }}
  uses: actions/setup-node@v1
  with:
    node-version: ${{ matrix.node-version }}
- name: npm install, build, and test
  run: |
    cd CHAP14/appdemo
    npm install
    npm run build --if-present
    npm test
  env:
    CI: true
```

6. Commit this file by clicking on the **Start commit** button at the top of the code editor. Once committed, the file will be present in the repository code and will trigger a new CI pipeline.

7. Finally, let's return to the **Actions** tab. We will see that a workflow has been triggered and successfully completed:

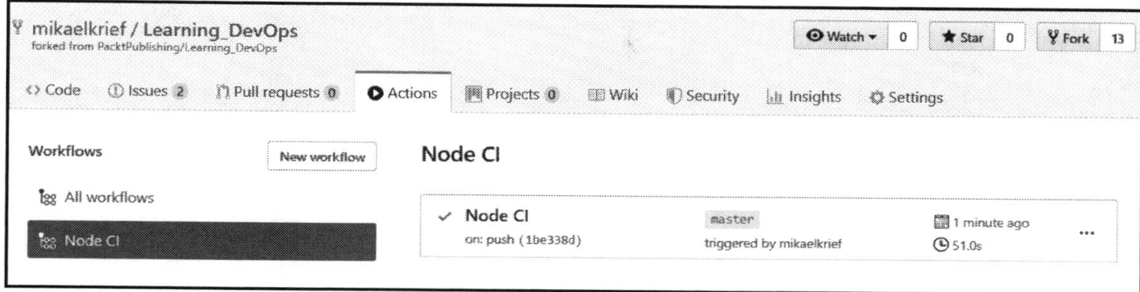

It is also possible to edit this workflow by clicking on the **Edit workflow** button at the top. From here, we can add a new workflow by clicking on the **Add a new workflow** button in the bottom-left panel. This is where we can view the list of workflows in this repository.

As for Travis CI, which we looked at in the previous section, this workflow is triggered during any commit to this repository, which happens either directly or during the merge of a pull request.

The great advantage of GitHub Actions is that it natively provides a very extensive catalog of actions in its Marketplace (`https://github.com/marketplace?type=actions`) and that it can also develop and publish its own actions (`https://help.github.com/en/articles/development-tools-for-github-actions`).

In this section, we discussed implementing a CI pipeline in GitHub using the new GitHub Actions feature, which allows for a complete DevOps integration for open source projects. So far in this chapter, we have focused on code management in GitHub and implementing CI/CD pipelines for open source projects using Travis CI and GitHub Actions.

In the upcoming sections, we will talk about open source code security. We will start with static code analysis with SonarCloud.

Analyzing code with SonarCloud

In `Chapter 10`, *Static Code Analysis with SonarQube*, we explained the importance of implementing static code analysis practices. For open source projects, code analysis is more important because the source code and its binaries are published publicly.

One of the roles of open source is to provide code and components that can be used in enterprise applications, so this code must be written correctly and without any security fails.

In this book, we have discussed that SonarQube, with its installations and uses, is one of the major tools that allows code analysis for enterprise applications. However, it requires the installation of an on-premises infrastructure, which is more expensive for a company.

For open source project code analysis, it is possible to use **SonarCloud** (`https://sonarcloud.io/`), which is exactly the same product as SonarQube but comes in a cloud solution that requires no installation.

SonarCloud has a free plan that allows us to analyze the code of open source public repository projects from GitHub, BitBucket, or even Azure Repos. For more information on its price plans, take a look at the following link: `https://sonarcloud.io/about/pricing`.

Let's look at how quick it is to set up code analysis on an open source project hosted on GitHub. Before implementing the analysis itself, we will connect to our GitHub repository in SonarCloud, and for that, we need to access the `https://sonarcloud.io/` page. From here, click on the **GitHub** button, as shown in the following screenshot:

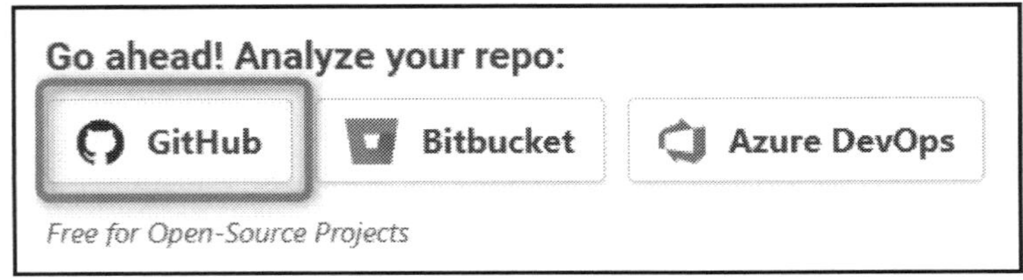

Then, we will configure SonarCloud so that we can create a project that will contain the analysis of our GitHub project, which is available at `https://github.com/mikaelkrief/Learning_DevOps/tree/master/CHAP14/appdemo`. To do this, follow these steps:

1. Once connected to SonarCloud with our GitHub account, click on the **Analyze new project** button on the home page:

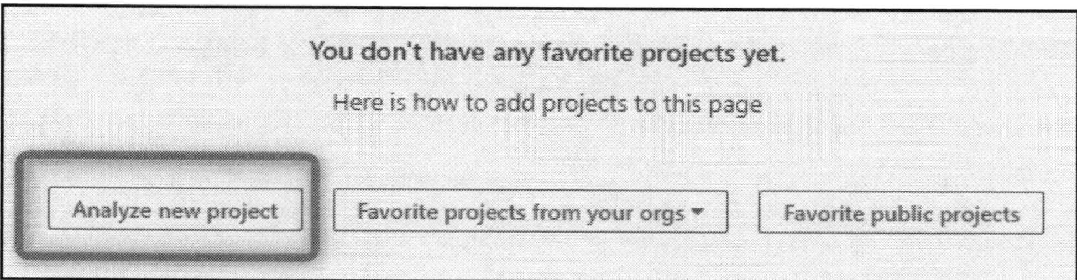

2. SonarCloud proposes a few steps for selecting the target GitHub repository for analysis. Let's choose the **Only select repositories** option and select the target repository:

3. Now, let's create a new organization based on our GitHub account by clicking on the **Create new SonarCloud organization from it** button. This will allow us to collaborate as a team on code analysis. We need to associate it with a free plan:

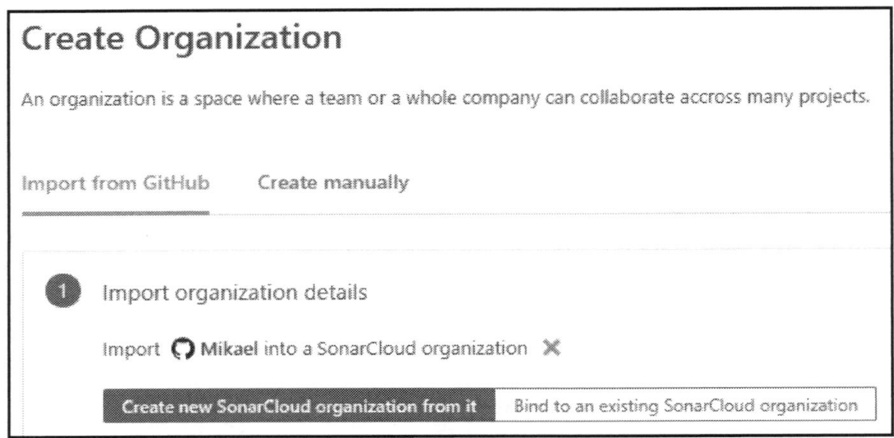

4. Next, mark the checkbox of our repository and associate it with our organization by clicking on the **Set Up** button:

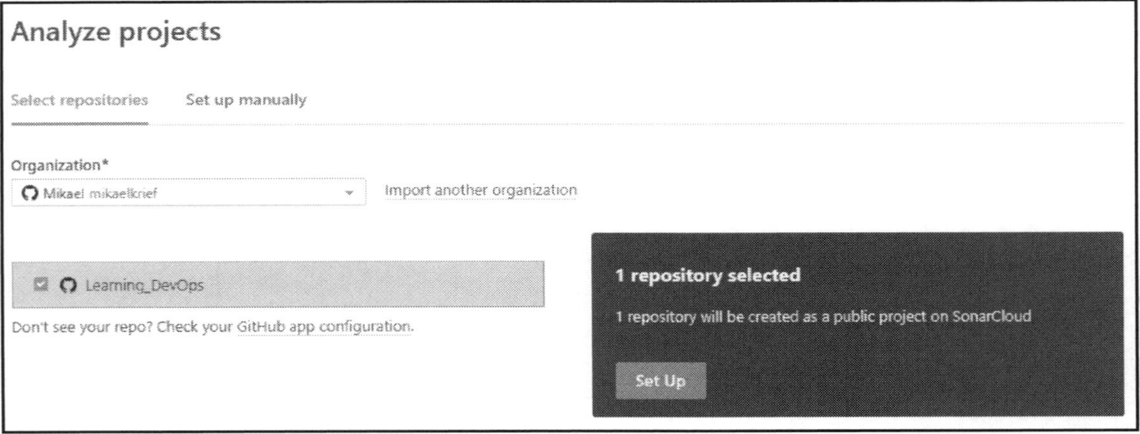

5. The project analysis dashboard will be displayed and indicating that the code has not yet been analyzed.

We have successfully configured the SonarCloud project. Now, let's proceed with the analysis of the project in the most basic way by following these steps:

1. At the root of our GitHub repository, we will create a `.sonarcloud.properties` file without any content that we'll archive with a commit. This `.sonarcloud.properties` file contains the SonarCloud analysis configuration of our project. By default, we leave it empty, without any particular configuration. To find out about the configuration of this file, take a look at its documentation: `https://sonarcloud.io/documentation/analysis/analysis-parameters/`.
2. The commit of this file will automatically trigger the SonarCloud analysis of this project.
3. After a few minutes, we will see the result of the analysis in the SonarCloud dashboard, as shown in the following screenshot:

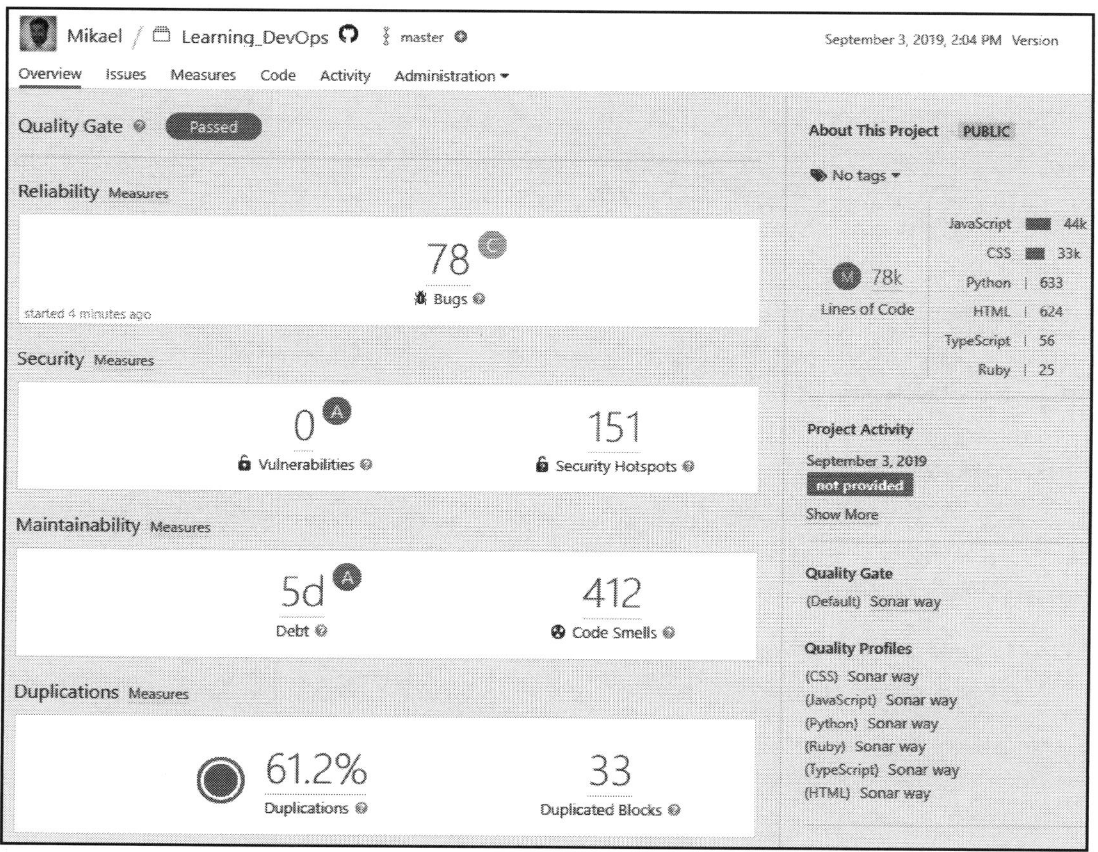

So, with each new code commit on this repository, either directly or via merging a pull request, the code analysis will be triggered and the SonarCloud dashboard will be updated.

It is clear that our end goal is to integrate SonarCloud analysis into a CI/CD pipeline, so here are some resources to help us integrate it:

- If you're using Azure DevOps, here is a complete tutorial that will guide you through the integration of SonarCloud into the pipeline: `https://docs.microsoft.com/en-us/labs/devops/sonarcloudlab/`.
- If you're using Travis CI, which we looked at in this chapter, take a look at the following documentation: `https://docs.travis-ci.com/user/sonarcloud/`.
- If you're using GitHub Actions, here is the Action that allows you to integrate the code analysis into a GitHub Actions workflow: `https://github.com/SonarSource/sonarcloud-github-action`.

In this section, we discussed how to configure SonarCloud, which is the cloud platform that analyzes static code. We did this to analyze the source code of an open source project on GitHub via a continuous integration process. Then, we looked at the result of this analysis on the dashboard.

In the next section, we will look at another aspect of open source code security, which is the analysis of code vulnerabilities using WhiteSource Bolt.

Detecting security vulnerabilities with WhiteSource Bolt

Due to their public visibility, open source projects or components are highly exposed to security vulnerabilities because it is easier to unintentionally inject a component (a package or one of its dependencies) containing a security vulnerability into them.

In addition to static source code analysis, it is also very important to continuously check the security of packages that are referenced or used in our open source projects.

There are many tools available that we can use to analyze the security of referenced packages in applications, such as SonaType AppScan (`https://www.sonatype.com/appscan`), Snyk (`https://snyk.io/`), and WhiteSource Bolt (`https://bolt.whitesourcesoftware.com/`).

For more information on open source vulnerability scanning tools, take a look at the following article, which lists 13 tools that analyze the security of open source dependencies: `https://techbeacon.com/app-dev-testing/13-tools-checking-security-risk-open-source-dependencies`.

Among all these tools, we will look at the use of **WhiteSource Bolt** (`https://bolt.whitesourcesoftware.com/`), which is free, able to analyze the package code of many development languages, and allows for direct integration into GitHub and Azure DevOps.

In our case, we will use it directly in GitHub to analyze the security of an application whose sources are available here: `https://github.com/PacktPublishing/Learning_DevOps/tree/master/CHAP14/appdemo`.

To do this security analysis, we will install and configure WhiteSource Bolt on our GitHub account and trigger a code analysis. Follow these steps to learn how to do this:

1. In a web browser, go to `https://bolt.whitesourcesoftware.com/github/` and click on the **DOWNLOAD FREE** button to install WhiteSource Bolt on your GitHub account.
2. We'll be redirected to the WhiteSource Bolt application, which can be found on the GitHub Marketplace (`https://github.com/marketplace/whitesource-bolt`). To install it using the free plan, click on the **Install it for free** button at the very bottom of the page, as shown in the following screenshot:

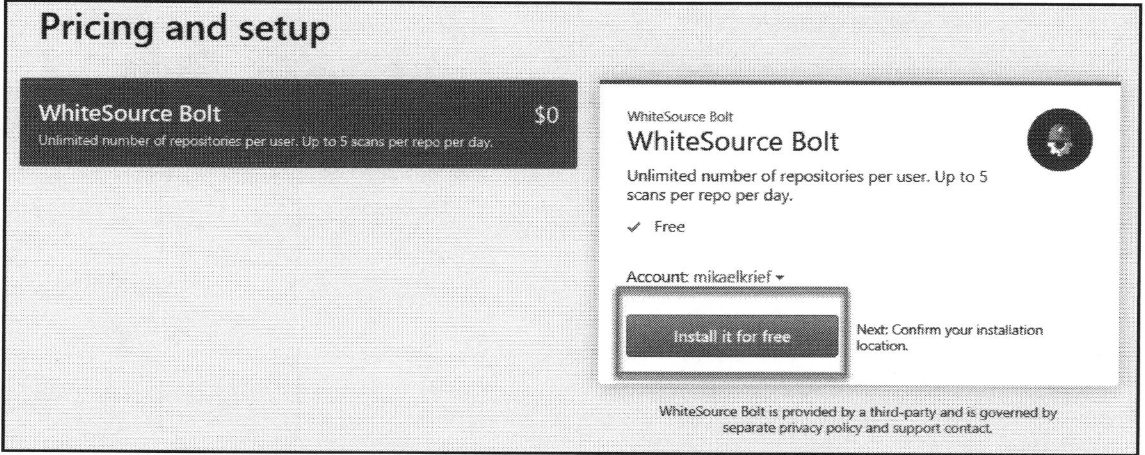

3. We confirm the purchase of the application for $0 by clicking on the **Complete order and begin installation** button. Then, on the next page, we confirm the installation of WhiteSource Bolt on the GitHub account.
4. After the installation has finished, we'll be redirected to the WhiteSource Bolt account creation page, where we will fill in our full name and country and then validate the form.
5. Now, let's activate the **Issues** features in GitHub by going to the **Settings** tab of the repository, which contains the code to scan, and checking the **Issues** checkbox, as shown in the following screenshot:

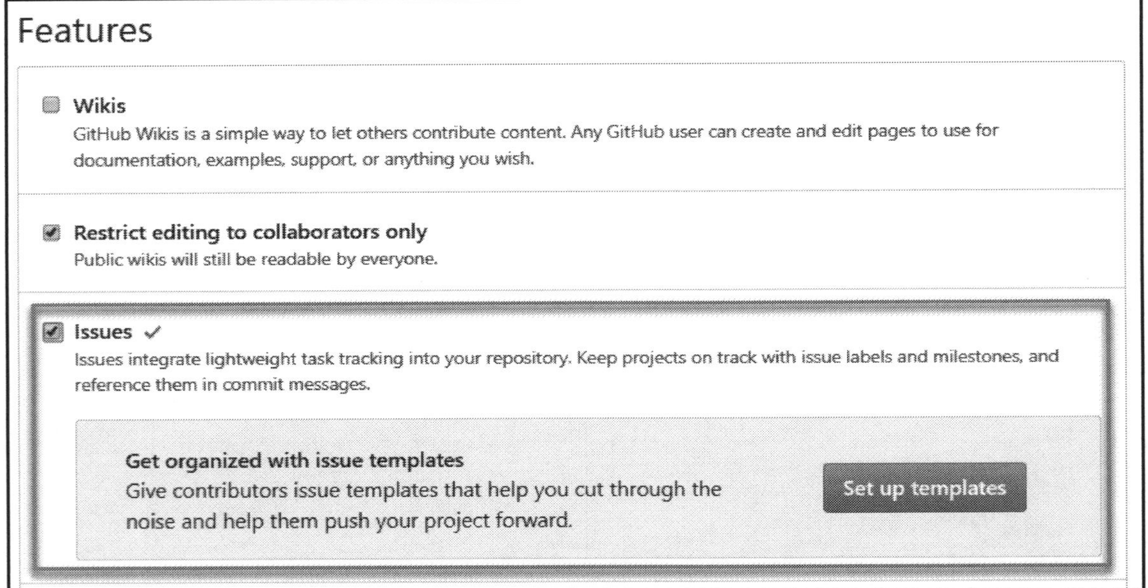

6. To configure the WhiteSource analysis on this repository, it is necessary to validate and merge the new pull request that we automatically created during the installation of WhiteSource Bolt:

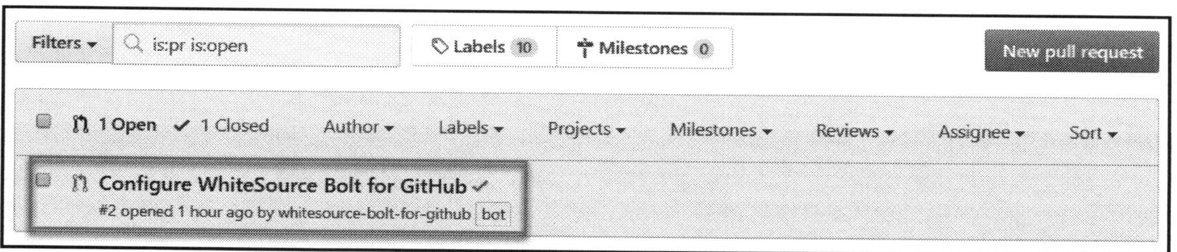

This pull request adds a `.whitesource` file that is used for configuration at the root of the repository.

7. Finally, we will trigger a code analysis by committing the code of this application in the GitHub repository.
8. After a few minutes, we will see a list of security issues in the **Issues** tab of the repository:

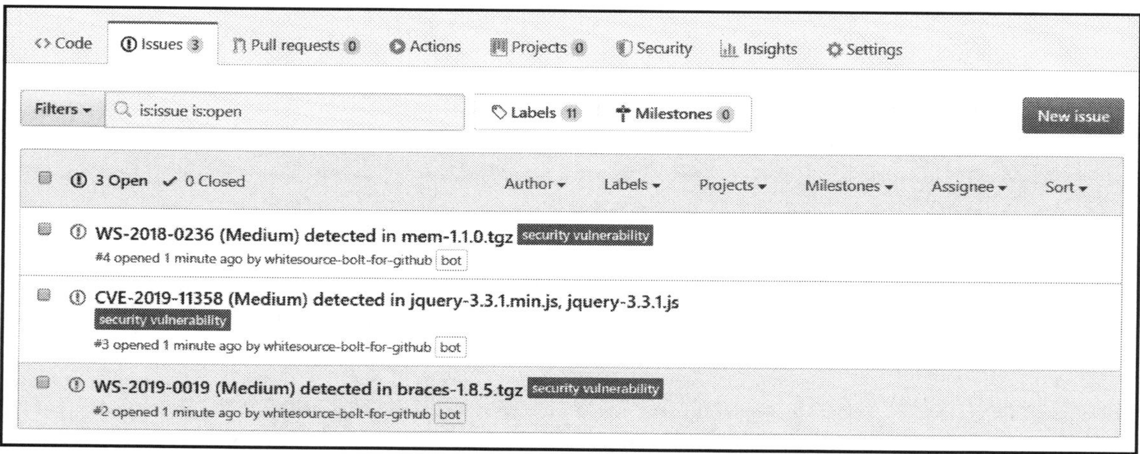

9. To find out the full details of a security issue, simply click on the desired issue, read it, and take the information provided in the description of the issue into consideration:

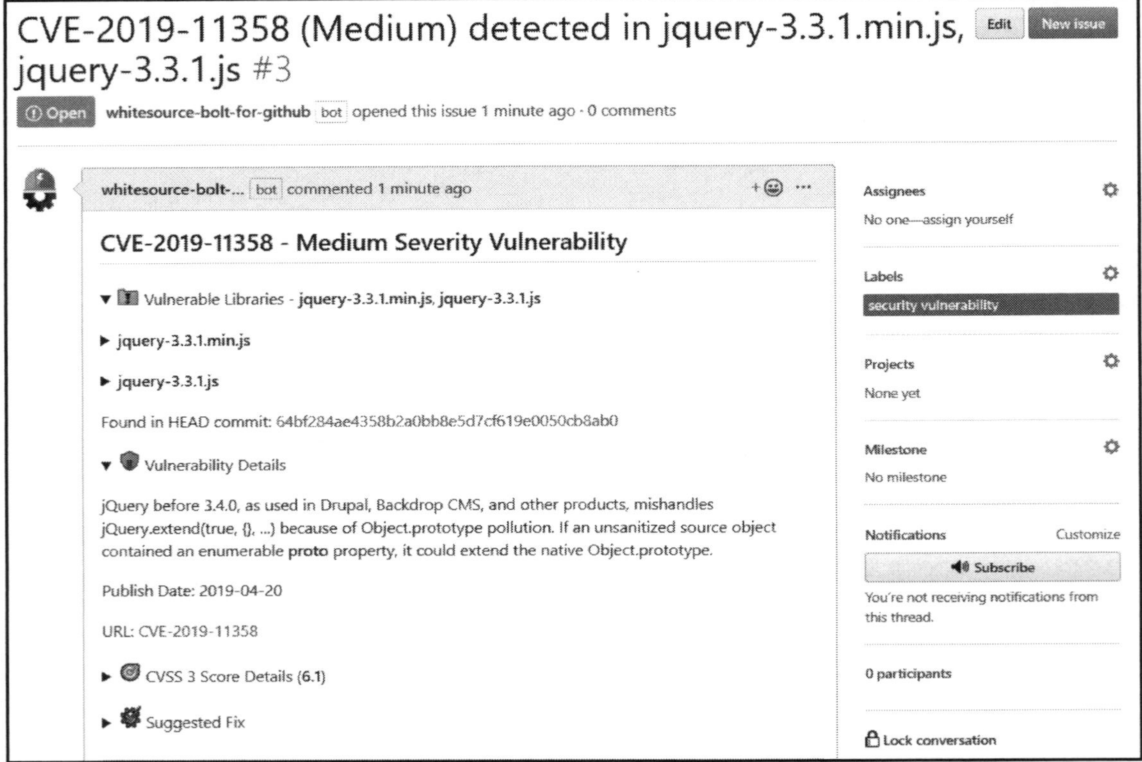

We will have to fix all of these problems and redo code commits to trigger new WhiteSource Bolt scans and ensure that we have a secure application for those who will use it.

In this section, we have seen how to analyze the code of this open source project using WhiteSource Bolt. We installed it and triggered a code analysis that revealed security issues in our demo application.

Summary

This chapter was dedicated to the DevOps best practices that can be applied on an open source project and especially on GitHub. In this chapter, we learned how to collaborate on open source code by starting with repository creation using GitHub and forks, and then looked at the usage of pull requests and how we can share binaries in GitHub Releases.

Then, we implemented continuous integration processes with two tools that integrate with GitHub, that is, Travis CI, which is a third-party cloud platform, and GitHub Actions, which is fully integrated with GitHub.

Finally, in the last part of this chapter, we looked at how to analyze open source code for static code analysis with SonarCloud and for security vulnerability analysis with WhiteSource Bolt.

In the next chapter, we will summarize every DevOps best practice we have talked about in this book.

Questions

1. In GitHub, can I modify the code of a repository of another user?
2. In GitHub, which element allows us to merge code changes between two repositories?
3. Which element allows us to simply display the history of code changes in an open source project?
4. In GitHub, which feature mentioned in this chapter allow to share binaries?
5. In Travis CI, where do we put the configuration of the CI pipeline?
6. What two tools that we have looked at in this chapter allow us to analyze the source code of an open source project?
7. In which GitHub tab are the security issues detected by WhiteSource Bolt listed?

Further reading

If you want to find out more about using DevOps practices on open source projects, take a look at the following resources:

- *GitHub Essentials, Achilleas Pipinellis* published by *Packt Publishing*: `https://www.packtpub.com/in/web-development/github-essentials-second-edition`
- Travis CI complete documentation: `https://docs.travis-ci.com/`

15
DevOps Best Practices

We have reached the last chapter of this book and, finally, after reading everything, you are probably asking yourself, what are the best practices to apply to effectively implement a DevOps culture?

This chapter, entitled *DevOps Best Practices*, is a great overview of the DevOps good practices that we have already seen and that will allow you to practice all the elements of what we have seen in this book.

We will discuss best practices in automation, tooling choice, Infrastructure as Code, application architecture, and infrastructure design. We will also discuss good practices to be applied in project management to facilitate the implementation of DevOps culture and practices. Then, we will review best practices about CI/CD pipelines, test automation, and the integration of security into your DevOps processes.

Finally, we will end this chapter with some of the best practices of monitoring in DevOps culture.

This chapter covers the following topics:

- Automating everything
- Choosing the right tool
- Writing all your configuration in code
- Designing the system architecture
- Building a good CI/CD pipeline
- Integrating tests
- Applying security with DevSecOps
- Monitoring your system
- Evolving project management

Automating everything

When you want to implement DevOps practices within a company, it is important to remember the purpose of the DevOps culture: it delivers new releases of an application faster, in shorter cycles.

To do this, the first good practice to apply is to automate all tasks that deploy, test, and secure the application and its infrastructure. Indeed, when a task is done manually, there is a high risk of error in its execution. The fact that these tasks are performed manually increases the deployment cycles of applications.

In addition, once these tasks are automated in scripts, they can be easily integrated and executed in CI/CD pipelines. Another advantage of automation is that developers and the operational team can spend more time and focus their work on the functionality of their business.

It is also important to start the automation of the delivery process at the beginning of project development; this allows us to provide feedback faster and earlier.

Finally, automation makes it possible to improve the monitoring of deployments by putting traces on each action and allows you to make a backup and restore very quickly in case of a problem.

Automating deployments will, therefore, reduce deployment cycles and the teams can now afford to work in smaller iterations. Thus, the time to market will be improved, with the added benefit of better-quality applications.

However, automation and orchestration require tools to be implemented, and the choice of these tools is an important element to consider in the implementation of a DevOps culture.

Choosing the right tool

One of the challenges a company faces when it wants to apply a DevOps culture is the choice of tools.

Indeed, there are many tools that are either paid for or free and open source and that allow you to version the source code of applications, process automation, implement CI/CD pipelines, and test and monitor applications.

Along with these tools, scripting languages are also added, such as PowerShell, Bash, and Python, which are also part of the DevOps suite of tools to integrate.

So it's a question I'm often asked: how do I choose the right DevOps tools that are useful for my company and business?

In fact, to answer this question, we must remember the definition of DevOps culture provided by Donovan Brown, which was mentioned in `Chapter 1`, *DevOps Culture and Practices*:

> *"The DevOps culture is the union of people, process, and products to enable continuous delivery of value to our end users."*

The important point of this definition is that DevOps culture is the *union* of Dev, Ops, processes, and also *tools*.

That is to say, the tools used must to be shared and usable by both Devs and Ops, and should be integrated into the same process.

In other words, the choice of tools depends on the teams and the company model. It is also necessary to take into account the financial system, by choosing open source tools, which are often free of charge; this is easier to use them at the beginning of the DevOps transformation of the company. That is not the case of paid tools, which are certainly richer in features and support, but require a significant investment.

Concerning scripting languages, I would say that the choice of language must be made according to the knowledge of the teams. For example, Ops teams that are more trained on Linux systems will be able to make automation scripts better with Bash than with PowerShell.

In this book, we have introduced you to several tools, some of which are open source and free, such as Terraform, Packer, Vault, and Ansible, and others that are paid for, such as Azure DevOps (for more than five users) or LaunchDarkly, and this is to help you better choose the tools that best suit you.

After this reflection on the choice of tools in the implementation of DevOps practice, we will look at another good practice, which is the fact of putting everything in code.

Writing all your configuration in code

We have seen throughout the book, especially with the first three chapters on Infrastructure as Code, that writing the desired infrastructure configuration in code offers many advantages both for teams and for the company's productivity.

It is, therefore, a very good practice to put everything related to infrastructure configuration in code. We have seen this in practice with Terraform, Ansible, and Packer, but there are many other tools that may be better suited to your needs and your organization. Among these tools, we are not only talking about the major editors, but also about the use of JSON files, Bash, PowerShell, and Python scripts that we'll apply to this configuration. The key is to have a description of your infrastructure in code, that is easy for a human to read, and tools that are adapted to you, as discussed in the previous section.

Moreover, this practice continues to evolve in other fields, as we have seen in `Chapter 12`, *Security in the DevOps Process with DevSecOps*, with the use of Inspec, which allows us to describe the compliance rules of the infrastructure in code.

We have also seen this Infrastructure as Code practice in several chapters of this book with what is called **Pipeline as Code**, with GitLab CI, Travis CI, GitHub Actions, and also Azure Pipelines, which also allows a YAML Pipeline mode (this mode has not been discussed in this book).

Putting any configuration in code is a key practice of the DevOps culture to take into account from the beginning of projects for both Ops and developers. For developers, there are also good practices about the designing of the application and infrastructure architecture that we'll discuss in the next section.

Designing the system architecture

A few years ago, all the services of the same application were *coded* in the same application block. This architecture design was legitimate since the application was managed in waterfall mode (`https://activecollab.com/blog/project-management/waterfall-project-management-methodology`), so new versions of the application were deployed in very long cycles. Since then, many changes have taken place in software engineering practices, starting with the adoption of the agile method and DevOps culture, then continuing with the arrival of the cloud.

This evolution has brought many improvements, not only in applications but also in their infrastructure.

However, in order to take advantage of an effective DevOps culture to deploy an application in the cloud, there are good practices to consider when designing software architecture and also when designing infrastructure.

First of all, cloud architects must work hand in hand with developers (or solution architects) to ensure that the application developed is in line with the different components of the architecture and that the architecture also takes into consideration the different constraints of the application.

In addition to this collaboration, security teams must also provide their specifications, which will be implemented by developers and cloud architects.

In order to be able to deploy a new version of the application more frequently without having to impact all of its features, it is good practice to separate the different areas of the application into separate code at first, then into different departments at a later stage. Thus, the separate code will be much more maintainable and scalable, and can be deployed faster without having to redeploy everything.

This method of separating the code into several services is part of the architecture pattern called **microservices**; to learn more, read the following comprehensive article: `https://microservices.io/patterns/microservices.html`.

Once decoupled, however, there is still a need to control dependency in order to implement a CI/CD pipeline that takes into account all the dependencies of the application.

In `Chapter 13`, *Reducing Deployment Downtime*, we discussed another good practice that allows deploying in production more frequently and it consists of encapsulating the functionalities of the application in feature flags. These feature flags must also be taken into account when designing the application, as they allow the application to be deployed in the production stage, enabling/disabling its features dynamically without having to redeploy it.

Finally, the implementation of unit tests and the log mechanism must be taken into account as soon as possible in the development of the application, because they allow feedback on the state of the application to be shared very quickly in its deployment cycle.

DevOps culture involves the implementation of CI/CD pipelines; as we have just seen, this requires changes in the design of applications, with the separation of functionalities to create less monolithic applications, the implementation of tests, and the addition of a log system.

After considering good practices for application design, we will explore some good practices for the implementation of CI/CD pipelines.

Building a good CI/CD pipeline

In this book, we have dedicated a complete chapter, `Chapter 6`, *Continuous Integration and Continuous Delivery*, to the creation of CI/CD pipelines using different tools such as GitLab CI, Jenkins, and Azure Pipelines, in which we have already mentioned the prerequisites for the implementation of CI/CD pipelines.

We also discussed the CI/CD process in `Chapter 14`, *DevOps for Open Source Projects*, with some examples of a CI pipeline for open source projects such as Travis CI and GitHub Actions.

Building a good CI/CD pipeline is indeed an essential practice in DevOps culture and, together with the correct choice of tools, allows faster deployment and better-quality applications.

One of the best practices for CI/CD pipelines is to set them up as early as the project launch stage. This is especially true for the CI pipeline, which will allow the code (at least the compilation step) to be verified when writing the first lines of the code. Then, as soon as the first environment is provisioned, immediately create a deployment pipeline, which will allow the application to be deployed and tested in this environment. The rest of the CI/CD pipeline process's tasks, such as unit test execution, can be performed as the project progresses.

In addition, it is also important to optimize the processes of the CI/CD pipeline by having pipelines that run quickly. This is used to provide quick feedback to team members (especially for CI) and also to avoid blocking the execution queue of other pipelines that may be in the queue.

Thus, if some pipelines take too long to run, such as integration tests, which can be long, it may be a good idea to schedule their execution for hours with less activity, such as at night.

Finally, it is also important to protect sensitive data embedded in CI/CD pipelines. So if you use a configuration manager tool in your pipelines, do not leave information such as passwords, connection strings, and tokens visible to all users.

To protect this data, use centralized secret management tools such as Vault, which we saw in `Chapter 12`, *Security in the DevOps Process with DevSecOps*, or use Azure Key Vault if you have an Azure subscription.

These are some of the best practices for the implementation of CI/CD pipelines; we have mentioned other good practices for CI/CD pipelines that you can study in the different chapters of this book and in other books dedicated to DevOps culture.

As a follow-up to the good practices for CI/CD pipelines, let's review good practices for the integration of tests into DevOps processes.

Integrating tests

Testing is, in today's world, a major part of the DevOps process, but also of development practices. Indeed, it is possible to have the best DevOps pipeline that automates all delivery phases, but without the integration of tests, it loses almost all its efficiency. For my part, I think that the minimum requirement for a DevOps process is to integrate at least the execution of the unit tests of the application. In addition, these unit tests must be written from the first line of code of the application using testing practices such as **Test-Driven Development** (**TDD**; `https://hackernoon.com/introduction-to-test-driven-development-tdd-61a13bc92d92`) and **Behavior-Driven Development** (**BDD**) and, in this way, the automatic execution of these tests can be integrated into the CI pipeline.

However, it is important to integrate other types of tests, such as functional tests or integration tests, that allow the application to be tested functionally from start to finish with the other components of its ecosystem.

It is certainly true that the execution of these tests can take time; in this case, it is possible to schedule their execution at night. But these integration tests are the ones that will guarantee the quality of the application's smooth operation during all stages of delivery until deployment to production.

However, there is often a very bad practice of disabling the execution of tests in the CI pipeline in case their execution fails. This is often done to avoid the blocking of the complete CI process and thus deliver faster in production. But keep in mind that the errors detected by the unit tests, including the ones that would have been detected by tests that have been disabled, will be detected in the production stage at some point and the correction of the failed code will cost more time than if the tests had been enabled during CI.

We have also learned about other types of tests in this book, such as code analysis tests or security tests, which are not to be ignored. The sooner they are integrated into CI/CD pipelines, the more value will be added for maintaining code and securing our application.

To summarize, you should not ignore the implementation of tests in your applications and their integration in CI/CD pipelines, as they guarantee the quality of your application.

After the good practices for test integration, I suggest you see what the good practices for the integration of security in your CI/CD processes are.

Applying security with DevSecOps

As we discussed in `Chapter 12`, *Security in the DevOps Process with DevSecOps*, security and compliance analyses must be part of DevOps processes. However, in companies, there is often a lack of awareness among development teams about security rules and this is why security is implemented too late in DevOps processes.

To integrate security into processes, it is, therefore, necessary to raise awareness among developers of aspects of application code security, but also of the protection of CI/CD pipeline configuration.

In addition, it is also necessary to eliminate the barrier between DevOps and security, by integrating security teams more often into the various meetings that bring together Developer and Operational teams, thus ensuring better consistency between developers, operational team, and also security. Regarding the choice of tools, do not use too many different tools, because the goal is for these tools to also be used by developers and be integrated into CI/CD pipelines. It is, therefore, necessary to select a few tools that are automated, do not require great knowledge of security, and provide reports for better analysis.

Finally, if you don't know where to start when it comes to analyzing the security of the application, work with simple security rules that are recognized by communities, such as the top 10 OWASP rules, `https://www.owasp.org/index.php/Category:OWASP_Top_Ten_Project`, which we saw in `Chapter 11`, *Security and Performance Tests*. You can use the ZAP tool, which uses these 10 rules, to perform security tests on a web application. You can also use the very practical checklist at `https://www.sqreen.com/checklists/devsecops-security-checklist` to check security points in a DevOps process.

These are some good practices for integrating security into DevOps culture to in order to achieve DevSecOps culture. We will now view some good practices for monitoring.

Monitoring your system

One of the main elements for the success of DevOps culture is the implementation of tools that will continuously monitor the state of a system and applications. The monitoring must be implemented at all levels of the system by involving all teams, with the aim of having applications with real added value for the end user.

Indeed, the first component that can be monitored is the application itself, by implementing, as soon as possible, a log or trace system that will serve to gather information on the use of the application. Then, we will measure and monitor the state of the infrastructure, such as the RAM and CPU level of the VMs or the network bandwidth. Finally, the last element that must be monitored is the status of DevOps processes. It is therefore important to have metrics on the execution of CI/CD pipelines, such as information on the execution time of pipelines, or the number of pipelines that are in success or failure. With this data, for example, we can determine the deployment speed of an application.

There are many monitoring tools, such as Graphana, New Relic, Prometheus, and Nagios, and others are integrated into the various cloud providers, such as Azure Application Insight or Azure Log Monitor.

Concerning good practices for monitoring, I would say that it is important to target the KPIs that are necessary for you and that are easy to analyze. It is useless to have a monitoring system that captures a lot of data or an application that writes a lot of logs, as it's too time-consuming when it comes to analyzing this information. In addition, on the volume of the captured data, we need to take care in ensuring its retention. The retention time of the data must be evaluated and different teams should be consulted. Too much retention can cause capacity saturation on VMs or high costs for managed components in the cloud, and with too little retention, the log history is shorter and therefore you can lose track of any problems.

Finally, when choosing the tool, you must make sure that it protects all the data that is captured, that the dashboards the tools present are understandable enough by all team members, and that it is integrated into a DevOps process.

We have explained that monitoring is a practice that must be integrated into DevOps culture, taking into account some points of good practice that can improve communication between Dev and Ops and improve product quality for end users.

After reviewing DevOps best practices with automation, CI/CD pipelines, and monitoring, we will take an overview of DevOps practices for project management and team organization.

Evolving project management

We have previously discussed some good DevOps practices to apply to projects, but all this can only be implemented and realized with a change in the way that projects are managed and teams are organized.

Here are some good practices that can facilitate the implementation of DevOps culture in project management within companies.

First of all, it should be remembered that DevOps culture only makes sense with the implementation of development and delivery practices that will allow applications to be delivered in short deployment cycles. Therefore, in order to be applicable, projects must also be managed with short cycles. To achieve this, one of the most suitable project management methods to apply DevOps culture and has proven its worth in recent years is the agile method, which uses sprints (short cycles of 2 to 3 weeks) with incremental, iterative deployments and strong collaboration between developers.

DevOps culture just extends the agile methodology by promoting collaboration between several domains (Dev/Ops/Security/Testers).

To learn more about the agile method and its different frameworks (Scrum and XP, for example), I recommend `http://agilemethodology.org/`, which provides a lot of documentation.

In addition, for a better application of DevOps implementations, it is important to change your organization by no longer having teams organized by areas of expertise, such as having a team of developers, another team of Ops, and a team of testers. The problem with this organizational model is that the teams are compartmentalized, resulting in a lack of communication (which we saw in `Chapter 1`, *DevOps Culture and Practices*, with the *wall of confusion*). This means that different teams have different objectives, which slams the brakes on applying good practices of the DevOps culture.

One of the models that allows for better communication is *feature team* organization with multidisciplinary project teams that are composed of people from all fields. In a team, we have developers, operational staff, and testers, and all these people work with the same objective.

If you want to know more about Microsoft's DevOps transformation, I suggest you watch the presentation by Donovan Brown at `https://www.agilealliance.org/resources/sessions/microsoft-devops-transformation-donovan-brown/`, which explains how Microsoft has changed its organization to adapt to DevOps culture to continuously improve its products while taking into account user needs.

We have just seen that to implement DevOps culture in companies, organizational changes are required, including agile project management and the composition of multidisciplinary teams.

Summary

In this final chapter, we have seen that the implementation of DevOps culture within projects requires the use of best practices regarding the automation of all manual tasks, the proper choice of tools, a less monolithic project architecture, and the implementation of monitoring.

On a large scale, for the organization of teams and the company as a whole, we have seen that the agile method, as well as multidisciplinary teams, contributes strongly to the implementation of DevOps culture.

To finish this book, my advice to all you readers who are adopting DevOps practices is to implement and monitor them on small projects and to start by using the tools that are most familiar and accessible to you. Then, as soon as your DevOps process is working properly, you can extend it to larger projects.

Questions

1. What are the advantages of deployment automation?
2. Is it necessary to use Terraform to do Infrastructure as Code?
3. What needs to be done to improve security in DevOps processes?
4. Does monitoring only concern the monitoring of the condition of the infrastructure?
5. What good practice should be implemented in the application architecture?
6. In a DevOps organization, how are teams constituted?

Further reading

If you want to know more about DevOps best practices, here are some articles:

- 16 Best Practices Of CI/CD Pipeline To Speed Test Automation: `https://www.lambdatest.com/blog/16-best-practices-of-ci-cd-pipeline-to-speed-test-automation/`
- Implementing Continuous Testing: `https://www.lambdatest.com/blog/how-to-implement-continuous-testing-in-devops-like-a-pro/`

- Secure DevOps: `https://www.microsoft.com/en-us/securityengineering/devsecops`
- 9 Pillars of Continuous Security Best Practices: `https://devops.com/9-pillars-of-continuous-security-best-practices/`
- The DevSecOps Security Checklist: `https://www.sqreen.com/checklists/devsecops-security-checklist`
- Top 5 Best Practices for DevOps Monitoring: `https://devops.com/top-5-best-practices-devops-monitoring/`
- DevOps success: 5 flow metrics to get you there: `https://devops.com/top-5-best-practices-devops-monitoring/`
- 10 pitfalls to avoid when implementing DevOps: `https://opensource.com/article/19/9/pitfalls-avoid-devops`
- Microsoft DevOps Journey: `http://stories.visualstudio.com/devops/`

Assessments

Chapter 1: DevOps Culture and Practices

1. DevOps is a contraction that is formed from the word *Developers* and *Operations*.
2. DevOps is a term that represents a culture.
3. The three axes of DevOps culture are collaboration, process, and tools.
4. The objective of continuous integration is to get quick feedback on the quality of the code archived by team members.
5. The difference between continuous delivery and continuous deployment is that the triggering of the deployment in production is done manually for continuous delivery, whereas it is automatic for continuous deployment.
6. Infrastructure as Code consists of writing the code of the resources that make up an infrastructure.

Chapter 2: Provisioning Cloud Infrastructure with Terraform

1. The language used by Terraform is **HashiCorp Configuration Language** (HCL).
2. Terraform's role is an Infrastructure as Code tool.
3. No. Terraform is not a scripting tool.
4. The command that allows you to display the installed version is `versionterraform version`.
5. The name of the Azure object that connects Terraform to Azure is the Azure service principal.
6. The three main orders of Terraform are `terraform init`, `terraform plan`, and `terraform apply`.
7. The Terraform command that allows us to destroy resources is `terraform destroy`.
8. We add the `--auto-approve` option to the `terraform apply` command.
9. The purpose of the `tfstate` file is to keep the resources and their properties throughout the execution of Terraform.

10. No, it is not a good practice to leave `tfstate` locally; it must be stored in a protected remote backend.

Chapter 3: Using Ansible for Configuring IaaS Infrastructure

1. The role of Ansible as detailed in this chapter is to automate the configuration of a VM.
2. No. We cannot install Ansible on a Windows OS.
3. The two artifacts studied in this chapter that Ansible needs to run are the inventory and the playbook.
4. The option is `--check`.
5. The name of the utility used to encrypt and decrypt Ansible data is Ansible Vault.
6. When using dynamic inventory in Azure, the script is based on VM tags that are used to return the list of VMs.

Chapter 4: Optimizing Infrastructure Deployment with Packer

1. The two ways to install Packer are manually or via a script.
2. The mandatory sections of a Packer template that are used to create a VM image in Azure are builders and provisioners.
3. The command used to validate a Packer template is `packer validate`.
4. The command that is used to generate a Packer image is `packer build`.

Chapter 5: Managing Your Source Code with Git

1. Git is a **distributed version control system** (DVCS).
2. The command to initialize a repository is `git init`.

3. The artifact is the commit that consists of saving part of the code.
4. The command that allows you to save your code in the local repository is `git commit`.
5. The command that allows you to send your code to the remote repository is `git push`.
6. The command that allows you to update your local repository from the remote repository is `git pull`.
7. The branch is the mechanism that allows you to isolate the code.
8. GitFlow is a branch management model in Git.

Chapter 6: Continuous Integration and Continuous Delivery

1. The prerequisite for setting up a CI pipeline is to have its code in a source control manager.
2. The CI pipeline is triggered every time a team member commits/pushes a code.
3. A package manager is a central repository used to centralize and share packages, development libraries, tools, or software.
4. The NuGet package manager allows you to store .NET libraries/Frameworks.
5. Azure Artifacts is integrated into Azure DevOps.
6. No, it's an on-premises tool that you have to install on a server.
7. In Azure DevOps, the service that manages CI/CD pipelines is Azure Pipelines.
8. The GitLab services consists of a source code manager, an CI/CD pipeline manager, and a board for project management.
9. In GitLabCI, a CI pipeline is built with a YAML file named `.gitlab-ci.yml`.

Chapter 7: Containerizing Your Application with Docker

1. Docker Hub is a public registry of Docker images.
2. The basic element is the Dockerfile.
3. The instruction is `FROM`.
4. The command to create a Docker image is `docker build`.

5. The instantiation command of a Docker container is `docker run`.
6. The Docker command to publish an image is `docker publish`.

Chapter 8: Managing Containers Effectively with Kubernetes

1. Kubernetes' role is to manage containers.
2. In Kubernetes, all objects are written in YAML specification files.
3. The Kubernetes CLI tool is called `kubectl`.
4. The command that applies a deployment in K8s is `kubectl apply`.
5. HELM is the package manager for Kubernetes.
6. Azure Kubernetes Services is a managed Kubernetes cluster in Azure.

Chapter 9: Testing APIs with Postman

1. Postman is a tool that allows you to perform API tests.
2. The first element to create is a collection.
3. In Postman, the API configuration is found in a request.
4. The runner collection allows you to execute all requests in a collection.
5. Newman is a command-line tool that performs Postman tests in a CI/CD pipeline.

Chapter 10: Static Code Analysis with SonarQube

1. SonarQube is developed in Java.
2. To install SonarQube, it is necessary to have Java installed.
3. SonarQube is a tool for static code analysis.
4. SonarLint allows developers to do code analysis while they write code.

Chapter 11: Security and Performance Tests

1. No. ZAP is not a tool to analyze the source code of an application.
2. In Postman, the performance metric is the execution time of each request.

Chapter 12: Security in the DevOps Process with DevSecOps

1. Its role is to test the compliance of a system or infrastructure.
2. The package manager is `Gem`.
3. The command is `inspec exec`.
4. Vault is edited by Hashicorp.
5. The command is `vault server -dev`.
6. No. When installed locally, it can only be used for development and testing.
7. In this mode, the data is stored in memory.
8. The prerequisites are to have a Windows OS with PowerShell (version 5 or later) installed on it.
9. The file format of the generated report is CSV.

Chapter 13: Reducing Deployment Downtime

1. The Terraform option that reduces downtime is `create_before_destroy`.
2. A blue-green deployment infrastructure is composed of one blue and one green environment and a router or load balancer.
3. Both patterns are canary release and dark launch.
4. The Azure components that allow blue-green deployment are the app service slots and the Azure Traffic Manager.
5. The role of feature flags is to enable or disable features of an application without having to redeploy it.
6. The `FeatureToogle` is a simple feature flag, open-source framework for .NET applications.
7. `LaunchDarkly` is a SaaS solution that frees you from any installation.

Chapter 14: DevOps for Open Source Projects

1. No, to modify the code of another repository, you will need to create a fork of this repository.
2. The element that allows the merge is the pull request.
3. The `CHANGELOG.md` file allows you to display the release notes.
4. Release is linked to a `git` tag.
5. In Travis CI, the configuration of a job is written to a YAML file.
6. The two tools are SonarCloud and WhiteSource Bolt.
7. Security vulnerabilities are listed in the **Issues** tab.

Chapter 15: DevOps Best Practices

1. Deployment automation eliminates manual errors and reduces deployment cycles.
2. No. Any tool that allows you to script the configuration of an infrastructure can be used.
3. Security teams must be integrated with Dev and Ops.
4. No, monitoring is about applications, infrastructure, and IC/CD pipelines.
5. In the application, it is a good practice to separate the features or domains of the application to have code that can be easily deployed
6. In the DevOps organizational structure the teams are multidisciplinary.

Other Books You May Enjoy

If you enjoyed this book, you may be interested in these other books by Packt:

Hands-On Security in DevOps
Tony Hsu

ISBN: 978-1-78899-550-4

- Understand DevSecOps culture and organization
- Learn security requirements, management, and metrics
- Secure your architecture design by looking at threat modeling, coding tools and practices
- Handle most common security issues and explore black and white-box testing tools and practices
- Work with security monitoring toolkits and online fraud detection rules
- Explore GDPR and PII handling case studies to understand the DevSecOps lifecycle

DevOps with Kubernetes - Second Edition
Hideto Saito, Hui-Chuan Chloe Lee, Cheng-Yang Wu

ISBN: 978-1-78953-399-6

- Learn fundamental and advanced DevOps skills and tools
- Get a comprehensive understanding of containers
- Dockerize an application
- Administrate and manage Kubernetes cluster
- Extend the cluster functionality with custom resources
- Understand Kubernetes network and service mesh
- Implement Kubernetes logging and monitoring
- Manage Kubernetes services in Amazon Web Services, Google Cloud Platform,and Microsoft Azure

Leave a review - let other readers know what you think

Please share your thoughts on this book with others by leaving a review on the site that you bought it from. If you purchased the book from Amazon, please leave us an honest review on this book's Amazon page. This is vital so that other potential readers can see and use your unbiased opinion to make purchasing decisions, we can understand what our customers think about our products, and our authors can see your feedback on the title that they have worked with Packt to create. It will only take a few minutes of your time, but is valuable to other potential customers, our authors, and Packt. Thank you!

Index

B

D

E

F

G

H

I

J

O

P

Q

R

S

T